Lecture Notes in Computer Science 14161

Founding Editors

Gerhard Goos
Juris Hartmanis

Editorial Board Members

The series Lecture Notes in Computer Science (LNCS), including its subseries Lecture Notes in Artificial Intelligence (LNAI) and Lecture Notes in Bioinformatics (LNBI), has established itself as a medium for the publication of new developments in computer science and information technology research, teaching, and education.

LNCS enjoys close cooperation with the computer science R & D community, the series counts many renowned academics among its volume editors and paper authors, and collaborates with prestigious societies. Its mission is to serve this international community by providing an invaluable service, mainly focused on the publication of conference and workshop proceedings and postproceedings. LNCS commenced publication in 1973.

Robert Thomson · Samer Al-khateeb ·
Annetta Burger · Patrick Park · Aryn A. Pyke
Editors

Social, Cultural, and Behavioral Modeling

16th International Conference, SBP-BRiMS 2023
Pittsburgh, PA, USA, September 20–22, 2023
Proceedings

 Springer

Editors
Robert Thomson ⓘ
Army Cyber Institute, United States Military
Academy
West Point, NY, USA

Annetta Burger ⓘ
Oak Ridge National Laboratory
Oak Ridge, TN, USA

Aryn A. Pyke ⓘ
Army Cyber Institute, United States Military
Academy
West Point, NY, USA

Samer Al-khateeb ⓘ
Creighton University
Omaha, NE, USA

Patrick Park ⓘ
Carnegie Mellon University
Pittsburg, PA, USA

ISSN 0302-9743 ISSN 1611-3349 (electronic)
Lecture Notes in Computer Science
ISBN 978-3-031-43128-9 ISBN 978-3-031-43129-6 (eBook)
https://doi.org/10.1007/978-3-031-43129-6

This Springer imprint is published by the registered company Springer Nature Switzerland AG
The registered company address is: Gewerbestrasse 11, 6330 Cham, Switzerland

Paper in this product is recyclable.

Preface

In the 16th year of the International Conference on Social Computing, Behavioral-Cultural Modeling and Prediction and Behavior Representation in Modeling and Simulation, SBP-BRiMS 2023, we highlighted the many advances in the computational social sciences. Improving the human condition requires understanding, forecasting, and impacting socio-cultural behavior in both the digital and non-digital world. Increasing amounts of digital data, embedded sensors that collect human information, rapidly changing communication media, changes in legislation concerning digital rights and privacy, spread of 4G technology to developing countries and development of 5G technology, the rise of large language models (e.g., ChatGPT), and other changes are creating a new cyber-mediated world in which the very precepts of why, when, and how people interact and make decisions is being called into question. For example, Uber understood human behaviors vis-à-vis commuting. It then developed software to support this behavior, which ended up saving time (and so capital) and reducing stress, which indirectly created the opportunity for people to evolve new behaviors.

Scientific and industrial pioneers in this area rely on both social science and computer science to help make sense of the impact of this new frontier. To be successful pioneers, a true merger of social science and computer science is needed. Solutions that rely only on the social sciences or computer science are doomed to failure. For example, Anonymous developed an approach for identifying members of terror groups, such as ISIS, on the Twitter social media platform using state-of-the-art computational techniques. These accounts were then suspended. This was a purely technical solution. The consequence was that those individuals with suspended Twitter accounts just moved to new platforms and resurfaced on Twitter under new IDs. In this case, failure to understand basic social behavior resulted in an ineffective solution.

The goal of this conference is to build this new community of social cyber scholars by bringing together and fostering interaction between members of the scientific, corporate, government, and military communities who are interested in understanding, forecasting, and impacting human socio-cultural behavior. It is the mission of this community to build this new field, its theories, methods, and scientific culture in a way that does not give priority to either social science or computer science, and to embrace change as the cornerstone of the community. Despite decades of work in this area, this scientific field is still in its infancy. To meet this charge, to move this science to the next level, this community must meet the following three challenges: deep understanding, socio-cognitive reasoning, and reusable computational technology. Fortunately, as the papers in this volume illustrate, this community is poised to answer these challenges. But what does meeting these challenges entail?

Deep understanding refers to the ability to make operational decisions and theoretical arguments based on an empirically-based deep and broad understanding of the complex socio-cultural phenomena of interest. Today, although more data is available digitally than ever before, we are still plagued by anecdotally based arguments. For example, in social media, despite the wealth of information available, most analysts focus on small samples, which are typically biased and cover only a small time period, to explain all events and make future predictions. The analyst finds a magic tweet or an unusual tweeter and uses that to prove their point. Tools that can provide more data or less biased data are not widely used, as they are often more complex or time-consuming than what the average analyst would use for generating results. Not only are more scalable technologies needed, but also a better understanding of the biases in the data and ways to overcome them, as well as a cultural change to not accept anecdotes as evidence.

Socio-cognitive reasoning refers to the ability of individuals to make sense of the world and to interact with it in terms of groups and not just individuals. Currently, most social-behavioral models either focus on (1) strong cognitive models of individuals engaged in tasks that produce a small number of agents with high levels of cognitive accuracy but with little if any social context, or (2) light cognitive models and strong interaction models, which results in a model depicting massive numbers of agents with high levels of social realism and little cognitive realism. In both cases, as realism is increased in the other dimension, the scalability of the models falls, while their predictive accuracy on one of the two dimensions remains low. In contrast, as agent models are built where the agents are not just cognitive but socially cognitive, we find that the scalability increases and the predictive accuracy increases. Not only are agent models with socio-cognitive reasoning capabilities needed, but so too is a better understanding of how individuals form and use these social-cognitions.

More software solutions that support behavioral representation, modeling, data collection, bias identification, analysis, and visualization are available to support human socio-cultural behavioral modeling and prediction than ever before. However, this software is generally just piling up in giant black holes on the web. Part of the problem is the fallacy of open source, namely the idea that if you just make code open source, others will use it. In contrast, most of the tools and methods available in Git or R are only used by the developer, if at all. Reasons for its unpopularity with analysts include the lack of documentation, interfaces, interoperability with other tools, difficulty of linking to data, and increased demands on the analyst's time due to a lack of tool-chain and workflow optimization. A part of the problem is the "not invented here" syndrome. For social scientists and computer scientists alike, it is just more fun to build a quick and dirty tool for your own use than to study and learn tools built by others. Another issue is the insensitivity of people from one scientific or corporate culture to the reward and demand structures of other cultures that impact what information can or should be shared and when. A related problem is double standards in sharing: universities are expected to share and companies are not, but increasingly universities are relying on intellectual property as a source of funding just like other companies. While common standards and representations would help, a cultural shift from a focus on sharing to a focus on re-use is as or more critical for moving this area to the next scientific level.

In this volume, and in all the work presented at the SBP-BRiMS 2023 conference, you will see suggestions of how to address the challenges just described. The SBP-BRiMS 2023 conference carried on the scholarly tradition of the past conferences out of which it emerged like a phoenix: the Social Computing, Behavioral-Cultural Modeling, and Prediction (SBP) Conference and the Behavioral Representation in Modeling and Simulation (BRiMS) Society's conference.

A total of 73 papers were submitted as regular track submissions. Of these, 31 were accepted to these proceedings, resulting in an acceptance rate of 42%. All papers were single-blind reviewed by 2 to 3 non-chair reviewers. Additionally, any papers from conference committee members were independently reviewed by 2–3 non-committee members. Furthermore, committee members' students were excluded from reviewing their papers.

Overall, the accepted papers come from an international group with authors from many countries, showing the broad reach of the computational social sciences community. The conference has a strong multi-disciplinary heritage. As the papers in this volume show, people, theories, methods, and data from a wide number of disciplines are represented, including computer science, psychology, sociology, communication science, public health, bioinformatics, political science, and organizational science. Numerous types of computational methods are used including, but not limited to, machine learning, large language models, language technology, social network analysis and visualization, agent-based simulation, and statistics.

This exciting program could not have been put together without the hard work of several dedicated and forward-thinking researchers serving as the organizing committees, listed on the following pages. Members of the Program Committee, the scholarship committee, publication, advertising, and local arrangements chairs worked tirelessly to put together this event. They were supported by the government sponsors, the area chairs, and the reviewers. Please join me in thanking them for their efforts on behalf of the community. In addition, we gratefully acknowledge the support of our sponsor – the Army Research Office (W911NF-21-1-0102). We hope that you enjoy the conference, and welcome to the community.

September 2023

Robert Thomson
Samer Al-khateeb
Annetta Burger
Patrick Park
Aryn A. Pyke

Organization

General Chairs

Kathleen M. Carley   Carnegie Mellon University, USA
Nitin Agarwal   University of Arkansas at Little Rock, USA

Program Committee Chairs

Robert Thomson   United States Military Academy, USA
Samer Al-khateeb   Creighton University, USA
Annetta Burger   Oak Ridge National Laboratory, USA
Patrick Park   Carnegie Mellon University, USA
Aryn A. Pyke   United States Military Academy, USA

Proceedings Co-chairs

Robert Thomson   United States Military Academy, USA
Samer Al-khateeb   Creighton University, USA

Agenda Co-chairs

Robert Thomson   United States Military Academy, USA
Kathleen M. Carley   Carnegie Mellon University, USA

Journal Special Issue Chair

Kathleen M. Carley   Carnegie Mellon University, USA

Tutorial Chair

Kathleen M. Carley   Carnegie Mellon University, USA

Publicity Chair

Donald Adjeroh   West Virginia University, USA

Graduate Program Chair

Aryn A. Pyke   United States Military Academy, USA

Challenge Problem Committee

Samer Al-khateeb   Creighton University, USA
Annetta Burger   Oak Ridge National Laboratory, USA
Patrick Park   Carnegie Mellon University, USA

Emeritus Steering Committee

Fahmida Chowdhury   National Science Foundation, USA
Rebecca Goolsby   Office of Naval Research, USA
Paul Tandy   Defense Threat Reduction Agency, USA
Edward T. Palazzolo   Army Research Office, USA
Patricia Mabry   Indiana University, USA
John Lavery   Army Research Office, USA
Tisha Wiley   National Institutes of Health, USA

Additional Reviewers

Palvi Aggarwal   Andrew Crooks
Nasrin Akhter   Jessica Dawson
Antonio Luca Alfeo   Vito D'Orazio
Walaa Alnasser   Peng Fang
Akshay Aravamudan   Laurie Fenstermacher
Kiran Kumar Bandeli   Juan Fernandez-Gracia
Mihovil Bartulovic   Erika Frydenlund
Emanuel Ben-David   Gayane Grigoryan
Amrita Bhattacharjee   Yuzi He
Nathan Bos   Adriana Iamnitchi
Keith Burghardt   Aruna Jammalamadaka
Edward Cranford   Jasser Jasser

Ujun Jeong
Bohan Jiang
Prakruthi Karuna
Hamdi Kavak
James Kennedy
Nayoung Kim
Charity King
Srijan Kumar
Ugur Kursuncu
Christian Lebiere
Stephan Leitner
Huan Liu
Chitaranjan Mahapatra
Murali Mani
Raha Moraffah
Jonathan Morgan
David Mortimore
Md. Saddam Mukta

Salem Othman
Jose Padilla
Samantha Phillips
Scott Renshaw
Patrick Rice
Rey Rodrigueza
Juliette Rouge
Jordan Richard Schoenherr
Mainuddin Shaik
Paras Sheth
Martin Smyth
Billy Spann
Farnaz Tehranchi
Joshua Uyheng
Alina Vereshchaka
Friederike Wall
Xupin Zhang

Contents

Detecting Malign Influence

Misogyny, Women in Power, and Patterns of Social Media Harassment   3
   Jennifer Golbeck

Stereotype Content Dictionary: A Semantic Space of 3 Million Words
and Phrases Using Google News Word2Vec Embeddings   12
   Xuanlong Qin and Tony Tam

Social Cybersecurity Analysis of the Telegram Information Environment
During the 2022 Invasion of Ukraine   23
   Ian Kloo and Kathleen M. Carley

The Dynamics of Political Narratives During the Russian Invasion
of Ukraine   33
   Ahana Biswas, Tim Niven, and Yu-Ru Lin

Anger Breeds Controversy: Analyzing Controversy and Emotions
on Reddit   44
   *Kai Chen, Zihao He, Rong-Ching Chang, Jonathan May,
   and Kristina Lerman*

Building a Healthier Feed: Private Location Trace Intersection Driven
Feed Recommendations   54
   Tobin South, Nick Lothian, Takahiro Yabe, and Alex 'Sandy' Pentland

Classifying Policy Issue Frame Bias in Philippine Online News   64
   Jose Mari Luis M. Dela Cruz and Maria Regina Justina E. Estuar

Dismantling Hate: Understanding Hate Speech Trends Against NBA
Athletes   74
   *Edinam Kofi Klutse, Samuel Nuamah-Amoabeng, Hanjia Lyu,
   and Jiebo Luo*

Feedback Loops and Complex Dynamics of Harmful Speech in Online
Discussions   85
   Rong-Ching Chang, Jonathan May, and Kristina Lerman

Chirping Diplomacy: Analyzing Chinese State Social-Cyber Maneuvers
on Twitter 95
  Samantha C. Phillips, Joshua Uyheng, Charity S. Jacobs,
  and Kathleen M. Carley

Vulnerability Dictionary: Language Use During Times of Crisis
and Uncertainty 105
  Wenjia Hu, Zhifei Jin, and Kathleen M. Carley

Tracking China's Cross-Strait Bot Networks Against Taiwan 115
  Charity S. Jacobs, Lynnette Hui Xian Ng, and Kathleen M. Carley

Assessing Media's Representation of Frustration Towards Venezuelan
Migrants in Colombia 126
  Brian Llinas, Guljannat Huseynli, Erika Frydenlund,
  Katherine Palacio, Humberto Llinas, and Jose J. Padilla

Human Behavior Modeling

Agent-Based Moral Interaction Simulations in Imbalanced Polarized
Settings 139
  Evan M. Williams and Kathleen M. Carley

Investigating the Use of Belief-Bias to Measure Acceptance of False
Information 149
  Robert Thomson and William Frangia

Simulation of Stance Perturbations 159
  Peter Carragher, Lynnette Hui Xian Ng, and Kathleen M. Carley

Integrating Human Factors into Agent-Based Simulation for Dynamic
Phishing Susceptibility 169
  Jeongkeun Shin, Kathleen M. Carley, and L. Richard Carley

Designing Organizations of Human and Non-Human Knowledge Workers 179
  David Mortimore, Raymond R. Buettner Jr., and Eugene Chabot

Modeling Human Actions in the Cart-Pole Game Using Cognitive
and Deep Reinforcement Learning Approach 189
  Aadhar Gupta, Mahavir Dabas, Shashank Uttrani, Sakshi Sharma,
  and Varun Dutt

CCTFv1: Computational Modeling of Cyber Team Formation Strategies 199
  Tristan J. Calay, Basheer Qolomany, Aos Mulahuwaish,
  Liaquat Hossain, and Jacques Bou Abdo

Simulating Transport Mode Choices in Developing Countries 209
   Kathleen Salazar-Serna, Lorena Cadavid, Carlos J. Franco,
   and Kathleen M. Carley

User Identity Modeling to Characterize Communication Patterns
of Domestic Extremists Behavior on Social Media 219
   Falah Amro and Hemant Purohit

Understanding Clique Formation in Social Networks - An Agent-Based
Model of Social Preferences in Fixed and Dynamic Networks 231
   Pratyush Arya and Nisheeth Srivastava

A Bayesian Approach of Predicting the Movement of Internally Displaced
Persons 241
   Obed Domson, Jose J. Padilla, Guohui Song, and Erika Frydenlund

Social-Cyber Behavior Modeling

Few-Shot Information Operation Detection Using Active Learning
Approach 253
   Meysam Alizadeh and Jacob N. Shapiro

Dynamic Modeling and Forecasting of Epidemics Incorporating Age
and Vaccination Status 263
   Nitin Kulkarni, Chunming Qiao, and Alina Vereshchaka

Inductive Linear Probing for Few-Shot Node Classification 274
   Hirthik Mathavan, Zhen Tan, Nivedh Mudiam, and Huan Liu

Regression Chain Model for Predicting Epidemic Variables 285
   Kirti Jain, Vasudha Bhatnagar, and Sharanjit Kaur

Physical Distancing and Mask Wearing Behavior Dataset Generator
from CCTV Footages Using YOLOv8 295
   Roland P. Abao, Maria Regina Justina E. Estuar,
   and Patricia Angela R. Abu

Moving Beyond Stance Detection in Cross-Cutting Communication
Analysis 305
   Rezvaneh Rezapour, Daniela Delgado Ramos, Sullam Jeoung,
   and Jana Diesner

Pandemic Personas: Analyzing Identity Signals in COVID-19 Discourse
on Twitter 316
   *Scott Leo Renshaw, Samantha C. Phillips, Michael Miller Yoder,
   and Kathleen M. Carley*

Poster Abstract

Selected Poster Presentation Abstracts 329
   Robert Thomson

Author Index 337

Detecting Malign Influence

Misogyny, Women in Power, and Patterns of Social Media Harassment

Jennifer Golbeck[✉]

College of Information Studies, University of Maryland,
College Park, MD 20740, USA
jgolbeck@umd.edu

Abstract. Online harassment is a well-documented and studied problem on social media. Who does this harassing, how, and to what degree are important questions that can inform platform policies and automated controls as well as helping understand harassers more broadly. This study investigates users who were discovered because they created a post that harassed a women in power using misogynistic slurs. Do these tend to be isolated incidents, or do such users engage in higher rates of harassment more generally? Findings from both Twitter and Parler suggest that this population uses offensive slurs at over 3 times the rate of control groups. We break down these findings and discuss the implications for moderation, automation, user well-being, and platform success.

Keywords: social media · online harassment

NB: This paper deals with abusive language. To accurately describe our methods, we describe slurs that many may find offensive.

1 Introduction

Harassment has become an all too common aspect of online life, with 41% of US adults reporting that they have been targets of it [15]. Social media platforms employ a variety of strategies to keep harassment in check, from automated filtering to forced post removal to account suspension. While this may improve things, enforcement is spotty and harassing posts often remain up, even when they violate a platform's terms and conditions [5].

Researchers have studied harassment from may perspectives, including understanding the motivations of trolls and harassers (e.g. [1]), the impact on frequently targeted groups (e.g. [14]), the efficacy of behavioral interventions (e.g. [12]), and automated techniques for identifying and filtering harassment (e.g. [11]). However, much remains unknown about harassers, their behavior, and their role in social media ecosystems.

This study focuses on understanding if certain harassing behaviors predict higher levels of harassment. Specifically, we were interested in people who direct sexist harassment at women in positions of power. Would users who author such

R. Thomson et al. (Eds.): SBP-BRiMS 2023, LNCS 14161, pp. 3–11, 2023.
https://doi.org/10.1007/978-3-031-43129-6_1

posts be more likely to harass more broadly? For the purposes of this study, we operationalize harassment as use of offensive slurs in social media posts. Greater slur use reflects more harassing behavior.

In samples pulled from both Twitter and Parler, we searched for users who used one of five common misogynistic slurs ("bitch", "whore", "hoe", "cunt", "slut") and the name of one of four prominent US women legislators: Vice President Kamala Harris, House Speaker Nancy Pelosi, Rep. Lauren Boebert, and Rep. Marjorie Taylor Greene. These users were our Harassing Sample. We compared their overall slur usage to control groups and found significant differences. We discuss the results of this analysis, the limitations and future directions for this work, and the implications for platform moderation, harasser detection, and user well-being.

2 Related Work

The research questions of this study focus primarily on harassers with a goal of better understanding their behavior and if certain behaviors make them easier to identify.

There have been many studies looking at the impact of online harassment but relatively few that study the harassers themselves. Work on online trolls [1] found that, compared to other groups, people who liked trolling scored significantly higher on the dark tetrad of personality traits: psychopathy, sadism, Machiavellianism, and narcissism. More recent work that looked at online harassers [9] found that impulsivity, reactive aggression, and premeditated aggression were distinguishing characteristics of harassers.

Data on online harassment instigation is harder pin down since definitions of "harassment" itself vary widely. A study of college students found 92% of subjects reported participating in some sort of cyber-harassment [16]. Studies of slightly younger participants generally found lower self-reported levels of perpetration. More importantly, these studies looked at various correlations with other anti-social behavior and found that higher levels of harassment predicted higher levels of other anti-social behavior.

One study of people aged 10–17 years found that 12% reported frequent or occasional perpetration of online harassment, while 17% said they had limited participation [18]. As the frequency of harassment increased, so did other anti-social behaviors including aggression and rule breaking. Another study of youth in Thailand found about half of participants had perpetrated online harassment in the last year, and that this was positively correlated with committing offline violence [17].

A large study of over 1,500 young people [18] found that perpetration of online harassment and unwanted sexual advances online was associated with a litany of problems, including "substance use; involvement in offline victimization and perpetration of relational, physical, and sexual aggression; delinquent peers; a propensity to respond to stimuli with anger; poor emotional bond with caregivers; and poor caregiver monitoring as compared with youth with little to no involvement."

These results, suggesting that higher rates of harassment are associated with both anti-social personality traits and anti-social behavior, emphasize the importance of understanding patterns of behavior in online harassment perpetration and identifying characteristics that make frequent perpetrators easier to identify.

3   Methods

3.1   Datasets

Bot Sentinel is a research firm that focuses on identifying mis- and disinformation, harassment, and malicious behavior on social media. In May 2022, they released a report, "Twitter's Response to Abuse and Bigotry Directed at Vice President Kamala Harris". Between January and May 2022, they identified 4,265 tweets that called Harris "bitch", "whore", "hoe", "cunt", "slut", and "nigger". As a test, Bot Sentinel staff reported 40 tweets, yet Twitter only removed two, leaving many aggressively harassing tweets up and available (see Fig. 1 for an example of some of the less graphic tweets; the full set is available in the cited Bot Sentinel report).

Their work put a spotlight on this type of harassment and the ineffectiveness of platform policies in curtailing even extreme violations of community guidelines. It also served as a seed dataset for this project.

We began with the Bot Sentinel data set of abusive or bigoted tweets directed at Vice President Kamala Harris. Each tweet was posted by a different user [7]. Since the focus of our work in this paper is misogyny, we included tweets using one of the five misogynistic terms and dropped those that used racist language (as discussed in the Future Work section below, we believe it is important to do in-depth work both on racist harassment and intersectional harassment, but it is beyond the scope of this project).

The authors of the remaining tweets became our initial set of harassers. Many of these accounts were removed after Bot Sentinel published their report. Some accounts also went private which prevented us from accessing their tweets. This left 910 accounts from the original dataset who were still active with tweets we could access. For each of these users, we collected their most recent tweets, with a maximum of 200, for a total of 171,137 tweets.

As a control, we used the Twitter API to search for users who tweeted "Kamala Harris". We randomly selected 100 users and collected their most recent tweets (up to 200), for a total of 18,302 tweets.

To expand beyond data about Vice President Harris, we collected data from Twitter for House Speaker Nancy Pelosi. Using the API, we searched for people who used one of the Bot Sentinel five misogynist slurs along with "Pelosi". As a baseline control, we searched for any tweets mentioning "Pelosi" that did not include those slurs. Again, we randomly selected 100 users from each group and collected their 200 most recent tweets, for a total of 14,119 tweets in the slur group and 15,364 tweets in the control.

Since Pelosi and Harris are both Democrats, we followed the same protocol as used on Nancy Pelosi to collect data sets mentioning Rep. Lauren Boebert and Rep. Marjorie Taylor-Greene, both higher profile Republicans. This allowed us to compare the use of slurs targeting both major US political parties. For Boebert, there were 21,825 tweets in the slur group and 18,784 in the control. For Taylor-Greene, there were 7,664 in the slur group and 18,094 in the control.

We also compared across platforms. Parler is a microblogging platform similar to Twitter but that was mostly unmoderated in the run up to the January 6, 2021 insurrection. It was a platform that attracted many right wing users who had either been banned from Twitter or who were seeking a platform with a stronger right-wing voice.

We used a dataset of 1.8 million text posts from Parler posted between January 6 and 10, 2021, available at https://mirrors.deadops.de/parler/. We repeated the process above, collecting posts that mentioned Kamala Harris, Nancy Pelosi, Lauren Boebert, and Marjorie Taylor-Greene with one the five misogynistic slurs originally used by Bot Sentinel. There were no posts in our dataset that used these slurs alongside either Republican legislator, so we can only compare slurs used towards Harris and Pelosi. We found 86 unique users in this dataset who had used the misogynistic slurs directed towards them. The control included all users who had mentioned Harris and Pelosi without slurs in that five day window, totaling 9,154.

Across all samples, we dropped users who had fewer than 10 posts since their slur frequency values could skew our results. The final number of accounts for each target and group is shown in Table 1.

Table 1. Number of accounts for each sample

	Harassers	Control
Twitter		
Kamala Harris	909	100
Nancy Pelosi	73	89
Lauren Boebert	112	98
Marjorie Taylor-Greene	39	91
Parler		
Kamala Harris + Nancy Pelosi	86	9,154

3.2 Slur Detection

To measure the frequency of slurs use among users in each group, we used the list of slurs provided at https://gate-socmedia.sites.sheffield.ac.uk/topics/elections-hate-speech/ge2019-supplementary-materials [4]. Two of the original five misogynistic slurs, "bitch" and "hoe", were not part of this list, so we added them for our analysis.

For each user selected, we tallied the number of slurs used in their posts and calculated the average number of slurs per post.

4   Results

We first analyzed the Twitter data. For the four targets of misogynistic harassment - Kamala Harris, Nancy Pelosi, Lauren Boebert, and Marjorie Taylor-Greene - we compared the average number of slurs per tweet between the harassing group and control group. As shown in Table 2, for all targets, harassers all used slurs at significantly higher rates than the control group for $p < 0.001$.

Table 2. Average number of slurs per 100 tweets for harassing and control populations for each of the four Twitter target accounts. Student's t-test shows that for every target, the Harassing Sample uses significantly more slurs than the control sample ($p < 0.001$)

Target	Harassing	Control
Kamala Harris	3.1	1.0
Nancy Pelosi	2.4	1.2
Lauren Boebert	3.0	1.2
Marjorie Taylor-Greene	4.0	1.3

Given the general similarity in slur per tweet rates seen in Table 2, we investigated whether or not there was any significant difference in the slurs per tweet among the Harassing Samples across targets or in the control samples across targets. An ANOVA shows that there is no significant difference in the slur per tweet average among the targets in either the Harassing Sample (Table 3) or the control sample (Table 4).

These results suggest that, for these samples, it does not matter if the target of the initial harassment is liberal or conservative. The people who harass them use slurs in their tweets at the same rate.

Because there was no significant difference in the rates of slur use by target among the control or Harassing Samples, we pooled the data across all four targeted women for the remainder of the analysis in this section. Note that because our sample of harassers was much larger for Kamala Harris than for the other three targets due to how the data was collected, we sample down to use only 100 randomly selected harassers from the Harris data so as not to have the statistics dominated by her group.

Overall, the slur frequency in the Harassing Sample was over 3 times higher than in the control (as shown in Table 5).

Table 3. There is no statistically significant difference in the rate at which slurs are used in the Harassing Samples for the four Twitter target accounts.

Source	DF	Sum of Square	Mean Square	F Statistic	P-value
Groups (between groups)	3	0.006779	0.00226	1.2812	0.2794
Error (within groups)	1129	1.9912	0.001764		
Total	1132	1.998	0.001765		

Table 4. There is no statistically significant difference in the rate at which slurs are used by control populations who tweet about the four Twitter target accounts.

Source	DF	Sum of Square	Mean Square	F Statistic	P-value
Groups (between groups)	3	0.0005524	0.0001841	0.787	0.5018
Error (within groups)	374	0.08751	0.000234		
Total	377	0.08806	0.0002336		

On Parler, there were no posts harassing the conservative targets that we selected in our Twitter analysis. Thus, we only considered the question of whether people posting misogynistic harassment toward either of our liberal targets used more slurs than a control group who posted slur-free messages about these targets.

As included in Table 5, the Harassing Sample used significantly more slurs than the control. Like on Twitter, the rate of slur usage was over three times higher in the Harassing Sample than the control.

However, for both Parler and Twitter, a simple harasser sample vs. control sample may not paint a full picture. Accounts in the Harassing Sample all used at least one slur (which is how they were selected). On the other hand, accounts in the control may have never used any slurs. Indeed, on Parler, 5,020 of 9,154 control accounts (54.8%) used zero slurs. On Twitter, 128 / 379 (33.8%) of control accounts used zero slurs. Thus, the Harassing Sample may have a significantly higher harassment rate and it may have nothing to do with the misogynistic targeting.

To account for this, we compared the Harassing Sample to controls who (1) had used at least one slur somewhere else in their body of tweets and (2) who had used at least one *misogynistic* slur in their body of tweets. As shown in Table 5, Harassers still used significantly more slurs in their posts compared to the Controls. This was true on both Twitter and Parler.

Finally, we were concerned that using one of the five misogynistic slurs that defined our Harassing Sample may itself predict greater slur usage than the control, independent of who was targeted. To control for this, we selected a subset of each control group that had used at least 1 of the five misogynistic slurs. The difference is that harassers targeted these slurs at one of the women in power, while the control used them without targeting these women. There were 896 control accounts using misogynistic slurs on Parler and 45 on Twitter. Again, on both platforms, accounts in the Harassing Sample used significantly more slurs than control accounts who used at least one of the misogynistic slurs.

5   Limitations

There are a number of limitations to this work. First, we focused only on the United States, and only on four prominent women legislators to choose our samples. Three of these women are white. While we are focused on misogyny, we

Table 5. Slur use among the harassing and various subsets of control groups on Parler and Twitter. Users in Harassing Sample use significantly more slurs than every other group ($*p < 0.05, **p < 0.01, ***p < 0.001$).

Platform	Harassers	Control	Controls that use at least 1 slur	Controls who use misogynistic slurs
Twitter	3.92	1.16***	1.75***	2.79**
Parler	9.47	2.6***1	5.77***	6.60*

hypothesize that both misogyny and other harassment would likely be higher for women of color. Our results here suggest there is a lot more to understand about which specific one-time behaviors might predict broader harassing patterns, and future work should consider a larger and more diverse set of accounts. One study found that women of color were 34% more likely to receive harassment than white women, and black women in particular faced 84% more harassment than white women [8]. The intersection of harassment based on gender, race, religion, and sexual orientation are all especially interesting and complex areas that need more exploration.

There is also much more work to do on understanding the identity of harassers. While one might infer that accounts that harass liberal women are mostly conservative, and accounts that harass conservative women are mostly liberal, this is not necessarily the case. It could be that the harassers are primarily misogynists, who do not care about the party of their targets, or they may be trolls more broadly who will harass anyone with whatever language is likely to get a reaction. Understanding who is doing the harassment and why would add more depth to our understanding of the harassment phenomenon.

6   Discussion and Future Work

If slur use is a proxy for harassing behavior, these results suggest that users who harass women in power with misogynistic slurs tend to be more prolific harassers. There was no significant difference in slur use based on which person was targeted, but on both Twitter and Parler, accounts in the Harassing Sample used more slurs than the control group, even if that control group used at least one slur, and even if that was one misogynistic slur. This suggests that there is something predictive specifically because powerful women are the target of the misogynistic harassment.

As Bot Sentinel reported, on Twitter these posts are violations of the community standards, and yet many are not removed even when reported. Figure 1 shows three examples from the Bot Sentinel report of tweets that were reported and that Twitter declined to remove.

If posting this type of misogynistic harassment directed at women in power is indeed predictive of greater harassing behavior - over three times greater in our samples - it becomes a feature to identify accounts that should be more closely scrutinized for moderation. As discussed above, online harassment is not only a violation of most platforms' terms of use, but also creates real impacts

●𝒪𝒮 ♗ ♦𝒽𝒪𝒮 ☠
@evilgirlboss

Replying to @baldguyfag

i m going to kill kamala harris

6:01 PM · Jan 14, 2022 · Twitter Web App

zero
@BuddhaCyberpunk

Kamala, you are a faggot cunt.

9:41 PM · Mar 8, 2022 · Twitter for Android

FuckTankZero
@JohnGar71651102

Kamala is a nigger Bitch

9:10 PM · Jan 6, 2022 · Twitter for iPhone

Fig. 1. Example tweets from Bot Sentinel that Twitter said did not violate their safety policies.

for the targets. Harassment targets are more likely to self-censor [2], experience emotional distress [13], and fear for their safety [10]. It is also bad for the business of the platform, since it leads to reduced engagement [3] and even platform abandonment [6].

If further studies replicate the results presented here, these easy-to-detect targeted harassments could be used as features for algorithms designed to detect online harassment. These results also suggest platforms could consider less-lenient policies for accounts who engage in this type of harassment since it is likely to predict greater, broader harassment from those accounts.

References

1. Buckels, E.E., Trapnell, P.D., Paulhus, D.L.: Trolls just want to have fun. Pers. Individ. Differ. **67**, 97–102 (2014)
2. Citron, D.K.: Hate Crimes in Cyberspace. Harvard University Press, Cambridge (2014)
3. Fox, J., Tang, W.Y.: Women's experiences with harassment in online video games: rumination, organizational responsiveness, withdrawal, and coping strategies. In: 65th Annual Conference of the International Communication Association, San Juan, Puerto Rico (2015)
4. Gorrell, G., Bakir, M.E., Roberts, I., Greenwood, M.A., Bontcheva, K.: Which politicians receive abuse? four factors illuminated in the UK general election 2019. EPJ Data Sci. **9**(1), 18 (2020)
5. Herring, S.C., Stoerger, S.: Gender and (a) nonymity in computer-mediated communication. Handbook Lang. Gender Sexuality **2**, 567–586 (2014)
6. Hess, A.: Why women aren't welcome on the internet. Pacific Standard (2014)
7. Bot Sentinel Inc., Twitter's response to abuse and bigotry directed at vice president kamala harris (2020)
8. Amnesty International and Element AI. Troll patrol findings: Using crowdsourcing, data science & machine learning to measure violence and abuse against women on twitter
9. Lee, S.M., Lampe, C., Prescott, J.J., Schoenebeck, S.: Characteristics of people who engage in online harassing behavior. In: CHI Conference on Human Factors in Computing Systems Extended Abstracts, pp. 1–7 (2022)
10. Maple, C., Short, E., Brown, A.: Cyberstalking in the United Kingdom: an analysis of the echo pilot survey. Technical report, University of Bedfordshire (2011)
11. Mishra, P., Yannakoudakis, H., Shutova, E.: Tackling online abuse: a survey of automated abuse detection methods. arXiv preprint arXiv:1908.06024 (2019)
12. Munger, K.: Tweetment effects on the tweeted: experimentally reducing racist harassment. Polit. Behav. **39**(3), 629–649 (2017)
13. Van Laer, T.: The means to justify the end: combating cyber harassment in social media. J. Bus. Ethics **123**(1), 85–98 (2014)
14. Vitak, J., Chadha, K., Steiner, L., Ashktorab, Z.: Identifying women's experiences with and strategies for mitigating negative effects of online harassment. In: Proceedings of the 2017 ACM Conference on Computer Supported Cooperative Work and Social Computing, pp. 1231–1245 (2017)
15. Vogels, E.A.: The state of online harassment. Pew Research Center, 13 (2021)
16. Wick, S.E., et al.: Patterns of cyber harassment and perpetration among college students in the United States: a test of routine activities theory. Int. J. Cyber Criminol. **11**(1), 24–38 (2017)
17. Ybarra, M.L., Espelage, D.L., Mitchell, K.J.: The co-occurrence of internet harassment and unwanted sexual solicitation victimization and perpetration: Associations with psychosocial indicators. Journal of adolescent health **41**(6), S31–S41 (2007)
18. Ybarra, M.L., Mitchell, K.J.: Prevalence and frequency of internet harassment instigation: implications for adolescent health. J. Adolescent Health **41**(2), 189–195 (2007)

Stereotype Content Dictionary: A Semantic Space of 3 Million Words and Phrases Using Google News Word2Vec Embeddings

Xuanlong Qin$^{(\boxtimes)}$ ⓘ and Tony Tam ⓘ

The Chinese University of Hong Kong, Shatin, Hong Kong
`xuanlong@link.cuhk.edu.hk`, `tony.tam@cuhk.edu.hk`

Abstract. This paper (Our data and codes are available at https:// github.com/XuanlongQ/Stereotype_Content_Dictionary) suggests an efficient procedure to feasibly extend the specification of stereotype content of millions of text strings without using the traditional costly approach of questionnaire surveys. Using Google News Word2Vec embeddings, this paper develops a new semantic differential model of social perception to construct a Stereotype Content Dictionary for 3 million English words and phrases by reducing the semantic space of 300 dimensions to the two dimensions of the influential Stereotype Content Model (SCM). The new procedure is based on the associations among 100 antonymous pairs we developed based on selected seed words from the list of words to illustrate the two theoretical dimensions of stereotype content. This study utilizes Nicholas et al.'s gold-standard classification of Rosenberg et al.'s 64 personality traits to assess and compare the validity and performance of our model and Fraser et al.'s word embedding model. The results reveal that the trait classification by the Stereotype Content Dictionary correctly predicts 75% of the gold standard and significantly improves accuracy by over 40% compared with the random chance. Our model achieves 1.6 times higher in predicated performance than Fraser et at.'s model. Our model generates not only the first-ever Stereotype Content Dictionary for public use but also an efficient and feasible tool for stereotype research based on big data across societies and contexts.

Keywords: Measuring stereotype content · Stereotype content dictionary · Computational stereotype model

1 Introduction

1.1 Background

The term "stereotype" was first used in the modern psychological sense by Lippmann (1922). He viewed stereotypes as knowledge structures that function as mental "pictures" of the groups in question [17]. Not all stereotypes are alike [7].

R. Thomson et al. (Eds.): SBP-BRiMS 2023, LNCS 14161, pp. 12–22, 2023.
https://doi.org/10.1007/978-3-031-43129-6_2

Stereotype content refers to the attributes that people think characterize a group. Studies of stereotype content examine what people think about others, rather than the reasons and mechanisms involved in stereotyping [3]. However, measuring stereotype content is problematic. Scholars are seeking methods for subjecting stereotype content to systematic principles to comprehensively understand groups within societies.

Nevertheless, stereotypes are subject to change over time. Tessa et al. (2022) provide a nuanced picture of change and persistence in stereotypes across 200 years [2]. It implies that if researchers want to accurately discern the patterns of stereotypes associated with a specific group, they need to conduct repeated survey-based investigations and undertake meticulous analyses at different time periods. Computational methods have the potential to overcome the limitations of traditional methods. Word embeddings and natural language processing models can be instrumental in conducting stereotype analyses and comparisons, thereby facilitating a more comprehensive sampling of groups and a random selection of respondents. Notably, upon implementing computational models in practical contexts such as social media and e-libraries, we could gain more in-depth insights into group stereotypes from diverse perspectives.

2   Related Work

The current research basically divides previous studies into two distinct categories, traditional and computational approaches to measuring stereotype content.

2.1   Traditional Measurement of Stereotype Content

There are two main theories that measure stereotypes with traditional approaches, Stereotype Content Model and the Semantic Differential theory.

The Stereotype Content Model (SCM) [7] is widely employed in the field of social psychology, positing that the assessment of social groups is fundamentally influenced by appraisals across two dimensions: "warmth" and "competence". These two dimensions yield four types of stereotypes: high warmth-high competence (HW-HC), high warmth-low competence (HW-LC), low warmth-high competence (LW-HC), and low warmth-low competence (LW-LC)[1]. Combinations of warmth and competence generate distinct emotions of admiration, contempt, envy, and pity [3]. Numerous in-depth analyses of stereotypes of social groups revealed "warmth" and "competence" as universal dimensions. These dimensions are evident in stereotypes of older people [3], Asian Americans [18], immigrants [16], various groups [10,11], cultures [4] and so forth, making it one of the most well-known and influential theories for measuring stereotype content.

Semantic Differential (SD) theory [22] utilized the constructs of "evaluation", "potency", and "activity" to measure social perception, and has been suggested

---

[1] HW-HC and LW-LC are known as extreme stereotypes.
HW-LC and LW-HC are known as ambivalent stereotypes.

as a tool for cross-cultural comparisons. Meanwhile, Kervyn et al. (2013) demonstrated that a positive correlation exists between warmth and evaluation, as well as between competence and potency [13]. It supports that stereotypes are influenced by perceptions of warmth and competence. Specifically, these two semantic differential dimensions, "warmth-coldness" and "competence-incompetence", could represent the meaning of stereotype content.

Notwithstanding its widespread employment within the realm of social psychology, the SCM theory remains subject to various challenges. Firstly, due to its reliance on questionnaire-based data collection and analysis, SCM theory is inherently limited in capturing a comprehensive understanding of stereotyping across entire societies, given its reliance on a narrow range of information sources. Secondly, SCM theory evaluates group stereotypes by studying individuals rather than textual data. Thus, the potential for selection bias is heightened by the overwhelming majority of respondents being drawn from college-educated subpopulations. Moreover, Lee et al. (2006) conceded the limited scope of the current study's participants [16]. The acknowledged limitations of participants in studies of this nature further compound the accuracy issue. Thirdly, the high costs associated with conducting questionnaire research, present significant challenges to obtaining a complete picture of stereotyping over time.

2.2   Computational Measurement of Stereotype Content

Researchers also explored the probability of measuring stereotypes by using computational methods. The literature suggests that two major categories of computational analysis of stereotypical work are being utilized to study stereotypes. 1) Word-embedding-based approach, word embedding model has been widely used in measuring nonmaterial traits, like culture [14], ethnic stereotypes [15], etc. Durrheim et al. (2023) illustrate embeddings yield valid and reliable estimates of bias and that they can identify subtle biases that may not be communicated explicitly [6]. Fraser et al. (2021) [9] developed the first computational model that utilizes word embeddings to effectively measure stereotypes and anti-stereotypes through combining the POLAR[2] framework with the SCM theory. 2) Language model-based approach, Nadeem et al. (2020) tried to measure stereotypical bias in pre-trained language models, he contrasted both stereotypical bias and language modeling ability of popular models like BERT, GPT2, and XLNET [20]. Nadeem provided a dataset, StereoSet, that serves as a valuable resource for researchers analyzing stereotypes bias. An et al. (2022) [1] used 37K questions as seed input to uncover biased groups in finetuning the BERT model and conducting social commonsense reasoning. Fraser et al. (2022) [8] expanded their model from word-level to sentence-level with the RoBERTa-STS model to evaluate stereotype content. We consider that is crucial to analyze stereotypes at the sentence level as most text data exists in the form of sentences. However, the sentence-level model is based on the preliminary work done by Fraser et al.

---

[2] POLAR framework using opposites to enable interpretability of pre-trained word embeddings, more details please refer to Mathew's (2020) work [19].

(2021) [9] at the word level. The effectiveness of sentence-level embedding is largely influenced by word-level embedding.

Whilst acknowledging the ingenuity and merit of their approaches, it must be noted that certain constraints persist in Fraser et al.'s (2021) [9] word-level model. Firstly, Fraser imported all "warmth" and "competence" words of stereotype content dictionaries (Nicolas [21]) to the POLAR framework directly. Without constructing a suitable semantic dimension with antonym pairs, there is a risk of meaning shifts and aggregation biases, as indicated by semantic differential theory [22] and bias of word embedding [6]. Secondly, the validation of Fraser's paper is contingent upon established stereotypes derived from psychological surveys, such as those pertaining to nurses, males, Africans, etc. However, it should be emphasized stereotype content for a given group varies across different societies and contexts. For instance, Japanese are perceived as cold-competent in Fraser et al.'s (2021) paper, whereas Cuddy et al. (2009) [4] and Lee et al. (2006) [16] suggested they are warm-incompetent. Therefore, relying on such a method for validation is not a viable approach. Thirdly, Fraser used some seed words from StereoSet [20] which is irrelevant to stereotype content but stereotype bias.

Consequently, this study aims to achieve three main objectives. Firstly, it aims to construct a stereotype content dictionary that can address the limitations of traditional methods and serve as a valuable tool for researchers. Secondly, it tries to utilize a rigorous and scientific methodology for evaluating computational models of stereotypes. Finally, it seeks to enhance the performance of existing computational methods for measuring stereotype content.

3   Datasets and Methods

This section introduces a new procedure for constructing the stereotype content dictionary with SCM theory. We delineate the three primary steps to build our model and expound upon the methods of model validation.

- Determining antonym pairs based on existing stereotype dictionaries [21].
- Indetifying stereotype dimensions in Google News Word Embeddings.
- Projecting word embeddings on stereotype semantic space.

We utilize the widely used Google News Word2Vec embeddings as original word embedding. It features pre-trained vectors that were specifically trained on a subset of the Google News dataset, encompassing approximately 100 billion words. The embedding consists of 300-dimensional vectors for over 3 million words and phrases. Meanwhile, we selected seed words from Nicolas's (2021) stereotype dictionary [21], where almost 14,000 words were assigned to seven types of stereotypes. The detailed steps are provided in the following sections.

3.1   Determining Antonym Pairs from Stereotype Dictionaries

Osgood (1964) [22] pointed the meaning of a sign can be conceived as some point in an n-dimensional space, and the evaluation of a group stereotype can be conducted through the examination of various dimensions. Once one dimension is fixed, it would reduce the uncertainty about location X in the n-dimensional space. Kozlowski et al. (2019) [14] proposed that word embeddings have relationships with complex semantics. Past work with word embeddings shows that semantically meaningful relations can be found between words not directly proximate in space [12]. Durrheim et al. (2023) [6] pointed out that associations between word meanings could decrease bias evaluations in the word embeddings. Thus, We utilize seed words of stereotype content to create antonym pairs, which are employed to capture the meaning of stereotype content in word embeddings.

Table 1. Examples of seed words from stereotype content dictionary(Nicolas [21]) and antonym pairs we built, for each of the components comprising the warmth and competence dimensions.

Dictionary	Component	Sign	Seed words	Antonym pairs
Warmth	Sociability	pos	sociability; friendliness;	friendliness-unfriendliness
		neg	unsociability; unfriendliness;	sociability-unsociability
	Morality	pos	moral; sincere; egoistic	moral-immoral
		neg	immoral; insincere; altruistic	sincere-insincere...
Competence	Agency	pos	competent; competitive;	competent-incompetent
		neg	incompetent; uncompetitive;	competitive-uncompetitive
	Ability	pos	ability; fearlessness;	ability-inability
		neg	inability; fear; insecure;	fearlessness-fear

An approximation of the warm-competent dimension of stereotype content is captured not only by single "warm-cold" and "competent-incompetent" pair but also by other pairs of words whose semantic difference corresponds to that SCM dimension, such as "intelligent-unintelligent", "wise-unwise", "sociable-unsociable", etc. To create a comprehensive and representative set of antonyms, this paper selects all seed words from Nicolas's Stereotype Content Dictionaries [21]. Nicolas et al. (2021) provided a list of English seed words, captured from the psychological literature, associated with the warmth and competence dimensions; specifically, associated with sociability and morality (warmth), and ability and agency (competence). The examples of seed words are shown in Table 1.

3.2   Identifying Stereotype Dimensions in Google News Embeddings

To identify stereotype dimensions in word embedding models, we average numerous pairs of antonymous words. Stereotype dimensions are calculated by simply taking the mean of all warmth and competence pair differences that approximate a given dimension, $\vec{dir} = \sum_{n=p}^{|P|} \frac{\vec{p_1} - \vec{p_2}}{|P|}$, where p is all antonym pairs in relevant set P, $\vec{p_1}$ and $\vec{p_1}$ are the first and second word-embedding of each pair. In the context of SCM theory, $\vec{dir}$ represents the warmth and competence dimensions, the examples are shown in the antonym pairs of Table 1.

This method refers to Kozlowski et al.'s (2019) [14] work he constructed cultural dimensions with antonymous pairs and used cosine similarity as an indicator. We selected antonymous pairs whose cosine similarity is less than 0.5 as an index.

3.3 Projecting Word Embeddings on Stereotype Semantic Space

Note that $\overrightarrow{dir}$ represents the change of basis matrix for stereotype content embedding subspace. We projected all 300-dimensional Google News Word2vec embeddings,$\overrightarrow{W_v}$, onto this 2-dimensional stereotype semantic space.

Let a word v in the embedding subspace be denoted by $\overrightarrow{E_v}$. So for word v we have by the rules of linear transformation as Eqs. (1) and (2) shows.

$$dir^T\overrightarrow{E_v} = \overrightarrow{W_v} \tag{1}$$

$$\overrightarrow{E_v} = (dir^T)^{-1}\overrightarrow{W_v} \tag{2}$$

After transforming all words, we obtain a new word embedding space E, and each dimension in E can be interpreted in terms of the warmth-competent opposites(The inverse of the matrix direction is accomplished through Moore-Penrose generalized inverse). Then, we get a stereotype content dictionary.

3.4 Model Validation

As our critical review in Sect. 2.2 suggests, there is currently no established validation procedure for any computational model of stereotype content. So we propose an intuitive approach to validity based on Nicholas et al.'s (2013) [13] gold-standard classification of 64 well-studied personality traits.

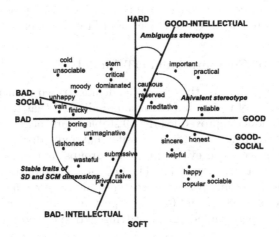

Fig. 1. The thumbnail of superimposition of the two figures from Rosenberg et al. [23], showing the relation between SD and SCM dimensions (SD dimensions are Evaluation (good-bad) and Potency (hard-soft). SCM dimensions are Warmth (intellectual good-bad) and Competence (social good-bad).).

Rosenberg et al. (1968) [23] used an empirical approach to investigate how the wealth of personality traits in language can array in social perception. Using a card-sorting task in which participants grouped traits often encountered in the same person, they derived distance ratings among the 64 most frequently used personality traits. Then, he analyzed the matrix of distances between these 64 traits using multidimensional scaling in order to find a two-dimensional solution that fits the data. Rosenberg et al.'s (1968) work can provide guidance for our objective to validate our model, especially SCM and SD theory had employed the classification of these 64 personality traits to test their theory [13, 23], thus we posit that it should be employed as the benchmark

Moreover, we found certain personality traits are difficult to accurately place in the SD and SCM dimensions, particularly those traits that appeared near the boundary lines. For instance, in Fig. 1, "unhappy", "vain", and "finicky" were defined as "LW-HC" in the SD dimension, but "LC-LW" in the SCM dimension. These ambiguous traits were then excluded from the experiment for testing the robustness of the models.

4   Result

Out of the 64 traits initially considered for our test, "high-strung" and "good-natured" were not an index in Google News embedding, "inventive", "liar" and "egotistical" were not appear in Fig. 1, and were consequently excluded from our analysis. As a result, the present study employed the remaining 59 personality traits as the first group to assess the accuracy of stereotypical models. Additionally, 39 traits that manifested consistency in both the SD and SCM dimensions were chosen as the second group to examine the robustness of stereotypical models. This paper establishes the baseline for trait distribution within each quadrant by random chance, whereby each personality trait has a 25% probability of being correctly predicted to compare QT model[3] and Fraser model[4].

4.1   Validation of Computational QT Model

Within the group of 59 traits, the QT model demonstrated proficient capture of the majority (66.1%) of traits, and the results indicated that the overall performance of the QT model improved for the 39 stable traits, achieving an accuracy of 69.2%. Social compensation theory posits that individuals are more likely to exhibit ambivalence when evaluating other groups, rendering it challenging to identify ambivalent stereotypes [5]. The QT model demonstrated stable performance in identifying ambivalent stereotypes across both 59 and 39 traits, achieving accuracies of 63.6% and 62.5%, respectively. Moreover, the QT model exhibited improved performance in identifying extreme stereotypes, achieving

---

[3] We use the name QT model to distinguish our new model and Fraser's model.

[4] We used personality traits as indicators to examine the performance of Fraser's (2021) model [9].

Table 2. Per quadrant labeling over two dimensions, competence and warmth. Ambivalent means HW-LC and LW-HC. Extreme means HW-HC and LW-LC.

Model Comparisons	59 traits		39 traits	
	Fraser model	QT model	Fraser model	QT model
High Warmth-High Competence	10/21	6/8	9/19	5/6
High Warmth-Low Competence	1/10	10/16	0/3	9/13
Low Warmth-High Competence	1/10	4/6	0/4	1/3
Low Warmth-Low Competence	12/18	19/29	9/13	12/17
Accuracy (Overall)	0.406	0.661	0.462	0.692
Accuracy (Ambivalent)	0.200	0.636	0.000	0.625
Accuracy (Extreme)	0.564	0.675	0.563	0.739
Predictive Performance (Overall)[a]	0.156	0.411	0.212	0.442
Ratio Index (Overall)[b]	-	1.6	-	1.1

[a] $Predictive\_Performance(PP) = Accuracy_{overall} - Accuracy_{baseline}$
[b] $Ratio\_Index = (PP_{QTmodel}/PP_{Frasermodel}) - 1$

accuracies of 67.5% and 73.9%, respectively. The QT model's predicative performance showed a significant improvement over the baseline model, with an improvement of 41.1% and 44.2% for 59-trait and 39-trait situations, respectively. Overall, the QT model performed excellently across personality traits (Table 2).

4.2   Comparison Between QT Model and Fraser Model

We applied 59 personality traits to both the QT model and the Fraser model. The results indicated that Fraser's model achieved a mere overall accuracy of 40.6%, whereas the QT model demonstrated a significant improvement of 62.8% in overall accuracy compared to the Fraser model. Both computational models exhibited an improvement in accuracy when compared to the baseline model. In terms of predicative performance, both models proved their efficacy in capturing stereotype content, with the Fraser model and QT model exhibiting improvements of 15.6% and 41.1%, respectively, over the baseline model. In terms of the ratio index, the performance of the QT model was found to be 1.6 times higher than that of the Fraser model.

Fraser's model (2021) has yielded unsatisfactory outcomes, with a mere 20% accuracy rate in identifying ambivalent stereotypes. Nonetheless, the QT model has demonstrated a 63.6% precision rate in the same quadrants, which is 3 times higher than Fraser's model. Additionally, when dealing with extreme stereotypes, the QT model and Fraser's model (2021) exhibited accuracies of 67.5% and 56.4%, respectively. Both of them demonstrate a significant improvement in the accuracy of extreme stereotypes, which may be attributed to the extreme phrasing of these personality traits, rendering them more identifiable. Therefore, the QT model exhibits superior performance to Fraser's model in identifying ambivalent and extreme personality traits.

Regarding the situation of 39 personality traits, the results revealed that both the QT model and Fraser model exhibited enhanced overall accuracies, achieving 69.2% and 46.2%, respectively. In terms of the predicative performance, the Fraser model and QT model demonstrated improvements of 21.2% and 44.2%, respectively, over the baseline model. The results indicate that the performance of both models improved across multiple indicators after removing ambiguous traits. Even in this case, in terms of the ratio index, the performance of the QT model was found to be 1.1 times higher than that of the Fraser model.

Furthermore, our findings revealed that the accuracy of ambivalent stereotypes has a significant decrease in Fraser's model, while only slightly affecting the QT model (resulting in a decrease from 63.6% to 62.5%). Additionally, when dealing with extreme stereotypes, the QT model and Fraser model exhibited accuracies of 73.9% and 56.3%, respectively. Although Fraser's model has improved in overall accuracy, it exhibits poor performance in identifying ambivalent traits. This observation suggests that ambiguous traits have a notable impact on measuring ambivalent stereotypes, while extreme stereotypes are more easily identifiable. This result implies that the QT model is more robust and exhibits superior performance in identifying such personality traits.

To sum up, this paper employs a gold-standard classification of 64 personality traits to evaluate and validate the effectiveness of computational models in measuring stereotype content, and the QT model exhibits superior performance as a result of meticulous and comprehensive evaluation.

5   Discussion and Future Work

Our research demonstrates that measuring the stereotype content of millions of text strings through word embedding is not only effective but also much more feasible and efficient than the traditional questionnaire survey approach. It is possible that a lower cosine similarity than the conventional threshold of 0.5 or a greater number of representative antonyms than we used may yield more accurate results. Despite this limitation, our method provides a useful new approach to measuring stereotype content. Employing word embedding is a valuable tool for exploring stereotype content, particularly in certain contexts such as social media and news where our model can deconstruct textual data. Our upcoming work will utilize Stereotype Content Dictionary to address real-world problems, compare different social perception models, and expand it from the word to the sentence level, which will be more significant and relevant.

References

1. An, H., Li, Z., Zhao, J.: Sodapop: open-ended discovery of social biases in social commonsense reasoning models. arXiv preprint arXiv:2210.07269 (2022)

2. Charlesworth, T.E., Caliskan, A., Banaji, M.R.: Historical representations of social groups across 200 years of word embeddings from google books. Proc. Natl. Acad. Sci. **119**(28), e2121798119 (2022)
3. Cuddy, A.J., Fiske, S.T., Glick, P.: Warmth and competence as universal dimensions of social perception: the stereotype content model and the bias map. Adv. Exp. Soc. Psychol. **40**, 61–149 (2008)
4. Cuddy, A.J., et al.: Stereotype content model across cultures: towards universal similarities and some differences. Br. J. Soc. Psychol. **48**(1), 1–33 (2009)
5. Durante, F., Tablante, C.B., Fiske, S.T.: Poor but warm, rich but cold (and competent): social classes in the stereotype content model. J. Soc. Issues **73**(1), 138–157 (2017)
6. Durrheim, K., Schuld, M.: Mafunda: using word embeddings to investigate cultural biases. Br. J. Soc. Psychol. **62**(1), 617–629 (2023)
7. Fiske, S.T., Cuddy, A.J., Glick, P., Xu, J.: A model of (often mixed) stereotype content: competence and warmth respectively follow from perceived status and competition. J. Pers. Soc. Psychol. **82**(6), 878 (2002)
8. Fraser, K.C., Kiritchenko, S., Nejadgholi, I.: Computational modeling of stereotype content in text. Frontiers in Artificial Intelligence 5 (2022)
9. Fraser, K.C., Nejadgholi, I., Kiritchenko, S.: Understanding and countering stereotypes: A computational approach to the stereotype content model. arXiv preprint arXiv:2106.02596 (2021)
10. Grigoryev, D., Fiske, S.T., Batkhina, A.: Mapping ethnic stereotypes and their antecedents in Russia: the stereotype content model. Front. Psychol. **10**, 1643 (2019)
11. Guan, Y., Deng, H., Bond, M.H.: Examining stereotype content model in a Chinese context: Inter-group structural relations and mainland Chinese's stereotypes towards hong kong chinese. Int. J. Intercult. Relat. **34**(4), 393–399 (2010)
12. Ji, S., Satish, N., Li, S., Dubey, P.K.: Parallelizing word2vec in shared and distributed memory. IEEE Trans. Parallel Distrib. Syst. **30**(9), 2090–2100 (2019)
13. Kervyn, N., Fiske, S.T., Yzerbyt, V.Y.: Integrating the stereotype content model (warmth and competence) and the osgood semantic differential (evaluation, potency, and activity). Eur. J. Soc. Psychol. **43**(7), 673–681 (2013)
14. Kozlowski, A.C., Taddy, M., Evans, J.A.: The geometry of culture: analyzing the meanings of class through word embeddings. Am. Sociol. Rev. **84**(5), 905–949 (2019)
15. Kroon, A.C., Trilling, D., Raats, T.: Guilty by association: Using word embeddings to measure ethnic stereotypes in news coverage. Journalism Mass Commun. Quart. **98**(2), 451–477 (2021)
16. Lee, T.L., Fiske, S.T.: Not an outgroup, not yet an ingroup: immigrants in the stereotype content model. Int. J. Intercult. Relat. **30**(6), 751–768 (2006)
17. Lippmann, W.: Public opinion harcourt, brace and company, New York (1922)
18. Maddux, W.W., Galinsky, A.D., Cuddy, A.J., Polifroni, M.: When being a model minority is good... and bad: Realistic threat explains negativity toward asian Americans. Personality Soc. Psychol. Bull. **34**(1), 74–89 (2008)
19. Mathew, B., Sikdar, S., Lemmerich, F., Strohmaier, M.: The polar framework: Polar opposites enable interpretability of pre-trained word embeddings. In: Proceedings of the Web Conference 2020, pp. 1548–1558 (2020)
20. Nadeem, M., Bethke, A., Reddy, S.: Stereoset: measuring stereotypical bias in pretrained language models. arXiv preprint arXiv:2004.09456 (2020)
21. Nicolas, G., Bai, X., Fiske, S.T.: Comprehensive stereotype content dictionaries using a semi-automated method. Eur. J. Soc. Psychol. **51**(1), 178–196 (2021)

22. Osgood, C.E.: Semantic differential technique in the comparative study of cultures. Am. Anthropol. **66**(3), 171–200 (1964)
23. Rosenberg, S., Nelson, C., Vivekananthan, P.: A multidimensional approach to the structure of personality impressions. J. Pers. Soc. Psychol. **9**(4), 283 (1968)

Social Cybersecurity Analysis
of the Telegram Information Environment
During the 2022 Invasion of Ukraine

Ian Kloo$^{(\boxtimes)}$ and Kathleen M. Carley

Carnegie Mellon University, Pittsburgh, PA 15213, USA
iankloo@cmu.edu

Abstract. The 2022 Russian invasion of Ukraine is a war being fought both on the physical battlefield and online. This paper studies Telegram activity in the first weeks of the invasion, applying social cybersecurity methods to characterize the information environment on a platform that is popular in both Ukraine and Russia. In a study of over 4 million Telegram messages, we find a contentious environment where channel discussions often contain content with the opposite stance of their associated main channels. We apply the BEND framework to characterize potential disinformation maneuvers on the platform, finding that the English-language community is the most contested in the information space. In addition to the specific analysis of the Russian invasion, we demonstrate the utility of Telegram as a useful platform in social cybersecurity research.

Keywords: Russian Disinformation · Social Cybersecurity · Telegram · BEND Maneuvers

1 Introduction

Modern military action involves both kinetic activities as well as information warfare. As a result, it is critical to understand the platforms that comprise the information space. The field of social cybersecurity aims to characterize and defend against malicious manipulative activity online [5]. Social cybersecurity researchers have built a strong understanding of Twitter, Reddit, Facebook, and other platforms, but there has been much less focus on Telegram.

Previous social cybersecurity work on Telegram demonstrated that disinformation is regularly shared on the platform. Ng and Loke found misinformation related to COVID-19 on Telegram but also noted that misinformation was often contested by users [10]. Walther and McCoy similarly found COVID-19 conspiracy theories among a broader range of typically far-right disinformation topics on Telegram [13]. Finally, Willaert et al. found QAnon and associated conspiracy theories on Dutch-speaking Telegram channels [14].

While these studies reached important conclusions and demonstrated the prevalence of disinformation on Telegram, they all focused on relatively small

R. Thomson et al. (Eds.): SBP-BRiMS 2023, LNCS 14161, pp. 23–32, 2023.
https://doi.org/10.1007/978-3-031-43129-6_3

data sets. Ng and Loke concentrated on a single channel, Walther and McCoy examined 125 channels, and Willaert et al. used 215 channels. This study will build upon this research by applying social cybersecurity analytic methods to a much larger data set (1,323 channels). Additionally, we will enhance the existing research on Telegram by studying the information environment of an active military conflict: Russia's invasion of Ukraine.

Several recent papers analyzed Telegram during the Russian invasion. Karpchuk et al. used analysis of a single Telegram channel to detect structural changes in Ukrainian strategic communication during the invasion [6]. Solopova et al. used a larger data set (approximately 69,000 Telegram messages) in a study that aimed to label pro-Kremlin propaganda in both news and Telegram data [11]. We seek to build upon this existing research by studying the information environment as a whole and by using a much larger data set (over 4 million messages).

The remainder of this paper will describe a large Telegram data set collected during the Russian invasion, provide an overview of the information space using network analysis, present a (pro- and anti-Russia) stance analysis on the data, and finally apply the BEND framework to identify potential disinformation maneuvers in the data.

2   Data

2.1   Telegram Overview

Telegram is a social media application developed by Russian entrepreneur Pavel Durov. It is popular globally but has taken on an especially significant role in the Russian invasion of Ukraine, where it has been used by military-focused bloggers to disseminate information during the war.

While Telegram has private chat capabilities, this analysis will focus on public-facing channels and groups. A channel is a microblog where a single user (or a small group of administrators) posts content. Some channels allow discussion in associated groups where users can post their own content.

2.2   Collection

The Telegram data was collected from public-facing channels and discussion groups using the official Telegram API. Telegram does not provide any way to search for content across public channels/groups, so we first built a list of relevant Telegram channels. We used Twitter's keyword search functionality to find tweets commenting on the Russian invasion that contained links to Telegram channels during the first month of the war (February - March 2022), which yielded 118 unique channels. Starting with these core channels, we performed a 1-hop snowball sample using forwards (i.e., adding a channel to the data set if a known channel forwarded content from that channel), resulting in 1,323 total channels from which we retrieved all message data.

2.3 Data Overview

The data set contains 4,026,648 messages from 478,971 users posted in 1,323 channels. All messages were posted between February 23, 2022, and March 15, 2022, covering the early stages of the Russian invasion of Ukraine. 55% of the channels are primarily in English, 26% are primarily in Russian, 13% are primarily in Ukrainian, and 6% are in other languages.

17% of messages were posted in channels and the remaining 83% were posted in discussion groups. One discussion was especially large, accounting for 16% of all messages in the data set. This discussion is associated with Vladimir Rudolfovich Solovyov who is a Russian TV commenter who has been vocally supportive of the invasion of Ukraine [12].

Excluding the Solovyov group, the average discussion contains 5,218 messages and the average channel contains 638 messages. The average user posted 6.24 messages, but the vast majority (88%) of users posted fewer than 10 messages. The top 1,000 users accounted for 17% of the total messages, all of which were in discussion groups.

3 Methodology

3.1 Stance Detection

Detecting stance in social media data is a well-researched problem with effective solutions in many circumstances. Classification models using supervised machine learning leveraging both linguistic and network features have been shown to be effective, but these methodologies require large training data sets that must be manually tagged and validated [1].

Weak supervision imposes a much smaller data-tagging burden and has also proven to be effective for stance detection [7]. This method takes a tagged set of concepts (e.g., hashtags) as input and uses network propagation to classify stances of documents and agents. While this is an effective strategy on Twitter, Telegram does not use hashtags (or other similar concepts) across channels which limits the potential of the network propagation procedure.

Several recent papers demonstrate the potential for using large language models (LLMs) to perform zero-shot stance labeling [8, 15]. In this approach, the researcher prompts an LLM to generate stance labels without additional model training. Zhang et al. demonstrate that the performance of zero-shot stance labeling using GPT-3 outperforms the existing machine learning approaches in a popular evaluation data set (SemEval Stance [9]) with F1 scores ranging from 78% to 59.3%, depending on the topic [15]. We validated these findings by performing our own zero-shot labeling on the SemEval data set and achieved similar F1 scores.

We used OpenAI's GPT-3.5-turbo model to assign stance labels at the channel level. Ideally, we would assign stances to every message, but this is cost and time prohibitive with over 4 million messages in the data set. To address this issue, we generated stance tags for 100 randomly sampled documents for each channel and determined the overall stance of the channel using the following formula:

$$stanceScore = \frac{proRussianDocs - antiRussianDocs}{proRussianDocs + antiRussianDocs} \qquad (1)$$

This results in stance scores close to -1 for channels with primarily anti-Russian content, close to 1 for channels with pro-Russian content, and 0 for channels with balanced content.

Note that we are using pro-Russian and anti-Russian as our stances. Because of the time period used in this analysis, we instructed the LLM to label pro-Ukrainian messages as anti-Russian. Likewise, anti-Ukrainian messages were tagged as pro-Russian. While stance is often a much more complicated issue, the fact that Russia and Ukraine were actively at war in the time period covered by this data set justifies this simplification.

3.2 BEND Analysis

The BEND framework is a methodology for labeling disinformation maneuvers on social media using 16 categories [3]. BEND has been validated in studies of COVID-19 disinformation [4] and Russian disinformation campaigns on Twitter [2]. While the framework was created using Twitter data, there is nothing Twitter-specific about BEND, and this paper demonstrates how it can be applied to other social media platforms.

The BEND framework can be used to study agents (typically users) and documents. This study will focus on document-level analysis because the Telegram API provides minimal information about users and does not provide a method to access users' posting history.

The document BEND analysis we employ labels documents according to 16 BEND categories: back, build, bridge, boost, engage, explain, excite, enhance, neutralize, nuke, narrow, neglect, dismiss, distort, dismay, and distract. We will use Netatonmic's ORA Pro software in combination with the NLP tools in Netanomic's Netmapper software to label BEND documents automatically.

4   Results and Discussion

4.1 Network Overview

Network analysis is a useful way to understand the complexities of social media data, and ORA Pro provides tools to parse and analyze Telegram data. The networks shown in Fig. 1 show the channel-to-channel connections where an edge between two channels signifies that content was forwarded from one channel to another. Network (a) colors the channel nodes by language, while network (b) colors them by stance towards Russia.

We can see three distinct communities in both networks. The right-most community is almost all English-speaking channels, with a few Russian channels. The left-most community is almost exclusively Russian-speaking. Finally, the top community contains mostly Ukranian-language channels, with some Russian channels also included. This intermingling makes sense as many Ukrainians speak Russian.

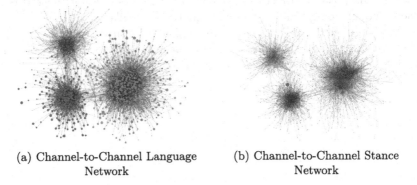

(a) Channel-to-Channel Language Network

(b) Channel-to-Channel Stance Network

Fig. 1. Both networks show connections between Telegram channels during the Russian invasion of Ukraine. Nodes correspond to channels, and links signify a channel forwarded content from another. In (a), color denotes language such that red is Russian, blue is Ukrainian, green is English, and gray is any other language. Node size correlates to a pro-Russian stance, where larger nodes are more pro-Russian. In (b), color denotes stance such that red nodes are pro-Russia, blue are anti-Russia, and gray are neutral. Nodes are sized by total degree centrality. (Color figure online)

The Russian-speaking community is almost exclusively pro-Russian. Similarly, the Ukrainian-speaking community is anti-Russian. Notably, the Russian-speaking channels included in the primarily Ukrainian-language community appear to have anti-Russian stances, suggesting these are Ukrainian communities that use the Russian language. English-speaking channels have a mix of pro-Russian and anti-Russian stances. Many channels in this group also have neutral status, which suggests they either host a balanced discussion or simply do not include posts about the invasion.

Overall, there are distinct language and stance communities in the Telegram data, but there are many connections between them. There is clearly an opportunity for information (and disinformation) to flow between these communities. Additionally, we see that the English-language channels contain diverse stances towards Russia and the pro- and anti-Russian English-speaking channels are not segregated into sub-communities. This suggests the English-speaking channels are the most contested information space in this data set.

4.2 Stance Analysis

After identifying the three main language communities, we performed a statistical analysis to further examine the stances within those groups. Figure 2(a) shows the distributions of stance for each of the three language communities.

An analysis of variance (ANOVA) test shows that the language groups had statistically significant stances (F = 284.6, p = 0.00), and a Tukey test confirms that all three group means were statistically significantly different (p = 0.00).

The Ukrainian channels had strongly anti-Russian views, with only a few channels in the positive stance space. The opposite was true for most Russian channels, though some Russian-language channels are likely run by Ukrainians due to the language diversity in the country. The English channels show a wide distribution that spans the entire stance space. The average stance in these channels is leaning slightly towards the pro-Russian stance (0.094).

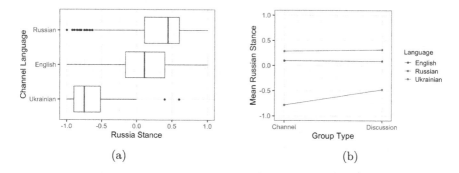

(a)   (b)

Fig. 2. (a) is a boxplot showing the stance distributions of channels by language. (b) is an interaction plot showing Ukrainian discussions were significantly less anti-Russian than channels.

The ANOVA test also showed a significant interaction effect between language and group type (channels vs. discussions). Figure 2(b) visualizes the interaction effect. The English and Russian channel and discussion stances were similar, but Ukrainian discussions were statistically significantly more pro-Russian than Ukrainian channels. While this effect is significant, it is important to note that the Ukrainian discussions were still firmly anti-Russian.

To further explore the differences between channel and group stances, Fig. 3 shows channels as blue points and their associated discussions on the same line in red. Here we can see that although the channels and discussions have similar stances on average, it is common for channels and their associated discussion groups to differ widely in stance. Notably, the most firmly anti- and pro-Russian channels have associated discussions with opposite stances. This supports the idea that post-invasion Telegram is a contentious information environment characterized by antagonistic activity between those with different stances toward Russia.

4.3   BEND Analysis of Maneuvers

BEND provides a framework for understanding and categorizing information maneuvers in social media. Figure 4 shows the total BEND maneuvers by stance and language.

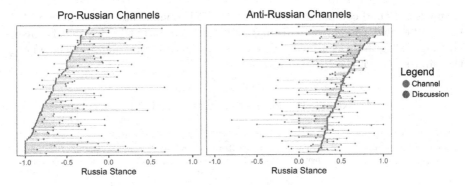

Fig. 3. A dot plot showing the difference between pro- anti-Russian stance in channels (blue) and their associated discussion groups (red). (Color figure online)

Partitioning by stance towards Russia, the pro-Russian channels contained the most documents that ORA labeled to be a part of a BEND maneuver. Distract and explain maneuvers were the most common, while there were few back, neutralize, dismiss, neglect, and enhance maneuvers. Anti-Russian channels saw less maneuvering in every BEND category except for dismiss (and this was a rare maneuver in general). Additionally, the neutral channels saw the highest number of maneuver documents in most categories. While there are some differences in the amount of maneuvering by stance, it is clear that many of the same maneuvers were present in all stance communities.

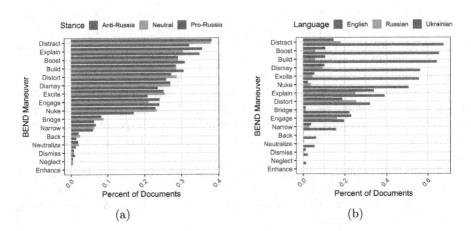

Fig. 4. Both plots show the frequency of BEND maneuvers in the Telegram data. (a) is split by channel stance, and (b) is split by channel language.

In contrast to the stance grouping, major differences exist in the amount of BEND maneuver documents in the different language groups. English language channels had many more BEND documents, while Russian and Ukrainian channels only had comparable numbers of BEND documents in the explain, distort, and engage categories. This finding further supports the idea that English-language Telegram was a much more contested information environment during the early stages of the Russian invasion.

Table 1. Example Documents Demonstrating BEND Maneuvers

BEND Category	Stance	Post	Context
Distract	Pro-Russian	i think they hate the Russians because they double crossed the globalist LGBTQI anti family agenda years ago, and decided to promote Christianity and having families	This is a reply to a comment that claims Russia will easily defeat Ukraine in combat with a mocking tone towards anti-Russian actors.
Boost	Pro-Russian	GREAT WORK BROTHER WWG1WGA here for you	This is a reply to a comment discussing a conspiracy theory that the Australian government is using the Russian invasion to consolidate power.
Dismay	Anti-Russian	THAT includes 40,000 civilians murdered by Russian forces??? DID those GODLY GOOD FOLK deserve to die? NO!! GOD warned Putin n this invasion n deaths GOD will avenge. ENOUGH murdering innocent folk!	This is a reply to a comment justifying the Russian invasion based on alleged corruption in the Ukrainian government.
Dismay/Build	Pro-Russian	World War 3 & Killer Vaccines: Welcome To The New World Order Join us @OPFREEDOM	This was original content in a typically pro-Russian channel

We present several examples of specific BEND maneuvers in Table 1. Distract is the most common pro-Russian maneuver in the data, and the first example in the table shows a common application. This message is a response to a comment about Russia's likely victory over Ukraine, but the response instead pivots away from the military to seemingly unrelated cultural issues.

The second example shows a boost maneuver that contains WWG1WGA, which stands for "where we go one, we go all". This is a common phrase/acronym in the QAnon community, and it is used here to give a sense of community in support of a pro-Russian conspiracy theory centering on the Australian government.

The third example is a dismay maneuver that references the high number of civilian casualties as a negative response to a message supporting the invasion. Many dismay maneuvers in the data set also contain pictures and videos showing bloody scenes from the war.

Finally, the last example shows a dismay and build maneuver combination. The message was posted to a firmly pro-Russian channel and sought to evoke strong negative emotions. This example also includes a call to join the channel (thus, building a community), showing how BEND maneuvers can be (and often are) used in combination.

5   Limitations and Future Work

First, we were unable to conduct a user-focused analysis given the limited information that Telegram makes available via the official API. It is also possible that our initial set of channels that we generated from Twitter is biased toward the types of things Twitter users find interesting. We expect, however, that our snowball sampling technique was sufficient to capture the key channels discussing the Russian invasion.

Next, the stance detection employed in this paper is only intended to work at the channel level due to the prohibitive cost and time required to label each document. Future studies should explore ways to scale this methodology to label a greater number of documents.

Finally, it is possible that the difference between the number of BEND maneuvers in English and other languages is exaggerated by the natural language processing (NLP) techniques used by Netmapper in our BEND detection pipeline. In particular, Netmapper has more robust idiom and slang detection techniques for English compared to other languages. It is possible that some of the differences in the number of BEND documents between the languages can be explained by the use of idioms and slang.

6   Conclusion

The analysis presented in this paper shows that the Telegram information environment during the early stages of the Russian invasion of Ukraine contained many pro- and anti-Russian sentiments in a generally contentious information environment. We found distinct language communities among Telegram channels, but there is substantial content forwarding between these communities. Additionally, we identified the English-language community as the most contentious in terms of stance towards Russia, but found that channel discussions often contain stances that run counter to the main channel stance. Applying the BEND framework in aggregate highlights the prevalence of disinformation maneuvers in English-speaking channels. Finally, exploring examples of popular BEND maneuvers in the data provides insight into the specific tactics employed in the data.

Beyond the analysis presented in the paper, this work motivates future work into less common social media platforms in the social cybersecurity field. Although Telegram does not support easy access to data, we show that it is feasible to collect and analyze millions of Telegram messages, extracting useful information about the information space. Future work should examine how to modify the existing social cyber security toolkit to work with Telegram data and other uncommon or regionally-specific social media platforms.

Acknowledgements. This material is based upon work supported by the U.S. Army Research Office and the U.S. Army Futures Command under Contract No. W911NF-20-D-0002. The content of the information does not necessarily reflect the position or the policy of the government and no official endorsement should be inferred.

References

1. ALDayel, A., Magdy, W.: Stance detection on social media: state of the art and trends. Inf. Process. Manage. **58**(4), 102597 (2021). https://doi.org/10.1016/j.ipm.2021.102597
2. Alieva, I., Moffitt, J., Carley, K.M.: How disinformation operations against Russian opposition leader alexei navalny influence the international audience on Twitter. Soc. Netw. Anal. Min. **12**(1), 80 (2022). https://doi.org/10.1007/s13278-022-00908-6
3. Blane, J.: Social-Cyber Maneuvers for Analyzing Online Influence Operations. Ph.D. thesis, Ph.D. thesis, Carnegie Mellon University (2023)
4. Blane, J.T., Bellutta, D., Carley, K.M.: Social-cyber maneuvers during the covid-19 vaccine initial rollout: content analysis of tweets. J. Med. Internet Res. **24**(3), e34040 (2022). https://doi.org/10.2196/34040
5. Carley, K.M.: Social cybersecurity: an emerging science. Comput. Math. Organ. Theory **26**(4), 365–381 (2020). https://doi.org/10.1007/s10588-020-09322-9
6. Karpchuk, N., Yuskiv, B., Pelekh, O.: The structure of strategic communications during the war: the case-study of the telegram channel insider ukraine. Politologija **107**(3), 90–119 (2022). https://doi.org/10.15388/Polit.2022.107.3
7. Kumar, S.: Social media analytics for stance mining a multi-modal approach with weak supervision. Ph.D. thesis, Carnegie Mellon University (2020)
8. Mets, M., Karjus, A., Ibrus, I., Schich, M.: Automated stance detection in complex topics and small languages: the challenging case of immigration in polarizing news media (2023). https://doi.org/10.48550/arXiv.2305.13047
9. Mohammad, S., Kiritchenko, S., Sobhani, P., Zhu, X., Cherry, C.: Semeval-2016 task 6: detecting stance in tweets. In: Proceedings of the 10th International Workshop on Semantic Evaluation (SemEval-2016), pp. 31–41 (2016). https://doi.org/10.18653/v1/S16-1003
10. Ng, L.H.X., Loke, J.Y.: Analyzing public opinion and misinformation in a covid-19 telegram group chat. IEEE Internet Comput. **25**(2), 84–91 (2020). https://doi.org/10.1109/MIC.2020.3040516
11. Solopova, V., Popescu, O.I., Benzmüller, C., Landgraf, T.: Automated multilingual detection of pro-kremlin propaganda in newspapers and telegram posts. Datenbank-Spektrum **23**(1), 5–14 (2023). https://doi.org/10.1007/s13222-023-00437-2
12. Vorobyov, N.: How are Russian media outlets portraying the Ukraine crisis? Al Jazeera, January 2022. https://www.aljazeera.com/news/2022/1/31/how-are-russian-media-outlets-portraying-the-ukraine-crisis
13. Walther, S., McCoy, A.: Us extremism on telegram: Fueling disinformation, conspiracy theories, and accelerationism. Perspectives Terrorism **15**(2), 100–124 (2021). https://www.jstor.org/stable/27007298
14. Willaert, T., Peeters, S., Seijbel, J., Van Raemdonck, N.: Disinformation networks: A quali-quantitative investigation of antagonistic dutch-speaking telegram channels. First Monday **27**(5), September 2022. https://doi.org/10.5210/fm.v27i5.12533
15. Zhang, B., Ding, D., Jing, L.: How would stance detection techniques evolve after the launch of chatgpt? (2023). https://doi.org/10.48550/arXiv.2212.14548

The Dynamics of Political Narratives During the Russian Invasion of Ukraine

Ahana Biswas[1]([envelope])[iD], Tim Niven[2][iD], and Yu-Ru Lin[1][iD]

[1] University of Pittsburgh, Pittsburgh, PA 15260, USA
{ahana.biswas,yurulin}@pitt.edu
[2] Doublethink Lab, Taipei, Taiwan
tim@doublethinklab.org

Abstract. The Russian invasion of Ukraine has elicited a diverse array of responses from nations around the globe. During a global conflict, polarized narratives are spread on social media to sway public opinion. We examine the dynamics of the political narratives surrounding the Russia-Ukraine war during the first two months of the Russian invasion of Ukraine (RU) using the Chinese Twitter space as a case study. Since the beginning of the RU, pro-Chinese-state and anti-Chinese-state users have spread divisive opinions, rumors, and conspiracy theories. We investigate how the pro- and anti-state camps contributed to the evolution of RU-related narratives, as well as how a few influential accounts drove the narrative evolution. We identify pro-state and anti-state actors on Twitter using network analysis and text-based classifiers, and we leverage text analysis, along with the users' social interactions (e.g., retweeting), to extract narrative coordination and evolution. We find evidence that both pro-state and anti-state camps spread propaganda narratives about RU. Our analysis illuminates how actors coordinate to advance particular viewpoints or act against one another in the context of global conflict.

Keywords: Computational propaganda · Russia-Ukraine conflict · Social network analysis

1 Introduction

Social media is often used as a means to manipulate public opinion during an international conflict, where polarized narratives are disseminated by different groups. Since the beginning of the Russian invasion of Ukraine, pro-Chinese-state (pro-state) and anti-Chinese-state (anti-state) accounts have spread divisive opinions, rumors, and conspiracy theories [10,24,29]. One of the propaganda narratives employed by the pro-state actors is the Ukraine biolabs conspiracy which states that the US has set up biolabs in Ukraine to manufacture bioweapons [1,10,29]. Such narratives were found to have originated from Russian social media accounts as a means to justify the invasion of Ukraine [10]. Existing research [1] found that the largest group of coordinated accounts tweeting Ukraine biolabs narratives also extensively retweeted Russian and Chinese

© The Author(s), under exclusive license to Springer Nature Switzerland AG 2023
R. Thomson et al. (Eds.): SBP-BRiMS 2023, LNCS 14161, pp. 33–43, 2023.
https://doi.org/10.1007/978-3-031-43129-6_4

state-funded media, suggesting certain coordinated efforts may exist between the Russian and Chinese users to sway public opinion on social media.

This study uses Chinese Twitter as a case study to examine political narratives surrounding the Russia-Ukraine war in the first two months of the Russian invasion of Ukraine (RU). Using text-based and network-driven approaches, we leverage a Twitter dataset to identify pro-state and anti-state actors. We investigate the subtopics introduced in the pro-state and anti-state networks, as well as how attention shifted toward the subtopics in subsequent weeks. We also look at how the pro- and anti-state camps contribute to the various narratives over time. We identify pro-state and anti-state influential actors based on their position in the retweet network and investigate how they influence the evolution of RU-related narratives. Our research questions (RQs) are:

RQ1 What were the dominant narratives in pro- and anti-state networks during the first two months of RU?

RQ2 How did the pro- and anti-state camps contribute to the evolution of the RU-related narratives? How did influential accounts shape the evolution of the RU-related narratives?

Main Contributions. The main contributions of this work are as follows:

(1) By combining text analysis and network-based approaches, we propose a method for identifying pro-state vs. anti-state leaning accounts, narratives, and influential actors in large social networks.

(2) This is the first large-scale analysis of narratives and interactions between pro- and anti-state camps on Chinese Twitter during the Russia-Ukraine war. We discover propagandist narratives from both the camps. Our study also reveals that anti-state voices dominate the Chinese Twitter space, which is consistent with previous studies [4].

2 Background and Related Work

Chinese Diplomatic Communication and Propaganda. Several works have studied the Chinese government's information censorship policies, online public opinion manipulation, and diplomatic communication strategies in digital/online space [4,18,28,30]. KING et al. [20] found that the Chinese government posts about 448M social media comments a year which are fabricated by government employees as part of their regular jobs. Media reports have suggested that the Chinese official Twitter accounts attempt to influence public opinion [7,19,27]. Bolsover and Howard [4] found that the anti-Chinese state groups predominantly manipulate information in the Chinese Twitter space. These observations regarding the distinct narratives promoted by pro-state and anti-state accounts inspired our study, which identifies the political orientations of the accounts from which the narratives originated.

Analysis of Propaganda. Social media manipulation to shape public opinion has become a matter of growing concern [5]. Existing works have explored the role of coordinated behavior [6,26] and bots/trolls to drive information campaigns [13,14,31]. González-Bailón and De Domenico [17] studied the visibility of bots, human accounts, and verified accounts and found that bots are less central in information campaigns than verified accounts during contentious political events. Research on propaganda detection from a text analysis perspective has focused on using distant supervision [23], and detecting the use of propaganda techniques [11,12]. It has been seen that 'malicious' actors change their behaviors to prevent detection by moderation tools [8]. Following the suggestion of Martino et al. [22], we leverage text-based and network analysis approaches to identify actors potentially involved in information campaigns.

Analysis of Evolving Networks, Communities and Topics. Previous works have explored content co-sharing networks to discover topic-based user communities and the engaging topics in these communities [15,16,25]. Arazzi et al. [2] studied the importance of adopted language in the formation of online communities around specific topics. In this work, we use n-grams co-occurrence in tweets to extract fine-grained narratives and analyze the evolution of narratives in the pro-state and anti-state camps.

3   Data

Twitter Dataset. We collect all tweets containing specific keywords related to Ukraine biolabs conspiracy (see Supplementary Materials[1]) from March 2022 until April 2022, i.e., during the first two months (8 weeks) of RU. This returns Ukraine biolabs-related tweets from 4494 Twitter accounts. We collect all the tweets for these 4494 accounts in the study period, resulting in a dataset containing over 4M tweets (including around 1.6M replies and 223K unique retweets).

Retweet Network. We construct a retweet network using the retweets in our Twitter dataset, where nodes are unique Twitter user accounts and a directed edge $A \rightarrow B$ indicates user A retweeted B. The edges are weighted by the number of times a source user retweeted a target user. Our retweet network comprises over 74K accounts and around 2M connections between them. We expand our data by extracting an additional 5000 accounts from the retweet network based on their impact on generating retweets, taking their immediate connections and network position into account. More specifically, we calculate an account's *influence* by the harmonic mean of its in-degree centrality and betweenness centrality in the retweet network.

---

[1] https://osf.io/xrb76/?view_only=a8439d2b7b8548d19b65dd4e56478aab.

4   Methodology

Tweet-Level Classification. We manually annotate around 12.6K tweets by the textual content as `pro` (pro-state stance), `anti` (anti-state stance), or `unsure` (neither pro-state nor anti-state). The inter-annotator agreement (Cohen's Kappa) is 0.7 (substantial agreement). We select a set of over 11K tweets containing specific topic-related keywords (see Supplementary Materials) for training our tweet-level classifier. We fine-tune the pre-trained Chinese RoBERTa model [9] provided by Huggingface to perform multi-class classification. A dense layer is added on top of RoBERTa while fine-tuning. We achieve an overall F1 score (imbalanced classes) of 0.77 over 5-fold cross-validation. The performance across classes is reported in Table 1.

Table 1. Performance summary of tweet-level classifier across classes.

Class	Precision	Recall	F1
anti	0.80	0.93	0.86
pro	0.69	0.69	0.69
unsure	0.81	0.51	0.63

Identifying Influential/core Accounts. We identify the `pro` and `anti` accounts that are most influential in driving the Chinese narratives around RU on Twitter, which we refer to as the *core* accounts. We manually identify 11 PRC (People's Republic of China) Official accounts in our dataset. We calculate an account's *influence* by the harmonic mean of its in-degree centrality and betweenness centrality in the retweet network and manually label 150 most central accounts by looking at 20 of their randomly selected tweets. This yields 87 `anti` core accounts and 10 `pro` core accounts which suggests that the `anti` actors are more central in our retweet network.

To expand our `pro` core we use a stratified sampling approach. We leverage the RoBERTa probabilities together with centrality measures to identify the `pro` core accounts. Specifically, we extract accounts with indegree centrality greater than 75% or closeness centrality greater than 70% and manually label 150 accounts with the highest upper quartile of pro probabilities for 20 topical tweets. This results in 87 `anti` core and 106 `pro` core (including PRC Official) accounts.

Account-Level Classification. To identify the political leanings of the narrative creators, we classify accounts as `pro` or `anti` leaning. Given the prevalent observations of users' homophilous retweeting behaviors, we consider the retweeting tendency as a proxy for political leaning, as calculated below:

$$leaning = \frac{\#retweet\ \texttt{pro}\ core}{\#retweet\ \texttt{pro}\ core + \#retweet\ \texttt{anti}\ core} \qquad (1)$$

Those accounts with a leaning greater than 0.5 are considered `pro` leaning, as they retweet the `pro` core accounts more often than the `anti` core accounts, while those with a leaning less than 0.5 are considered `anti` leaning. To account for temporal variation in our analysis, we divide our dataset into two time periods for account-level classification: March 2022 (T1) and April 2022 (T2). Thus, accounts can switch leaning from T1 to T2. We validated our leaning proxy assumption by manually labeling 25 `pro` leaning and 25 `anti` leaning samples from each T1 and T2. We train two BiLSTM classifiers in T1 and T2 using the leaning labels as ground truth to perform binary classification in T1 and T2 (details in Supplementary Materials). Table 2 shows our classifier's T1 and T2 class proportions.

Table 2. Proportion of classes using account-level classifier in T1 and T2

Setting	#pro	#anti	#unsure	#samples
T1 train, T1 test	488 (12%)	1743 (43%)	1781 (45%)	4012
T2 train, T2 test	1139 (21%)	1751 (33%)	2411 (46%)	5301

The majority (53%) of new accounts appearing in T2 are `anti` leaning. Most of the `anti` and `pro` accounts in T1 remain `anti` and `pro` in T2 with around 20% of accounts becoming `unsure` for both classes. We find that around 8% `pro` accounts shift their leaning to `anti` and 6% `anti` accounts shift their leaning to `pro` from T1 to T2 using this approach.

Identifying Narratives. We extract all the trigrams (having frequency more than two) from our tweets dataset. We create weekly n-gram ($n = 3$) co-occurrence networks based on n-grams co-occurring in a tweet. We further create n-gram co-occurrence networks for `pro` account, `pro` core account, `anti` account, and `anti` core account tweets (including retweets) for each week. Using the Louvain community detection [3] method, we construct n-gram clusters from our n-gram co-occurrence networks and manually identify 38 most popular subtopics related to RU from the n-gram clusters. These subtopics captured the diverse narratives surrounding RU on Chinese Twitter.

5   Results

5.1   RQ1: Chinese Political Narratives Around RU

Tweets with subtopic-related n-grams are mapped to the 38 subtopics. This yields 5209 (about 11%) subtopical tweets. Figure 1 shows subtopics by account leaning. The most common subtopic is "Ukraine (U) rescue and fund new China (C)"–funding the New China Federation to rescue Ukraine. This topic is driven

38   A. Biswas et al.

by `anti` accounts. The `anti` accounts tend to focus on topics that criticize the Chinese Communist Party (e.g., "GreatTranslationMovement," "Take down the CCP"), Russia and Putin, Ukraine, and China's relationships with the United States and Russia, respectively. The `pro` account narratives praised China's efforts (e.g., 'Chinese (C) in Ukraine (U) rescued', 'China (C) will play a constructive role'), criticized the US (e.g., 'West lecturing', 'gun violence in the US'), Sino-US relationship, and Russia-China partnership. Both the `pro` and `anti` accounts contribute to the narrative building for certain subtopics. We find polarized opinions on these subtopic-related narratives. For example, the `anti` accounts support US arms to Ukraine (U), while the `pro` accounts call it US imperialism.

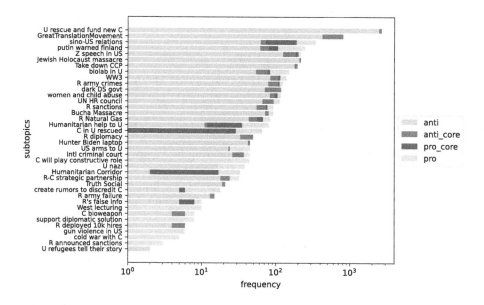

Fig. 1. Frequency of subtopics across account leaning

Figure 2(a) shows the `pro` and `anti` account activity (#tweets + #retweets) during the 8 weeks. Weeks 4–8 show increased `anti` account activity. From week 5 to week 8, `pro` account activity drops. The `anti` core activity however remains consistent over the weeks. We examine subtopical tweet account activity in Fig. 2(b) to see if a similar trend exists. For RU-related subtopics, `pro` accounts were more active in the first four weeks and less active in the later weeks, similar to the overall activity trend. Subtopic-related `anti` activity increases from week 5 to 8 with a sudden increase in week 8.

We examine narrative evolution using the top 12 subtopics. Other interesting subtopics include "US arms to U," "R-C strategic partnership," "UN Human Rights (HR) council," and "Bucha Massacre." Fig. 3 shows the 16 subtopics' distribution over time. Weeks 5 and 8 saw increased activity due to the subtopic

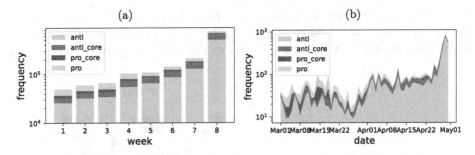

Fig. 2. (a) Overall activities (b) Activities with respect to the subtopical tweets

"U rescue and fund new C." We see a wider range of subtopics in the first 4 weeks, then fewer subtopics in the later weeks.

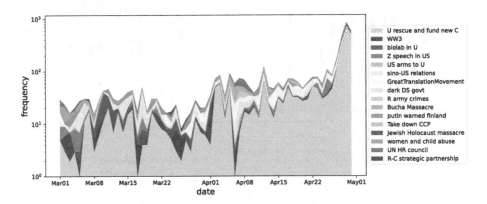

Fig. 3. Distribution of 16 selected subtopics

5.2   RQ2: Contribution of Camps to RU-Related Narrative Evolution

To investigate the `pro` and `anti` camp's contributions to narrative evolution, we look at the weekly attention received by the subtopics across these camps. Figure 4(a) shows the 16 subtopics' weekly frequencies in colors. Many subtopics (e.g., "U rescue and fund new C," "Take down CCP," "GreatTranslationMovement") appeared most in week 8. The subtopic "biolab in U" peaked during week 2 and disappeared after week 5, suggesting that the `pro` camp moved on from the Ukraine biolabs conspiracy.

We examine `pro` and `anti` camps's weekly contributions to these subtopics. Each subtopic's weekly leaning score is calculated by subtracting its `pro` activity from its `anti` activity and dividing by their sum (#total activity). A score near

1 indicates that `anti` accounts dominated the subtopic, while a score near -1 indicates `pro` accounts dominated. Only subtopics with more than three tweets in the week are considered. Figure 4(b) shows the weekly subtopic leanings.

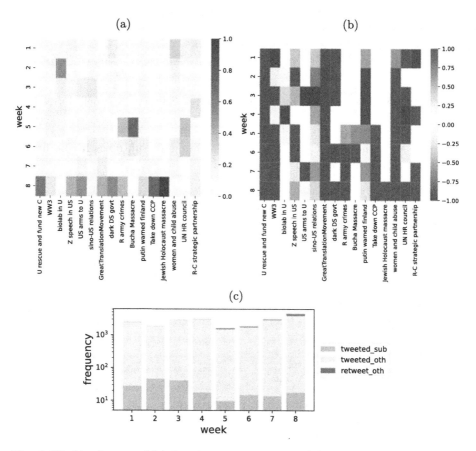

Fig. 4. Weekly changes of (a) distribution of subtopics, (b) subtopic leanings, and (c) PRC Official activity.

Throughout our study, the `anti` camp focused on subtopics like "U rescue and fund new C," "Take down CCP," "GreatTranslationMovement," and "dark Deep State (DS) govt." The `pro` camp focused on subtopics "biolab in U," "Sino-US relations," and "Putin warned Finland." The "Sino-US relations" subtopic is polarized, with both sides contributing but the `pro` side promoting it. "R army crimes," "Bucha massacre," and "Jewish Holocaust massacre" were introduced by `anti` camp while the `pro` camp introduced the subtopic "US arms to U". Week 2's "biolab in U" subtopic was equally attended by `pro` and `anti` camps.

The `anti` camp hijacked `pro` camp subtopics like "WW3", "R-C strategic partnership," and "US arms to U" in later weeks. When discussing these subtopics, the `anti` camp often flipped the `pro` camp's narratives. For the subtopic "WW3," the `pro` camp claimed that US-funded biolabs in Ukraine were making bioweapons for World War III: *"U.S.-funded Ukrainian biological weapons laboratory destroyed by Russia, the highest level of silent needle war against the people..."*. The `anti` camp narrative claimed that CCP started the WW3 by "releasing" the COVID-19 virus and *"Russia's invasion of Ukraine is just an extension of the virus war to the hot war..."*. This suggests that `anti` camp actively counters the narratives of `pro` camp, whereas no such observations were found from the `pro` camp. Consistent with our previous findings, `pro` camp activity decreased in weeks 5–8 while `anti` camp activity increased.

To understand the shift of focus of the `pro` camp, we look at how the 11 PRC Official accounts retweeted and were retweeted during this time period, shown in Fig. 4(c). During the first month, PRC accounts were retweeted more on the subtopical tweets ("tweeted_sub") as compared to the second month. The overall trend of PRC Official accounts being retweeted ("tweeted_oth") remained consistent. Interestingly, we find that the PRC Official accounts started retweeting ("retweet_oth") from week 5, most of which were related to Taiwan. This suggests that the `pro` core accounts started actively pursuing a different topic which may be the reason the `pro` camp shifted focus from RU-related narratives.

6   Discussion and Future Work

We studied the evolution of RU-related narratives in the Chinese Twitter space and how the anti-state and pro-state actors contributed to the narrative evolution. We found that propagandist narratives arise in both the pro-state and anti-state camps. Our study reveals that the Chinese Twitter space is dominated by anti-state voices. Both these findings are consistent with previous studies [4].

The results showed that a few influential accounts are responsible for introducing the propagandist narratives in both pro-state and anti-state camps which are then spread (through retweets) by the rest of the camp members. This is a core aspect of participatory propaganda where an information campaign is seeded by strategic actors and spread to the mainstream public by co-opting community members [21]. It was seen that the pro-state camp members shifted their focus from RU-related narratives when the pro-state core accounts actively started pursuing other topics (e.g., Taiwan). Nevertheless, understanding the exact strategies behind how a disinformation cascade unfolds requires more analysis and we leave that as a part of the future work.

We found that the pro-state influential accounts mostly focus on criticizing the US and anti-state influential accounts on criticizing the CCP. There were instances of polarized narratives by the pro-state and anti-state camps on the same subtopics to achieve their respective propaganda goals. Moreover, our analysis revealed that certain subtopics initiated by the pro-state camp were more likely to be appropriated by the anti-state camp in subsequent weeks, whereas the converse was less likely, suggesting that the pro-state camp may engage less actively in countering its critics, which resonates with previous studies [20].

This may suggest that the two groups have distinct offensive-defensive strategies, with one group focusing more on initiating novel propagandist narratives than the other. Future research may investigate the plausibility of the hypothesis in a different national or international context.

One of the limitations of our study is the short-term observation of the community activity—especially the lack of observation before the Russian invasion of Ukraine. Future work can look into the longer-term evolution of key communities in the pro-state and anti-state networks and community contribution to narrative building.

Acknowledgment. The authors gratefully acknowledge the support from AFOSR, ONR, Minerva, and Pitt Cyber Institute's PCAG. This research was supported in part by the resources provided by the University of Pittsburgh Center for Research Computing, RRID:SCR_022735, through support by NIH #S10OD028483. Any opinions, findings, and conclusions or recommendations expressed in this material do not necessarily reflect the views of the funding sources.

References

1. Alieva, I., Ng, L.H.X., Carley, K.M.: Investigating the spread of Russian disinformation about biolabs in Ukraine on Twitter using social network analysis. In: 2022 IEEE Big Data, pp. 1770–1775 (2022). https://doi.org/10.1109/BigData55660.2022.10020223
2. Arazzi, M., Nicolazzo, S., Nocera, A., Zippo, M.: The importance of the language for the evolution of online communities: an analysis based on twitter and reddit. Expert Syst. Appl. **222**, 119847 (2023)
3. Blondel, V.D., Guillaume, J.L., Lambiotte, R., Lefebvre, E.: Fast unfolding of communities in large networks. JSTAT **2008**(10), P10008 (2008)
4. Bolsover, G., Howard, P.: Chinese computational propaganda: automation, algorithms and the manipulation of information about Chinese politics on twitter and weibo. Inf. Commun. Soc. **22**(14), 2063–2080 (2019)
5. Bradshaw, S., Bailey, H., Howard, P.N.: Industrialized disinformation: 2020 global inventory of organized social media manipulation. Computational Propaganda Project at the Oxford Internet Institute (2021)
6. Chetan, A., Joshi, B., Dutta, H.S., Chakraborty, T.: Corerank: Ranking to detect users involved in blackmarket-based collusive retweeting activities. In: WSDM 2019, pp. 330–338 (2019)
7. Collins, B., Cox, J.: This twitter bot army is chasing down a Chinese dissident and mar-a-lago member. The Daily Beast 17 (2017)
8. Cresci, S., Di Pietro, R., Petrocchi, M., Spognardi, A., Tesconi, M.: The paradigm-shift of social spambots: evidence, theories, and tools for the arms race. In: WWW 2017, pp. 963–972 (2017)
9. Cui, Y., Che, W., Liu, T., Qin, B., Wang, S., Hu, G.: Revisiting pre-trained models for Chinese natural language processing. In: EMNLP 2020, pp. 657–668. Association for Computational Linguistics, Online, November 2020. https://www.aclweb.org/anthology/2020.findings-emnlp.58
10. Curet, M.: China repeats false claim that u.s has biolabs in Ukraine (2022). https://www.politifact.com/factchecks/2022/mar/10/instagram-posts/china-repeats-false-claim-us-has-biolabs-ukraine/

11. Da San Martino, G., Barron-Cedeno, A., Nakov, P.: Findings of the nlp4if-2019 shared task on fine-grained propaganda detection. In: Proceedings of the Second Workshop on Natural Language Processing for Internet Freedom: Censorship, Disinformation, and Propaganda, pp. 162–170 (2019)
12. Da San Martino, G., Yu, S., Barrón-Cedeno, A., Petrov, R., Nakov, P.: Fine-grained analysis of propaganda in news article. In: EMNLP-IJCNLP 2019, pp. 5636–5646 (2019)
13. Ferrara, E.: What types of covid-19 conspiracies are populated by twitter bots? arXiv preprint arXiv:2004.09531 (2020)
14. Ferrara, E., Varol, O., Davis, C., Menczer, F., Flammini, A.: The rise of social bots. Commun. ACM 59(7), 96–104 (2016)
15. Ferreira, C.H., et al.: On the dynamics of political discussions on Instagram: a network perspective. Online Soc. Networks Media 25, 100155 (2021)
16. Gomes Ferreira, C.H., et al.: Unveiling community dynamics on Instagram political network. In: 12th ACM Conference on Web Science, pp. 231–240 (2020)
17. González-Bailón, S., De Domenico, M.: Bots are less central than verified accounts during contentious political events. PNAS 2021 118(11), e2013443118 (2021)
18. Huang, Z.A., Wang, R.: Building a network to "tell china stories well": Chinese diplomatic communication strategies on twitter. Int. J. Commun. 13, 2984–3007 (2019)
19. Kaiman, J.: Free tibet exposes fake twitter accounts by china propagandists. The Guardian (2014)
20. King, G., Pan, J., Roberts, M.E.: How the Chinese government fabricates social media posts for strategic distraction, not engaged argument. Am. Political Sci. Rev. 111(3), 484–501 (2017). https://doi.org/10.1017/S0003055417000144
21. Lewandowsky, S.: 20 fake news and participatory propaganda. Cognitive Illusions: Intriguing Phenomena in Thinking, Judgment, and Memory, p. 324 (2022)
22. Martino, G.D.S., Cresci, S., Barrón-Cedeño, A., Yu, S., Di Pietro, R., Nakov, P.: A survey on computational propaganda detection. arXiv preprint arXiv:2007.08024 (2020)
23. Mintz, M., Bills, S., Snow, R., Jurafsky, D.: Distant supervision for relation extraction without labeled data. In: ACL-IJCNLP 2009, pp. 1003–1011 (2009)
24. Niven, T.: The spread of ukraine biolabs conspiracy content in chinese on twitter (2022). https://medium.com/doublethinklab/the-spread-of-ukraine-biolabs-conspiracy-content-in-chinese-on-twitter
25. Nobre, G.P., Ferreira, C.H., Almeida, J.M.: A hierarchical network-oriented analysis of user participation in misinformation spread on whatsapp. Inf. Process. Manage. 59(1), 102757 (2022)
26. Pacheco, D., Flammini, A., Menczer, F.: Unveiling coordinated groups behind white helmets disinformation. In: WWW 2020, pp. 611–616 (2020)
27. Phillips, T.: China's communist party takes online war to twitter. The Telegraph 8 (2014)
28. Schliebs, M., Bailey, H., Bright, J., Howard, P.: China's inauthentic UK Twitter diplomacy: a coordinated network amplifying PRC diplomats (2021)
29. Wong, E.: Us fights bioweapons disinformation pushed by Russia and China. The New York Times, March 2022. https://www.nytimes.com/2022/03/10/us/politics/russia-ukraine-china-bioweapons.html. Accessed on 19(6) (2022)
30. Wu, S., Mai, B.: Talking about and beyond censorship: mapping topic clusters in the Chinese twitter sphere. Int. J. Commun. 13, 23 (2019)
31. Zhang, J., Zhang, R., Zhang, Y., Yan, G.: The rise of social botnets: attacks and countermeasures. IEEE TDSC 15(6), 1068–1082 (2016)

Anger Breeds Controversy: Analyzing Controversy and Emotions on Reddit

Kai Chen[1(✉)], Zihao He[1], Rong-Ching Chang[2], Jonathan May[1], and Kristina Lerman[1]

[1] USC Information Sciences Institute, Los Angelos, CA, USA
{kchen035,zihaoh}@usc.edu, {jonmay,lerman}@isi.edu
[2] University of California, Davis, CA, USA
rocchang@ucdavis.edu

Abstract. Emotions play an important role in interpersonal interactions and social conflict, yet their function in the development of controversy and disagreement in online conversations has not been fully explored. To address this gap, we study controversy on Reddit, a popular network of online discussion forums. We collect discussions from various topical forums and use emotion detection to recognize a range of emotions from text, including anger, fear, joy, admiration, etc. (Code and dataset are publicly available at https://github.com/ChenK7166/controversy-emotion). We find controversial comments express more anger and less admiration, joy, and optimism than non-controversial comments. Moreover, controversial comments affect emotions of downstream comments, resulting in a long-term increase in anger and a decrease in positive emotions. The magnitude and direction of emotional change differ by forum. Finally, we show that emotions help better predict which comments will become controversial. Understanding the dynamics of emotions in online discussions can help communities to manage conversations better.

Keywords: Controversy · Emotion · Reddit

1 Introduction

The social web has linked millions of people worldwide, creating "digital town squares" for exchanging ideas, opinions, and beliefs. Unfortunately, the same mechanisms also create unique new vulnerabilities. Exchanging diverse viewpoints can lead to controversy and polarization, which amplifies hate speech and risks undermining collective trust in democracy [16].

Online communities have tried to mediate discussions to remove toxic messages that violate community norms [23]; however, manual moderation does not scale to the volume and speed of online conversations. Although machine moderation offers tools that automatically recognize harassment, hate speech, and other toxic speech [2,20,24], these methods treat the symptoms rather than the causes of the problem. We need to understand how controversy develops

© The Author(s), under exclusive license to Springer Nature Switzerland AG 2023
R. Thomson et al. (Eds.): SBP-BRiMS 2023, LNCS 14161, pp. 44–53, 2023.
https://doi.org/10.1007/978-3-031-43129-6_5

and derails online conversations before we can effectively—and automatically—moderate them.

The role of emotions in developing controversy and disagreement in online discussions has not been fully explored. Previous research mostly relied on network structures [15], user activities [18], or language cues [23,27] to identify controversial discussions. We argue that emotions can play a crucial role in controversial discussions as they often shape the social response to conflict [3]. Emotions also contribute to the viral spread of topics online [6,7,11].

To investigate the role of emotions in controversial discussions, we use online discussions on Reddit. Reddit is a popular network of online communities or 'subreddits,' where members post new topics for discussion and respond to the comments of others. Community members can up vote or downvote any comment, expressing their agreement or disagreement with it. Reddit automatically flags comments as *controversial* if there are large numbers of upvotes and downvotes. To study emotions in controversial discussions, we pose the following research questions:

RQ1 Are controversial comments more emotional than non-controversial ones?
RQ2 How do controversial comments change the emotional tone of discussions?
RQ3 Can we identify controversial comments at the time they are posted, i.e., before they become controversial?

We find that controversial comments express more anger and less joy than non-controversial comments. Although controversial comments are a small fraction of all comments in a discussion—typically 3%—they change the long-term emotional tone of discussions. Finally, we show that adding emotions to controversial comment classification leads to significant performance improvement. This enables us to predict whether a comment will become controversial, allowing a moderator to step in to keep the discussion on track.

2   Related Work

Previous work on controversy detection mostly relied on content, network structures, and user activities [15,18,27]. Others identified controversial posts on Reddit based on the ratio of upvotes to all votes [17]. In this section, we focus on introducing recent works that have laid the groundwork by combining emotions and controversy detection.

Emotions and Controversy Online. The role of emotions in controversy is an under-explored area of research. One study found a direct association between both positive and negative emotions and their spread by examining the diffusion of highly controversial topics on Facebook [6]. Similarly, others discover that emotional tweets related to political discussions on Twitter tend to spread widely and rapidly [8,25]. While another study observes controversial news contains more negative emotional words [21].

Building upon these prior studies, we focus on psycholinguistic indicators. Specifically, we study emotions expressed in comments and their relationship and dynamics with controversial comments at different levels of granularity. This includes individual comments, the impact of controversial comments on subsequent discussions, and the use of emotional cues for detecting controversial comments.

3   Methods

We use Pushshift API [5] to collect Reddit discussions. Pushshift has archived Reddit data back to 2005, including complete discussion comments and metadata. A comment is automatically flagged by Reddit as controversial when the numbers of upvotes and downvotes it has are both high. We further define a controversial discussion as the one that includes at least one controversial comment.

3.1   Data

We collected data from the 100 most popular subreddits based on the number of subscribers. For very large subreddits, we randomly under-sample discussions so that the number of discussions from all subreddits are roughly similar. We then filter out discussions with fewer than five comments, and discard ten subreddits that disallow users from commenting after 2022, such as *r/announcement*. The remaining 90 subreddits cover a large variety of topics, such as art (*r/Art*), music (*r/Music*), sports (*r/sports*), politics (*r/politics*), science (*r/science*), humor (*r/jokes*), gender (*r/TwoXChromosomes*), advice (*r/LifeProTips*), gaming (*r/PS4*), and emotional reactions (*r/wholesomememes*), among many others. The complete list of the 90 subreddits is shown in our GitHub repository. The ratio of controversial comments in subreddits ranges from 0.3% to 9.7%, with an average of 3%.

For controversy prediction, we sample six subreddits from the popular forums covering science (*r/science* and *r/technology*), Q&A (*r/AskScience* and *r/AskReddit*), news (*r/news* and *r/worldnews*). We add multilingual discussions from subreddits *r/france* (in French) and *r/de* (in German). We report the statistics of this multilingual forums dataset in Table 1.

Table 1. Statistics of the Multilingual Subreddits Dataset.

	Science	Technology	News	Worldnews	AskReddit	AskScience	france	de
# discussions	32,744	102,246	8,691	17,858	188,177	73,665	50,558	63,058
# comments	1.7M	1.5M	2.1M	2.7M	6.6M	1.7M	1.6M	1.8M
ave disc. length	51.3	14.4	243.5	150.8	35.2	5.7	32.2	29.2
contro. comms	4.4%	5.8%	7.4%	7.9%	1.2%	1.1%	5.9%	5.1%
removed comms	42.5%	10.8%	18.5%	11.2%	5.8%	47.3%	6.3%	6.3%

3.2   Emotion Detection

To recognize emotions expressed in text, we use a multilingual emotion detection model from [9]. The model is based on SpanEmo [1] that is the state-of-the-art in emotion detection. The original backbone language model BERT [13] is replaced with multiligual XLM-T [4,14], which is more suitable in a multilingual setting to handle text inputs in large number of languages. The model was finetuned on SemEval 2018 Task 1 E-c data [22] and GoEmotions [12], with ten simplified *emotion clusters*: "Anger, Hate, Contempt, Disgust," "Embarrassment, Guilt, Shame, Sadness," "Admiration, Love," "Optimism, Hope," "Joy, Happiness," "Pride, National Pride," "Fear, Pessimism," "Amusement," "Other Positive Emotions," and "Other Negative Emotions." Further details of the model and performance are described in [9].[1] Given input text, the model returns one scalar value per emotion, indicating the *confidence* that the emotion is present. Since it is a multi-label classification setting, the model can assign multiple (or no) emotions to text input. Therefore, for each input Reddit comment, we have a 10d confidence vector for the ten emotion clusters.

4   Results

We answer our research questions by analyzing emotions and controversy of online discussions. We define a discussion to be controversial if it has at least one comment that has been tagged as controversial by Reddit[2]. As a reminder, a comment is tagged as "controversial" if it has the same (or similar) number of upvotes as downvotes, and both numbers are large.

4.1   RQ1: Emotions in Controversial Comments

To answer **RQ1**, we compare emotions expressed in controversial comments to emotions in non-controversial comments. Let CC be the set of controversial comments and NC be the set of non-controversial comments in the same subreddit. We define emotion gap δ_e for a subreddit as the difference between the mean confidence of emotion e in CC and its mean confidence in NC comments:

$$\delta_e = \frac{1}{|CC|} \sum_{i \in CC} f_i(e) - \frac{1}{|NC|} \sum_{i \in NC} f_i(e).$$

A positive emotion gap δ_e indicates controversial comments express stronger emotion e compared to non-controversial comments. While the negative emotion gap δ_e suggests that non-controversial comments elicit stronger emotion e than controversial comments.

---

[1] https://github.com/gchochla/Demux-MEmo.
[2] Results do not differ qualitatively when using a higher threshold of the number of controversial comments to define controversial discussions.

We depict the emotion gaps δ_e for all emotion clusters across all 90 popular subreddits in Fig. 1. We find that some negative emotions, such as Anger/-Hate/Disgust, are consistently more common in controversial comments than in non-controversial comments, as indicated by the positive δ_e. However, not all negative emotions show strong differences between controversial and non-controversial comments. For example, the low arousal Embarrassment/Guilt/-Shame and Fear/Pessimism show only weak positive differences. We also find amusement appears to be stronger in controversial comments. This emotion is usually used to denote text that is funny, but sometimes also captures sarcasm.

In contrast, positive emotions more common in non-controversial comments. For example, Admiration/ Love, Joy/ Happiness, and Positive-other have shown a negative emotion gap δ_e.

There are variations in the strength of the emotion gap across subreddits. For example, compared to other forums, *r/Music, r/listentothis, r/Art* have strongest differences across all emotions. Controversial comments on these forums have much more Anger, but also much less Admiration and Joy than non-controversial comments. Surprisingly, the forums we expected to have more controversy, like *r/politics* and *r/AmItheAsshole*, show smaller emotional differences between controversial and non-controversial comments.

These results answer our **RQ1**. *Controversial comments are angrier than non-controversial comments and express less positive emotion, like love, joy, and optimism.*

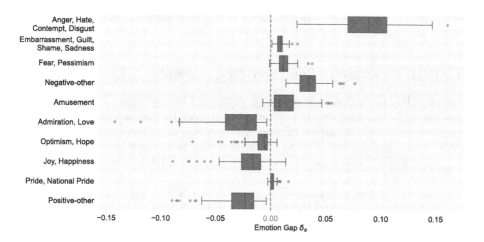

Fig. 1. Emotion gaps between controversial and non-controversial comments. Boxplots show the median, 25% and 75% of the emotion gaps across the 90 popular subreddits, with colors showing negative (red), neutral (brown), and positive (blue) emotions. Values to the right of $x = 0.0$ indicate higher emotion confidence for controversial comments. Conversely, values to the left of $x = 0.0$ indicate higher emotion confidence for non-controversial comments. (Color figure online)

4.2   RQ2: Controversial Comments Change Emotions in Discussions

To quantify the impact of a controversial comment on emotions in a discussion, we calculate the difference between the average confidence of the emotion in comments posted *after* that comment and comments that came *before* it. Let L be the length of a discussion, and i be the position of a controversial comment. We refer to comments in positions $\{1, \ldots, i-1\}$ as *predecessors* of the controversial comment i, and comments in positions $\{i+1, \ldots, L\}$ as its *successors*. The emotional impact θ_e^i of comment i is:

$$\theta_e^i = \frac{1}{(L-i)} \sum_{j>i} f_j(e) - \frac{1}{(i-1)} \sum_{j<i} f_j(e).$$

If $\theta_e^i > 0$, the controversial comment i leads to more emotion e in subsequent comments; otherwise, if $\theta_e^i < 0$, it reduces that emotion in the discussion.

To measure the overall impact of controversial comments on emotions, we average the impact of all controversial comments within each subreddit's discussions. Figure 2 shows this quantity across all 90 subreddits.

We find that most subreddits follow the same pattern: negative emotions like Anger/Hate/Contempt/ Disgust, Fear/Pessimism, and other negative emotions rise after a controversial comment. While positive emotions generally, though not always, fall after a controversial comment. There are a few exceptions to this trend. For example, anger decreases after a controversial comment in *r/nosleep* and *r/AmItheAsshole* subreddits. On top of that, we also find positive emotions, such as Admiration/Love and Joy/Happiness, in *r/photoshopbattle* and *r/WritingPrompts* increase after a controversial comment. Amusement rises systematically in almost all subreddits, on par with negative emotions. This suggests rising sarcasm in controversial discussions.

These findings answer our **RQ2** and suggest that *controversial comments lead to long-term changes in the emotional tone of discussions,* typically, not only raising anger of downstream comments but also reducing positive emotions of discussions. However, different communities respond differently to controversy, resulting in some deviations from this pattern. Therefore, automatic tools to moderate controversial comments will need to take the varying community context into account.

4.3   RQ3: Predicting Controversy from Emotions

To predict whether a comment i in a Reddit discussion is controversial, the author used a variety of lexical and user activity features [18]. They found that the most predictive features were related to user activity, which they calculated based on comments that preceded the comment i in a discussion (predecessors) and comments that followed it (successors):

- Predecessor features: number of comments before the current comment i, number of unique authors of preceding comments, elapsed time since the first predecessor, and the average elapsed time between predecessors.

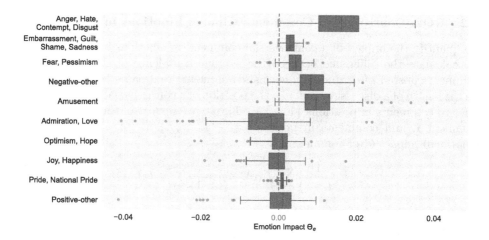

Fig. 2. Emotional impact of controversial comments. We compare the average emotion confidence of successors of a controversial comment to its predecessors. Boxplot shows the median emotion impact and its 25% and 75% values across 90 popular subreddits. Values to the right of $x = 0.0$ show that controversial comments increase emotions in downstream comments in a discussion, while values to the left of $x = 0.0$ indicate a long-term suppression of emotions.

- Successor features: number of comments posted after the comment i, number of unique authors, elapsed time since the current comment to the last successor, and average time between successors.

We supplement these features with emotions expressed in the current comment, and those expressed in predecessors or successors.

- All emotions: confidence of emotion clusters in predecessor comments, successor comments, or the comment i's own emotion.
- Positive emotions: subset of emotions that includes Admiration/Love, Optimism/Hope, Joy/Happiness, Pride, other positive emotions.
- Negative emotions: subset that includes Anger/Hate/Contempt/Disgust, Embarrassment/Sadness, Fear/Pessimism, and other negative emotions.

To better understand the impact of emotions on predicting controversial comments, we construct different feature sets for the classification model (as shown in Fig. 3). Following [18], we used Gradient Boosted Decision Trees with ten-fold cross-validation for the prediction task. We conducted a grid search to select hyperparameters that achieve the best performance on the validation set. We utilized the pretrained language model RoBERTa [19] and multilingual model XML-RoBERTa [10], using the original comments as inputs without any feature engineering, to predict comment controversy.

We applied the model to predict whether a comment is controversial in the multilingual dataset that includes discussions from six popular subreddits (in

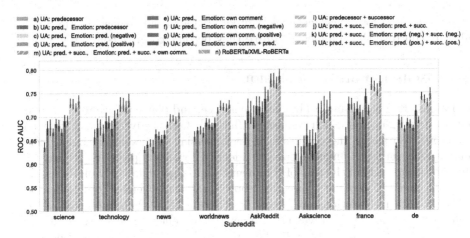

Fig. 3. Performance of controversial comment prediction in English-language, French and German subreddits. Ablation tests with user activity (UA) features and emotion features of the current comment, its predecessors, and its successors. Hatched bars show results using successor features. Overall the models that take in emotion features achieve better performance.

English) and also discussions in French and German. The data is highly imbalanced in that controversial comments make up only 5% of all comments; therefore, we under-sample non-controversial comments to create a balanced dataset for testing. We use ROC-AUC to measure classification performance.

Figure 3 shows the results of controversy classification separately for the eight subreddits. Overall, emotions help the model to consistently improve performance compared to using only activity features. Such non-trivial improvement demonstrates the effectiveness of emotions in controversial comment detection. Despite advancements in pretrained language models, detecting controversial comments remains challenging. The best performance is achieved on *r/AskReddit*, with AUC of 78.6%, followed by *r/france* and *r/de* with AUC of 77.3% and 74.7%. These scores represent 4%–6% improvement compared to using activity features alone. Interestingly, these three subreddits are also the least moderated forums in our data (see Table 1), with the lowest share of comments removed by moderators. This suggests that it is easier to identify controversial comments in less-moderated or unmoderated forums.

Overall, the model that uses a comment's own emotion (feature sets e, f, g) outperforms models that use emotions of predecessor comments (feature sets b, c, d). In addition, negative emotions are more effective than positive emotions in predicting controversy, suggesting that they better convey controversy. Features of successors (feature sets i, j, k, l, m) are more helpful in strongly moderated subreddits, such as *r/AskScience*. Finally, although using features of predecessors and the current comment does not perform as well as using all features (including successors), it represents a more realistic prediction scenario. On this task, adding emotions to the feature set significantly improves classification performance across all subreddits. Our results show emotions help identify controversial comments **before** they become controversial.

5 Discussion and Conclusion

Emotions are fundamental to human experience, and play a critical role in social interactions [26], and interpersonal conflict [3]. Our study demonstrates that emotions also shape the evolution of controversial discussions in online communities. Leveraging language cues, we identify a range of positive and negative emotions expressed in the text of comments. Our large-scale study of discussions on more than 90 popular subreddits shows that controversial comments have stronger negative emotions and fewer positive emotions than non-controversial comments. Although controversial comments represent a small share of all comments in a discussion, they shift the emotional tone of the entire discussion, leading to angrier and less positive subsequent comments. We also show that an emotionally aware classification model could better recognize comments that will become controversial, even in multilingual discussions.

Our work suggests that moderating controversial comments may help improve the emotional tone of discussions. It is possible to catch such comments at the time of their creation, and step in to help the author regulate the negative emotions such comments express. Reducing the long-term impact of controversial comments will help improve the overall quality of online discussions.

Acknowledgments. This material is based upon work supported in part by the Defense Advanced Research Projects Agency (DARPA) under Agreements No. HR00112290025 and HR001121C0168, and in part by the Air Force Office for Scientific Research (AFOSR) under contract FA9550-20-1-0224.

References

1. Alhuzali, H., Ananiadou, S.: SpanEmo: casting multi-label emotion classification as span-prediction. In: European ACL, pp. 1573–1584 (2021)
2. Plaza-del Arco, F.M., Molina-González, M.D., Urena-López, L.A., Martín-Valdivia, M.T.: Comparing pre-trained language models for Spanish hate speech detection. Expert Syst. Appl. **166**, 114120 (2021)
3. Bar-Tal, D., Halperin, E., De Rivera, J.: Collective emotions in conflict situations: Societal implications. J. Soc. Issues **63**(2), 441–460 (2007)
4. Barbieri, F., Anke, L., Camacho-Collados, J.: XLM-t: multilingual language models in twitter for sentiment analysis and beyond (2021)
5. Baumgartner, J., Zannettou, S., Keegan, B., Squire, M., Blackburn, J.: The pushshift reddit dataset. In: ICWSM, vol. 14, pp. 830–839 (2020)
6. Bi, N.C.: How emotions and issue controversy influence the diffusion of societal issues with imagined audience on facebook. Beh. Inf. Technol. **41**(6), 1245–1257 (2022)
7. Brady, W.J., McLoughlin, K., Doan, T.N., Crockett, M.J.: How social learning amplifies moral outrage expression in online social networks. Sci. Adv. **7**(33), eabe5641 (2021)

8. Brady, W.J., Wills, J.A., Jost, J.T., Tucker, J.A., Van Bavel, J.J.: Emotion shapes the diffusion of moralized content in social networks. PNAS **114**(28), 7313–7318 (2017)
9. Chochlakis, G., Mahajan, G., Baruah, S., Burghardt, K., Lerman, K., Narayanan, S.: Using emotion embeddings to transfer knowledge between emotions, languages, and annotation formats. In: ICASSP, pp. 1–5. IEEE (2023)
10. Conneau, A., et al.: Unsupervised cross-lingual representation learning at scale. arXiv preprint arXiv:1911.02116 (2019)
11. Coviello, L., et al.: Detecting emotional contagion in massive social networks. PLoS ONE **9**(3), e90315 (2014)
12. Demszky, D., et al.: Goemotions: a dataset of fine-grained emotions. arXiv preprint arXiv:2005.00547 (2020)
13. Devlin, J., Chang, M.W., Lee, K., Toutanova, K.: BERT: pre-training of deep bidirectional transformers for language understanding. In: ACL, pp. 4171–4186 (2019)
14. Wolf, T., et al.: Transformers: state-of-the-art natural language processing. In: EMNLP, pp. 38–45. ACL (2020)
15. Garimella, K., Morales, G.D.F., Gionis, A., Mathioudakis, M.: Quantifying controversy on social media. ACM Trans. Soc. Comput. **1**(1), 1–27 (2018)
16. Haidt, J.: Why the past 10 years of American life have been uniquely stupid. The Atlantic (2022)
17. Hessel, J., Lee, L.: Something's brewing! early prediction of controversy-causing posts from discussion features. In: ACL, pp. 1648–1659 (2019)
18. Koncar, P., Walk, S., Helic, D.: Analysis and prediction of multilingual controversy on reddit. In: WebScience. WebSci 2021, pp. 215–224 (2021)
19. Liu, Y., et al.: Roberta: a robustly optimized Bert pretraining approach. arXiv: abs/1907.11692 (2019)
20. MacAvaney, S., Yao, H.R., Yang, E., Russell, K., Goharian, N., Frieder, O.: Hate speech detection: challenges and solutions. PLoS ONE **14**(8), e0221152 (2019)
21. Mejova, Y., Zhang, A.X., Diakopoulos, N., Castillo, C.: Controversy and sentiment in online news. arXiv preprint arXiv:1409.8152 (2014)
22. Mohammad, S., Bravo-Marquez, F., Salameh, M., Kiritchenko, S.: Semeval-2018 task 1: Affect in tweets. In: SemEval, pp. 1–17 (2018)
23. Park, C.Y., et al.: Detecting community sensitive norm violations in online conversations. In: Findings of EMNLP, pp. 3386–3397 (2021)
24. Poletto, F., Basile, V., Sanguinetti, M., Bosco, C., Patti, V.: Resources and benchmark corpora for hate speech detection: a systematic review. Lang. Resour. Eval. **55**(2), 477–523 (2021)
25. Stieglitz, S., Dang-Xuan, L.: Emotions and information diffusion in social media-sentiment of microblogs and sharing behavior. J. Manag. Inf. Syst. **29**(4), 217–248 (2013)
26. Van Kleef, G.A., Cheshin, A., Fischer, A.H., Schneider, I.K.: The social nature of emotions. Front. Psychol. **7**, 896 (2016)
27. Zayats, V., Ostendorf, M.: Conversation modeling on Reddit using a graph-structured LSTM. Trans. ACL **6**, 121–132 (2018)

Building a Healthier Feed: Private Location Trace Intersection Driven Feed Recommendations

Tobin South[1,2]([✉]), Nick Lothian[3], Takahiro Yabe[1],
and Alex 'Sandy' Pentland[1,2]

[1] MIT Connection Science, Massachusetts Institute of Technology, Cambridge, USA
`tsouth@mit.edu`
[2] MIT Media Lab, Massachusetts Institute of Technology, Cambridge, USA
[3] Verida, Adelaide, Australia

Abstract. The physical environment you navigate strongly determines which communities and people matter most to individuals. These effects drive both personal access to opportunities and the social capital of communities, and can often be observed in the personal mobility traces of individuals. Traditional social media feeds underutilize these mobility-based features, or do so in a privacy exploitative manner. Here we propose a consent-first private information sharing paradigm for driving social feeds from users' personal private data, specifically using mobility traces. This approach designs the feed to explicitly optimize for integrating the user into the local community and for social capital building through leveraging mobility trace overlaps as a proxy for existing or potential real-world social connections, creating proportionality between whom a user sees in their feed, and whom the user is likely to see in person. These claims are validated against existing social-mobility data, and a reference implementation of the proposed algorithm is built for demonstration. In total, this work presents a novel technique for designing feeds that represent real offline social connections through private set intersections requiring no third party, or public data exposure.

Keywords: Social Media Feeds · Private Data Sharing · Personal Data Stores · Mobility Data

1 Introduction

Modern social media platforms design feeds to explicitly optimize for user attention, creating negative effects for both users and broader democratic discourse [10]. These feeds create a feedback loop of preferential attachment to already popular content and users, leading to disproportionally dominant agents [11]. To mitigate this rich-get-richer regime, feeds need to focus on promoting low-popularity content as well, but this is not sufficient - you can't just show people low-quality content or users stop engaging. Rather than promoting the content of the largest creators or random content pulled from across the social internet, we want to optimize social capital [19].

© The Author(s), under exclusive license to Springer Nature Switzerland AG 2023
R. Thomson et al. (Eds.): SBP-BRiMS 2023, LNCS 14161, pp. 54–63, 2023.
https://doi.org/10.1007/978-3-031-43129-6_6

Ideally, we want platforms that achieve sustainability and longer-term value by increasing real-world community connections, trust, capacity for collective action (known as bonding social capital), as well the number of trusted links to people in other communities (known as bridging capital). At its core, your feed should be representative of people that matter to you.

The physical spaces we navigate are strong determinants of many aspects of our lives and values, from our wellbeing [9] to our choice of collaborators [4] and the opportunities for economic growth we're exposed to [3]. Location history is a better predictor of your interests than the demographics used by current matching apps (e.g., dating app, content recommendation, etc.) [6], and is strongly predictive of which friendships you already have [7]. Leveraging GPS location data in social media has a long history demonstrating its value, but has been a series of privacy nightmares that require the sharing of this location data to a third party, often to be stored indefinitely and monetized without any visibility or accountability to the user.

We propose a new approach to building social media feeds that are optimized for social capital building via private matching over location histories, where friends are shown in proportion to their potential to build new within-community relationships and to reinforce community social capital, requiring no third-party interactions or capture of data.

All data sharing is done securely such that no parties ever see anyone else's location data, and only each pair of users can see the size of the location data overlap between them.

This approach is naturally extensible to a huge range of similar matching approaches ranging from shared photo matching, or shared friendship matching (including from existing social networks), to simple interest matching, all without needing trusted third parties or the exposure of any personal information. In doing so, this creates an in-built incentive for users to make personal data accessible (in an encrypted and private manner) to ensure they appear on others' feeds and can participate in the social ecosystem.

In summary, this paper:

- Presents a novel approach to building a social feed that represents real social relationships via calculations of set intersections between pairs of individuals without the need for a third party or transmission of *any* unencrypted data, ensuring no escape of any private information to anyone.
- Validates the value of location trace based matching as a measure of friendship on existing data using two different computational examinations, in turn motivating its use in building a feed optimized for social capital building.
- Prototypes the approach with a reference implementation and demo app to show that this can be done with extremely low computational cost.
- Proposes and discusses the extensibility of this private set intersection driven feed paradigm to other personal datasets and explores how this creates new social incentives around private data sharing for matching.

A longer, more verbose version of this publication with additional figures will be made available online via pre-print servers due to page limits.

2 A Social Recommendation Algorithm

For two friends Alice & Bob, the rate at which content is shown on a feed should be proportional to the physical-world relevancy of their ties (here the likelihood they'll be at the same location).

When Bob friends Alice, Bob can make a request to Alice to begin a multistep private set intersection on the mobility traces (location history) of both Bob and Alice. To make matching on this information easier, Alice can provide each point in her GPS trace, T_A^r, as a geohash [16] at an arbitrary resolution, r.

Geohash converts GPS coordinates to strings at a varying length of string to provide an arbitration resolution. For example, one can know that a GPS coordinate is inside the MIT Media Lab building using the string 'drt2yr7x'. If Alice desires, she can allow others to match on extremely detailed location traces by using a high-resolution geohash in the form of a long string ($r = 8$ characters). Or, if Alice decides her mobility trace should not inform her friend's feeds, she can share the geohash at a low resolution, $r = 5$, roughly city level. At the extreme, Alice can choose to deny sharing any data at all. However, sharing more information to match with Bob will allow Alice to be seen more often in Bob's feed, creating an incentive to share data at the level of detail one is comfortable with.

Bob can now perform a private set intersection using his original data, choosing the resolution of his mobility data as well. If Alice and Bob plan to share the exact same resolution of location, the task is simple to perform a private set intersection cardinality where Alice acts as a server and Bob acts as a client [17]. Alice encrypts each element of her mobility trace T_A^r using a *commutative* encryption scheme, $H(\cdot)^k$, where k is a users secret key, producing a sequence of encrypted geohash strings denoted $\{H(T(i)_A^r)^{k=A}\}_{\forall i}$. These elements are then inserted into a Bloom filter with a chosen false positive rate e and sent to Bob. Similarly, Bob encrypts each element of his sequence with his key to produce $\{H(T(i)_B^r)^{k=B}\}_{\forall i}$. Bob sends his encrypted sequence back to Alice who can then encrypt it again using her secret key, $\{H(H(T(i)_B^r)^{k=B})^{k=A}\}_{\forall i}$, to send back to Bob. In order to stop Bob from knowing what locations they match on, Alice can shuffle the mobility sequence to ensure that Bob can only see the size of the set intersection, rather than the actual matches. Once returned, Bob uses the commutative properties of the encryption, $\{H(H(T(i)_B^r)^{k=B})^{k=A}\}_{\forall i} = \{H(H(T(i)_B^r)^{k=A})^{k=B}\}_{\forall i}$, to decrypt the returned sequence using his secret key. This now allows Bob to see the size of the overlap between the elements of his sequence, $\{H(T(i)_B^r)^{k=A}\}$, and Alice's sequence, $\{H(T(i)_A^r)^{k=A}\}_{\forall i}$, both encrypted using Alice's secret.

This works simply when Alice & Bob operate at the same resolution r. However, there may be instances where either Alice or Bob want to share information at a lower resolution than the other. This adds difficulty, as the encrypted geohashes will not be able to match if $r_A \neq r_B$. To address this, Alice can share her mobility trace at their chosen resolution and below *simultaneously*. Bob should have a choice in how his algorithm matches on different resolutions. If Bob wants to match on only the best possible resolution, then he can perform the cardinal private set intersection repeatedly, starting from his highest resolution down to

$r = 1$ or until he finds a non-zero overlap. Alternatively, Bob can just perform a set intersection using all his available resolutions in one go, rewarding Alice for sharing more detailed data. To make this feasible, one should limit the resolution to below $r = 9$, since GPS's accuracy becomes untenable below the roughly one-meter size of a resolution 9 geohash.

Adding to this, we can incorporate time into the match. So far we've taken the locations to be a bag-of-words style collection, however matching on time of day is a simple extension by sharing tuples of location-time pairs, where time is binned into hourly segments.

Computing these private set intersection cardinalities is fast, requiring only three transmissions as serialized protocol buffers, and can be done with existing open source software [17]. Once these mobility overlaps are calculated, we can turn to how they can be used to populate a feed.

To echo the physical-world social capital we're hoping to reinforce, we want the time we see someone's content in the feed to be proportional to the time they're likely to see each other in person. At a first pass, it would seem ideal for the probability of Alice being shown in Bob's feed, $P(A)$, to be proportional to the simple overlap of their locations, $P(A) \propto |T_A \cap T_B|$. However, to avoid Alice gaming the system by sharing all, or a large number of possible geohashes, we need to normalize by the size of Alice's location set. Taking a simple proportion, $\frac{|T_A \cap T_B|}{|T_A|}$, would mean that Alice is incentivized to share few locations that are highly likely to match, and penalizes sharing high-quality data. As such, we normalize by $\sqrt{|T_A|}$ to balance these tradeoffs. This gives the simple formula of $P(A) \propto \frac{|T_A \cap T_B|}{\sqrt{|T_A|}}$.

In general, one benefit of this content ranking paradigm is its ability to incentivize people to share data in an attempt to appear on their friend's feeds, helping address the cold-start problem. However, to facilitate seeing friends who post little or no content, Laplace smoothing could be applied to the ranking weighting.

2.1 An Extendable Paradigm

This method, although grounded in mobility traces and its value for building social capital, is extremely extensible to other private data. Any common set of interests or activities could be used to build a recommendation system. In particular, where individuals may want to privately hold data and not share it with a wider community is ideal for this paradigm. Examples include: existing friendship ties (which many on social media choose to keep private), which could be used to over-emphasize showing the content of friends who have many mutual friends; a set of private interests (for example, the collection of movies on Netflix one has watched); or a stated set of shared goals (are you looking to find people on this service that are here to make jokes, discuss politics, or look for new friends outside your current circle), without having to publicly declare such a desire. Beyond simple set matching, intersection on more complex data (akin to the problem with GPS locations) can also be addressed.

As an example, take the case of wanting to see posts from friends who have the same (or very similar) photos from parties (often transmitted between friends via AirDrop, messaging platforms, or photo sharing platforms rather than posting publicly) or, in technically identical but somewhat different framing, those with similar memes stored as you. While transmitting these photos to perform set intersection is feasible, minor differences in photos (e.g., cropping, editing, etc.) and transmission costs from photo sizes make this suboptimal. Instead, we can use perceptual hashing [20] as a stand-in for the geohashing technique. This is an almost exactly analogous approach as the algorithm above, this time just optimizing for shared experiences and memories or shared cultural taste.

This in turn goes a long way to address a fundamental question: do users actually want this feature? In a bottom-up social media ecosystem, users have the choice to use, or not use, any given set of features so long as there is a plurality of interoperable interface providers. In general, users may desire not to optimize for social capital, preferring plain entertainment, or purely shared taste as in the above example. While no solution will be possible without any user demand, this approach works to incentivise user uptake of data availability for matching through the desire to appear on friends' feeds, in turn requiring you to share data.

2.2 A View Towards Web3

While altering the feeds of traditional social media platforms is challenging by design, the emergence of Web3 social protocols makes building this into a service significantly more feasible. Public social graphs such as the Lens protocol or the AT Protocol by Bluesky Labs allow for direct querying of friends' addresses/handles which could be used to request a private set intersection on mobility data. Tying into this, Web3 has presented new paradigms for storing private data [18] that can be requested on-chain through toolkits such as Verida Vault or Disco Data Backpacks, multi-chain protocols for interoperable database storage. In general, any system acting as a personal data store (such as open-PDS [12,13]) will enable this approach.

The composable nature of technology in Web3 is oriented towards allowing systems like feeds, content networks, and social graphs to be interoperable and interchangeable. While this is one possible tool in the plural toolkit of social feeds, as we will see in the validation section below, the tool developed here is surprisingly strong for enabling real-world connections and community social capital.

3 Validation and Demonstration

While, the relationship between mobility traces and friendships are well established [5,7], it is important to validate that even after geohashing the locations they remain a predictor of friendship. We draw on three datasets of real human behaviors to demonstrate the performance.

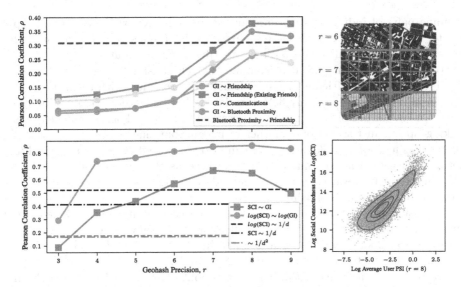

Fig. 1. Diverse measures of friendship correlate with set intersections of geohashed location histories between individuals. With sufficient resolution (r), the geohash intersection (GI) between members of a local community [1] is more predictive of self-reported strength of friendship for existing friends and almost as predictive of baseline friendship as bluetooth proximity (top left). More generally, in the bottom row, averages across zip code pairs of all pairwise GI are strongly correlated with Facebook's social connectedness index (SCI) [2] across multiple resolutions, outperforming a zip-zip gravity model.

To show that friendship can be predicted at the individual pair level, we used the Friends and Family experiment where data was collected on 130 residents of a young-family residential living community adjacent to a major research university in North America [1]. This data included GPS coordinates of individuals collected through their mobile phones in addition to Bluetooth proximity between participants' phones in the dataset, call and SMS records between participants, and surveys completed at regular intervals where participants were asked about their perceived strength of friendship with one another (on a scale of zero to seven, where zero is no friendship and 7 is the closest possible friend or partner).

As shown in the top left of Fig. 1, we can geohash these GPS locations at varying levels of resolution to compare how this geohash precision r maps onto reported friendships. To examine this relationship, we apply the same procedure as above, finding set intersection cardinalities between geohashes of individual mobility traces. As a baseline, we compare the relationship between Bluetooth proximity, a measure of who you're physically nearby at all times, and the reported friendship. At high resolution, the private set intersection cardinality of your geohashed mobility data is as powerful a predictor of your self-reported friendships as the baseline from Bluetooth proximity; and moreover, if we only examine pairs of individuals who are already friends (reporting any friendship greater than zero), we find that the strength of this friendship can be explained better by your location trace than by your Bluetooth proximity.

To help explain this, we can examine the relationship between your location intersection with friends and your Bluetooth proximity with them. Even at high resolutions, these are only 35% correlated. This is largely explained by the time window nature of the algorithm presented in Sect. 2. The mobility trace is picking up not just seeing one another in person, but sharing a common set of interests as measured by location. We can also see this effect in the relationship between location and communication (measured as the sum of total calls and SMS messages between each pair of individuals) where again, communication does not fully explain friendship.

To ensure this generalizes beyond this small community, we combined two large-scale recent datasets. First, a collection of privacy-preserving mobility traces of over 200K anonymized and opted-in individuals' mobile phones from five major US metropolitan areas over 2016 and 2017, provided by Cuebiq[1]. For each pair of zip codes within each metropolitan area, the location geohash intersection is calculated between every combination of individuals who sleep overnight in the zip codes. The average zip-to-zip location intersection is then compared to Meta's publicly available social connectedness index (SCI), a measure of how many Facebook friendships exist between two zip codes [2].

The results are broadly consistent across the metropolitan areas, with the bottom of Fig. 1 showing the relationship between SCI and the geohash intersection (GI) for Boston. The left subfigure shows the correlation of these zip-zip measures across geohash resolutions, with higher resolutions well outperforming a gravity model (both inverse of distance and inverse of distance squared). The rightmost subfigure shows this relationship for a high-resolution geohash with added 2D kernel density mapping.

3.1 Reference Architecture

To help demonstrate how this algorithm could be deployed we have implemented a reference architecture built on these principles and created a small demo app to accompany it[2]. At a basic level, this app allows a user to enter simple records on the client end and have them matched via a secure private set intersection cardinality with an existing prefilled server. This is done entirely through Javascript, and is an extremely low computational cost operation that can be done quickly on both the client and server side. This demo allows for users to match with not only mobility data, but other list-style data as well.

Much future work exists to build out this algorithm and toolkit into a broader product. In order to effectively interface this algorithm with its use case, you need a feed of content to leverage off of. This could be done by attempting to alter one's existing social feeds (e.g. Twitter) using their APIs or through manipulation of their webpages, but cleaner solutions exist using new Web3 social protocols.

---

[1] https://www.cuebiq.com/.
[2] https://github.com/tobinsouth/PSI-Social-Feed.

4   Discussion

4.1   Social Rankings Gamability

Many existing social and information feeds have resulted in content creators working to 'game' the feed through optimizing their content to match the algorithm. In contrast, the ranking proposed here deliberately makes it hard to promote oneself more widely than deserved.

If individuals try to game the system by sharing many location points, they're penalized. Popular influencers can't be everywhere at once to be seen on everyone's feed, and even if a user fabricates a large amount of location data, that would be penalized during normalization. Showing few high-resolution places is the best way to appear in the feeds of those you're most likely to bond with. As a result, the 'fitness' of any given person in the algorithm is fundamentally limited, stopping the possibility of dominant agents emerging [11].

4.2   Sharing Incentives

This protocol only works as users make their data available for private set matching, creating a cold start problem. Fortunately, as a protocol like this proliferates, users have an incentive to begin sharing their data as soon as their friends do to optimize for exposure on their friends' feeds.

This cold start problem is not insurmountable. Recent years have seen the rise in alternative social media platforms which have captured mass adoption through creating features that meaningfully consider the negative effects of previously dominant social media offerings (such as BeReal).

4.3   Echo-Chambers

By design, this approach to building a social feed reinforces and strengthens existing ties between people and communities. However, this same principle is inherently a driver of echo-chamber formation [15]. In a context where people have shared experiences, context, or location data with people of the same opinions, this algorithm would reinforce ideological segregation and contribute to political polarization. At a fundamental level, this is already at play in the ways we interact with our share physical spaces [14]; this algorithm may act to reinforce this tendency.

Ensuring broad exposure to ideas to mitigate political and social fracture is ideal. In general, this is far from a solved problem in any social media [8] but is an important consideration in the design and ultimate rollout of any feed approach.

The proposed Laplace smoothing goes some way to address the echo chamber problem, but we could also draw from existing literature and experience on avoiding polarization and echo-chamber formation. One such approach could be to periodically show content from non-similar users. This could be achieved by effectively adding noise to the recommendation feed; which, while potentially making a less desirable feed for the user, would ensure more diversity of exposure.

Finally, it is worth noting that the type of homophily being reinforced by this approach is different in character from echo-chamber formation in traditional social media. Traditional social media explicitly ideologically narrows your content based on engagement patterns, allowing some preferences and interests to be exemplified and radicalized as you are presented more concentrated agreeing opinions. In contrast, this approach does not immediately send individuals down specific content 'rabbit holes,' potentially leading toward radicalization based only on casual initial interest. In this case, any polarization that occurs happens in step with one's expressed real-world interests and beliefs, and is complementary to physical-world interactions and conversations.

5   Conclusion

This work presents a novel approach to constructing a social feed that represents relationships with friends in terms of real-world quantities beyond social media activity. This is achieved without exposing one's data publicly or relying upon third-party data brokers via private set intersection cardinalities applied to personal datastores. The system supports a variety of data for intersection, focusing firstly on the value of private location traces as tools to measure the likelihood of real-world interaction. Geohashing is used to map real-valued GPS points to strings of arbitrary resolution for matching. The value of location hashes for social capital is validated using existing data on friendships and location, and a demonstration app is built. This work presents an altogether different strategy for populating social feeds by focusing on private data sharing, user consent, and building social capital.

Acknowledgements. We thank Dinh Tuan Lu for his contributions to the code-base for the demonstration of this idea.

References

1. Aharony, N., Pan, W., Ip, C., Khayal, I., Pentland, A.: Social fMRI: investigating and shaping social mechanisms in the real world. Pervasive Mob. Comput. **7**(6), 643–659 (2011)
2. Bailey, M., Cao, R., Kuchler, T., Stroebel, J., Wong, A.: Social connectedness: measurement, determinants, and effects. J. Econ. Perspect. **32**(3), 259–80 (2018)
3. Chetty, R., Hendren, N., et al.: The impacts of neighborhoods on intergenerational mobility: childhood exposure effects and county-level estimates. Harvard Univ. NBER **133**(3), 1–145 (2015)
4. Claudel, M., Massaro, E., Santi, P., Murray, F., Ratti, C.: An exploration of collaborative scientific production at MIT through spatial organization and institutional affiliation. PLoS ONE **12**(6), e0179334 (2017)
5. Dong, W., Lepri, B., Pentland, A.: Modeling the co-evolution of behaviors and social relationships using mobile phone data. In: Proceedings of the 10th International Conference on Mobile and Ubiquitous Multimedia, pp. 134–143 (2011)

6. Dong, X., Suhara, Y., Bozkaya, B., Singh, V.K., Lepri, B., Pentland, A.S.: Social bridges in urban purchase behavior. ACM Trans. Intell. Syst. Technol. **9**(3), 1–29 (2017)

7. Eagle, N., Pentland, A.S., Lazer, D.: Inferring friendship network structure by using mobile phone data. Proc. Natl. Acad. Sci. **106**(36), 15274–15278 (2009)

8. Fernandez, M., Bellogin, A.: Recommender systems and misinformation: the problem or the solution? In: OHARS Workshop. In: 14th ACM Conference on Recommender Systems. OHARS Workshop. 14th ACM Conference on Recommender Systems (2020)

9. Jaques, N., Taylor, S., Azaria, A., Ghandeharioun, A., Sano, A., Picard, R.W.: Predicting students' happiness from physiology, phone, mobility, and behavioral data. In: 2015 International Conference on Affective Computing and Intelligent Interaction (ACII), pp. 222–228 (2015)

10. Kubin, E., von Sikorski, C.: The role of (social) media in political polarization: a systematic review. Ann. Int. Commun. Assoc. **45**, 188–206 (2021)

11. Lera, S.C., Pentland, A.S., Sornette, D.: Prediction and prevention of disproportionally dominant agents in complex networks. Proc. Natl. Acad. Sci. U.S.A. **117**, 27090–27095 (2020)

12. de Montjoye, Y.A., Shmueli, E., Wang, S.S., Pentland, A.S.: openPDS: protecting the privacy of metadata through SafeAnswers. PLoS ONE **9**(7), e98790 (2014)

13. de Montjoye, Y.A., Wang, S.S., Pentland, A., Anh, D.T.T., Datta, A., et al.: On the trusted use of large-scale personal data. IEEE Data Eng. Bull. **35**(4), 5–8 (2012)

14. Moro, E., Calacci, D., Dong, X., Pentland, A.: Mobility patterns are associated with experienced income segregation in large us cities. Nat. Commun. **12**(1), 1–10 (2021)

15. Nasim, M., et al.: Are we always in strife? a longitudinal study of the echo chamber effect in the Australian Twittersphere. arXiv preprint arXiv:2201.09161 (2022)

16. Niemeyer, G.: Geohash (2008). Accessed 6 Jun 2018

17. OpenMined: Private set intersection cardinality protocol based on ECDH and bloom filters. OpenMined (2022)

18. Pentland, A., Lipton, A., Hardjono, T.: Building the New Economy: Data as Capital. MIT Press (2021)

19. Putnam, R.D., et al.: Bowling Alone: The Collapse and Revival of American Community. Simon and Schuster, New York (2000)

20. Zauner, C.: Implementation and benchmarking of perceptual image hash functions. Ph.D. thesis, University of Applied Sciences Hagenberg (2010)

Classifying Policy Issue Frame Bias in Philippine Online News

Jose Mari Luis M. Dela Cruz[✉][iD] and Maria Regina Justina E. Estuar[iD]

Department of Information Systems and Computer Science,
Ateneo de Manila University, Quezon City 1108, Philippines
mari.delacruz@obf.ateneo.edu, restuar@ateneo.edu

Abstract. Media plays an important role in disseminating news to the public [1,14]. However, the selection of what information to write about, how information is presented, and when it is broadcasted are within the control of the media outlet. Media can therefore shape the consumers' opinion based on how publicly available news [1,9] contents are published. Framing bias occurs when the author selects and highlights aspects of the news [8]. This study developed a model to classify the policy issue frames in select Philippine online news articles. Media Frame Corpus [5] was tested on the classification of policy issue frames using supervised learning methods including Bidirectional Encoder Representations from Transformers (BERT), Long Short-Term Memory (LSTM), Gated Recurrent Unit (GRU), linear support vector machine (SVM), and logistic regression (LR). Results showed that the BERT model performed best with an accuracy of 74.73% with political frame (Frame 13) and economic frame (Frame 1) as leading policy issue frames. Implementing the MFC-BERT classification model on the Philippine dataset shows that there is a dominant policy issue frame across all selected media outlets. However, there is a significant difference in the usage of policy issue frames among these media outlets except for the Fairness and Equality frame (Frame 4) and the Quality of Life frame (Frame 10). Initially, ambiguous topics among media outlets when observed over time exhibit noticeable policy issue framing bias. However, a gradual transition to having similar policy issue frames among most media outlets occurred to these topics, similar to the unambiguous topics.

Keywords: Framing · Policy Issue Frames · Philippine Online News Media · Text Classification

1 Introduction

Media circulates news such as political affairs, economic circumstances, and other issues around the globe [1,14]. News dissemination comes in many forms and sources including print media, broadcast media, and online media. The public

Ateneo de Manila University - Social Computing Laboratory of Ateneo.

depends on news articles as one of the primary sources of information source [1,9]. With the advent of the Internet and the emergence of different online platforms, news articles are now available online. These online articles are now considered vital information sources that extend traditional media sources and compete with new sources like social media [7,9,14].

The media's most integral obligation in society is to communicate vital information and events to the people [10,13,15], dedicated to having informed citizenship [15]. However, different news media outlets have different ways of reporting the news. Given that news articles have a vital function in forming individuals' opinions [1,9], the variations in narratives on how the media presents an issue influences the consumers' recognition and understanding of the event [13]. This process of selecting and highlighting aspects of news is called framing, and the actual presentation of the news is called a frame [8].

Studies on policy issue frames, also known as issue definitions, have been a topic of research for over a decade [2,3]. Policy issues are usually one of the more contentious topics in news writing. Policy issues are defined as economic frames, morality frames, political frames, and quality of life frames [5,6] to name a few. These frames are used by politicians, the media, and the voting public to communicate policy issues [3]. The Policy Frames Codebook [4] is a non-issue specific framing schema, which cuts across all policy issues providing a system for categorizing framing cues across policy issues [3]. Media Frame Corpus (MFC) is an established dataset composed of 39,395 news articles primarily from Western News articles on various policy issues such as tobacco, same-sex marriage, immigration, gun control, and the death penalty, annotated in media framing that used the Policy Frame Codebook as a guide [5].

This study aims to test the feasibility of developing a policy issue frame classification model from the Media Frame Corpus that can be applied to Philippine online news.

2 Methodology

Bidirectional Encoder Representations from Transformers (BERT), Gated Recurrent Unit (GRU), Long short-term memory (LSTM), Linear Support Vector Machine (SVM), and Logistic Regression (LR) models were used on Media Frame Corpus (MFC) for policy issue frame classification. The best-performing model was used to classify policy issue frames using the dataset extracted from Philippine online news articles.

2.1 Selection of Media Frame Corpus (MFC) Training Dataset

MFC contains annotations of policy issue framing on various issues. The MFC dataset was narrowed down to sentences and phrases where agreement was found with at least two annotators. The final dataset resulted in a total of 46799 rows of sentences or phrases. The distribution of the 15 policy issue frames used in MFC is shown in Fig. 1.

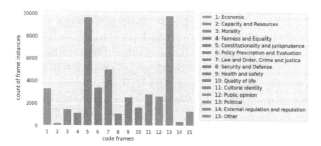

Fig. 1. Distribution of the different policy issue framing - code frames found in the MFC.

2.2   Selection of the Philippine Online News Dataset

GMA, Rappler, Manila Bulletin, Philippine News Agency (PNA), and Sunstar Philippines were used to represent Philippine online news. Selection criteria for online news included popularity, use of the English language, and availability of extracting news articles for a longer time period. The final Philippine dataset comprised of headlines and leads of the online news hailing from the different media outlets. A total of 21603 most recent news from online national news articles written in English were extracted from May 5, 2023, to May 7, 2023, using the first 2000 news articles from each media outlet.

Beautifulsoup and selenium packages were used for the extraction of online news articles. Standard data extraction methods were used. A Chrome driver was used to visit the media outlets' pages for national headline news. The HTML elements tags that lead to relevant article links and metadata (e.g. <p>, <li>) were identified for each media outlet. Using these identifiers, each news' URL, author, headline, and date were collected. The Chrome driver was used to scrape the metadata and the content of the article using predetermined HTML element tags of each media outlet from this list.

A dataframe is composed of rows of news articles, including metadata such as the source, headline, and date. The standard method including removing links, photo captions and references, and media outlets' specific terms (e.g. media outlet site name), and the separation of each paragraph of news articles into respective rows were used for pre-processing.

Four variations of the dataset were developed. The first dataset comprised of news articles from all media outlets. The second dataset comprised of news articles separated per media outlet. Criteria for the third dataset, comprised of nonambiguous articles narrowed down using publications from '2023-04-23' to '2023-04-29', and selected using top keywords: 'Sudan', 'Balikatan', 'SIM card registration'/'Sim registration'. The fourth dataset comprised of ambiguous articles using the following keywords, 'COVID'/'Covid', 'DepEd', and 'China'.

2.3   Developing Classification Models

Bidirectional Encoder Representations from Transformers (BERT), Gated Recurrent Unit (GRU), Long short-term memory (LSTM), and supervised machine learning classifiers (Linear Support Vector Machine and Logistic Regression) were used in training the model on the MFC dataset. Tokenization was used for BERT. Word embeddings were used for LSTM and GRU.term frequency-inverse document frequency (tf-idf) was used for Linear SVM and logistic regression. All models used 85–15 train-test split, stratified according to the 15 frames.

2.4   Classification of Policy Issue Frames on Philippine Dataset Using BERT

The resulting trained model on BERT was used to classify policy issue frames on all variations of the Philippine dataset, where each sentence or phrase is limited to only one policy issue frame. Accuracy score was used to test the performance of the model.

2.5   Testing the Difference of Policy Frames Among Media Outlet

Chi-square goodness of fit was used to test for the significance of differences in the use of policy issue frames labeled per article among the five media outlets. Actual frame counts were used for observed values while the mean score of observed values was used for expected values.

3   Results and Discussion

Accuracy scores resulting from the MFC training for all the classification methods are shown in Table 1. Results showed the accuracy scores, specifically for the BERT, LSTM, and GRU are similar to reference studies that were also trained using MFC [11,12] validating the use of MFC corpus for establishing ground truth.

Table 1. Performance Metric of the Models trained with MFC

Model	Accuracy(%)	Referenced Studies' Accuracy
BERT	74.73	71.50
LSTM	70.78	67.80
GRU	70.74	68.70
SVM	63.07	N/A
LR	68.25	N/A

BERT model was the highest performing model, followed by LSTM, and then GRU. On the other hand, the SVM and LR were the least accurate.

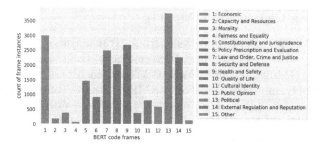

Fig. 2. Distribution of the policy issue frames on the aggregated dataset using BERT Model

3.1 Frame Classification on the Aggregated Philippine Dataset

Figure 2 shows the frequency count of resulting frame classification when BERT model was implemented on the Philippine dataset. The political frame (Frame 13) had the highest count followed by the economic frame (Frame 1). The next dominant frame is the health and safety frame (Frame 9).

3.2 Comparison of Frame Classification per Media Outlet

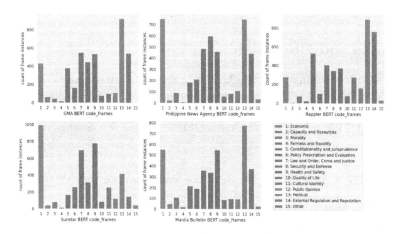

Fig. 3. Distribution of the policy issue frames on the dataset per Media Outlet using BERT Model

Figure 3 shows the frequency count of frame instances of each policy issue frame in each media outlet. Specifically for the BERT model, when the dataset is divided into different media outlets the general observation is that the prevalent frames vary for each media outlet. The political frame (Frame 13) is the frame with the highest count from most outlets however, the distribution of the remaining frames varies per outlet. Furthermore, the test for goodness of fit showed that there is a significant difference among preferences on policy issue frame per media outlet for all frames [p < 0.05] except for the Fairness and Equality frame (Frame 4) [p = 0.14] and the Quality of Life frame (Frame 10) [p = 0.15].

3.3 Comparison of Frame Classification per Media Outlet on Unambiguous Topics

Table 2. Comparison of Top 3 Policy Issue Frames where topic = Sudan.

| | Sudan Top Frame | | |
	Top 1	Top 2	Top 3
GMA	Economic	Security and Defense	Political
PNA	Security and Defense	External Regulation and Reputation	Health and Safety
Rappler	Security and Defense	Quality of Life	Capacity and Resources/External Regulation and Reputation
Sunstar	Security and Defense	Economic	External Regulation and Reputation
Manila Bulletin	Security and Defense	External Regulation and Reputation	Economic

Table 3. Comparison of Top 3 Policy Issue Frames where topic = Balikatan.

| | Balikatan Frame | | |
	Top 1	Top 2	Top 3
GMA	Security and Defense	External Regulation and Reputation	Cultural Identity/Law and Order, Crime and Justice
PNA	Security and Defense	External Regulation and Reputation	Cultural Identity
Rappler	Security and Defense	External Regulation and Reputation	Political
Sunstar	Security and Defense	External Regulation and Reputation	Cultural Identity
Manila Bulletin	Security and Defense	External Regulation and Reputation	Cultural Identity

Table 4. Comparison of Top 3 Policy Issue Frames where topic = SIM Registration.

| | SIM Registration Top Frame | | |
	Top 1	Top 2	Top 3
GMA	Policy Prescription and Evaluation	Political	Constitutionality and Jurisprudence
PNA	Policy Prescription and Evaluation	Economic	Political
Rappler	Constitutionality and Jurisprudence	Policy Prescription and Evaluation	Political
Sunstar	Policy Prescription and Evaluation	Economic	Constitutionality and Jurisprudence
Manila Bulletin	Policy Prescription and Evaluation	Economic	Political

Selected topics include the Repatriation of Filipinos from Sudan, the Balikatan Exercise between the United States and the Philippines, and Sim Card Registration.

Frames per Issue: Tables 2, 3, and 4 show the top 3 frames with the highest count per media outlet. The Sudan topic, illustrated in Table 2, is about the repatriation of Filipino from strife-torn Sudan and the actions of the Philippine government in helping safeguard the welfare of the Filipinos. All media outlets, except GMA, used the security and defense frame (Frame 8) as the top frame for the issue. GMA, however, used the economic frame (Frame 1) as its number one frame, followed by the security and defense frame (Frame 8). It is observable, however, that media outlets do not have the same succeeding frames.

The Balikatan topic, depicted by Table 3, is about the annual joint military exercise between the United States Army and the Philippine Army. It is observable that the security and defense frame (Frame 8) is the consensus top 1 frame for all media outlets with external regulation and reputation frame (Frame 14) as the top 2 frame. However, the top 3 frame differs for each outlet.

The SIM topic in Table 4 is about how the SIM registration ACT is being implemented in the Philippines and the challenges faced by policymakers as well as the public. The policy prescription and evaluation frame (Frame 6) is the dominant frame for all outlets except Rappler, which has the constitutionality and jurisprudence frame (Frame 5) as its top frame. Similarly, the succeeding secondary and tertiary frames vary per media outlet. From these observations, we can infer that each topic has a primary frame.

3.4 Comparison of Frame Classification per Media Outlet on Ambiguous Topic

Topics were selected to test if policy issue frame bias will be present when topics are ambiguous. Figure 4, 5, and 6 show the policy issue frame count distribution per Media Outlet from March of 2023 to May of 2023.

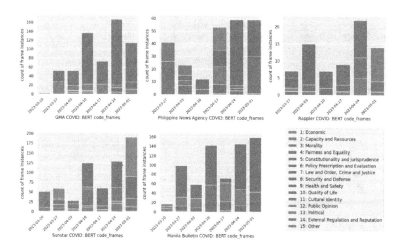

Fig. 4. Distribution of Policy issue frames over time per Media Outlet on keyword = COVID.

Figure 4 shows at first glance that the dominant frame is Health and Safety frame (Frame 9). However, it is also observed that there are changes in the use of policy issue frames over time which varies across media outlets. For example, there was a shift from Health and Safety frame (Frame 9) to the Economic frame (Frame 1) in Sunstar. Additionally, Rappler shifted its dominant frame to Political frames (Frame 13) from the Health and Safety frame (Frame 9).

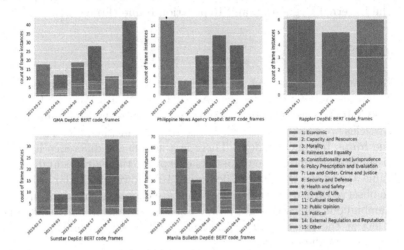

Fig. 5. Distribution of Policy issue frames over time per Media Outlet on keyword = DepEd.

Figure 5 shows that only three out of the five media outlets begin with the Political frame (Frame 13) and transition to different frames over time. Rappler, on the other hand, with limited stories on DepEd started with Policy Prescription and Evaluation frame (Frame 6) as its initial frame, then transitioned to the

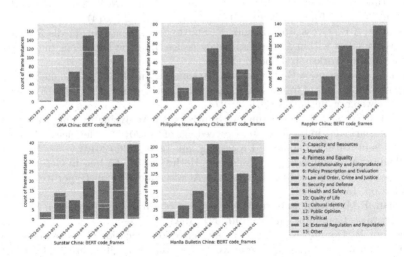

Fig. 6. Distribution of Policy issue frames over time per Media Outlet on keyword = China.

Political frame (Frame 13) as its final frame. Manila Bulletin shows two policy issue shifts from Policy Prescription and Evaluation frame (Frame 6), followed by the Political frame (Frame 13), then ended with the Cultural Identity frame (Frame 11).

On the other hand, Fig. 6 shows one dominant policy issue frame over time namely, the External Regulation and Reputation frame (Frame 14).

4   Conclusion

The Media Frame Corpus served as ground truth to predict policy issue frames for the Philippine online news dataset. Albeit trained on a Western-based corpus, the MFC-BERT classification model is applicable to the Philippine dataset. When all articles are combined as a whole regardless of media outlets, a dominant policy issue frame becomes evident. There is a significant difference in the use of policy issue frames among media outlets except for the Fairness and Equality frame (Frame 4) and the Quality of Life frame (10). However, similar policy issue frames are observed when unambiguous topics are chosen as an additional criterion. For ambiguous topics though, policy issue frame bias is noticeable among media outlets at the onset. Policy issue frame shift towards similar usage over time across most media outlets.

References

1. Al-Hindawi, F.H., Ali, A.H.: A pragmatic study of CNN and BBC news headlines covering the Syrian conflict. Adv. Lang. Literary Stud. **9**(3), 43–51 (2018)
2. Bogaerts, J., Carpentier, N.: The postmodern challenge to journalism: Strategies for constructing a trustworthy identity. In: Rethinking Journalism, pp. 60–71. Routledge (2013)
3. Boydstun, A.E., Card, D., Gross, J., Resnick, P., Smith, N.A.: Tracking the development of media frames within and across policy issues (2014)
4. Boydstun, A.E., Gross, J.H., Resnik, P., Smith, N.A.: Identifying media frames and frame dynamics within and across policy issues. In: New Directions in Analyzing Text as Data Workshop, London (2013)
5. Card, D., Boydstun, A.E., Gross, J.H., Resnik, P., Smith, N.A.: The media frames corpus: annotations of frames across issues. In: Proceedings of ACL (2015), https:// www.aclweb.org/anthology/P15-2072/
6. Card, D., Gross, J.H., Boydstun, A., Smith, N.A.: Analyzing framing through the casts of characters in the news. In: Proceedings of the 2016 Conference on Empirical Methods in Natural Language Processing, pp. 1410–1420 (2016)
7. Dallmann, A., Lemmerich, F., Zoller, D., Hotho, A.: Media bias in German online newspapers. In: Proceedings of the 26th ACM Conference on Hypertext & Social Media, pp. 133–137 (2015)
8. Entman, R.M.: Framing: towards clarification of a fractured paradigm. McQuail's Reader Mass Commun. Theory **390**, 397 (1993)
9. Hamborg, F., Donnay, K., Gipp, B.: Automated identification of media bias in news articles: an interdisciplinary literature review. Int. J. Digit. Libr. **20**(4), 391–415 (2019)

10. Jenkins, J., Nielsen, R.K.: Proximity, public service, and popularity: a comparative study of how local journalists view quality news. J. Stud. **21**(2), 236–253 (2020)
11. Liu, S., Guo, L., Mays, K., Betke, M., Wijaya, D.T.: Detecting frames in news headlines and its application to analyzing news framing trends surrounding us gun violence. In: Proceedings of the 23rd Conference on Computational Natural Language Learning (CoNLL) (2019)
12. Naderi, N., Hirst, G.: Classifying frames at the sentence level in news articles. Policy **9**, 4–233 (2017)
13. Pristianita, S., Marta, R.F., Amanda, M., Widiyanto, Y.N., Boer, R.F.: Comparative analysis of online news content objectivity on COVID-19 between detik. com and kompas. com. Informatologia 53 (2020)
14. Spinde, T., et al.: Automated identification of bias inducing words in news articles using linguistic and context-oriented features (2021). https://doi.org/10.1016/j.ipm.2021.102505, https://www.sciencedirect.com/science/article/pii/S0306457321000157
15. Tiffen, R., et al.: Sources in the news: a comparative study. J. Stud. **15**(4), 374–391 (2014)

Dismantling Hate: Understanding Hate Speech Trends Against NBA Athletes

Edinam Kofi Klutse, Samuel Nuamah-Amoabeng, Hanjia Lyu[✉],
and Jiebo Luo

University of Rochester, Rochester, NY 14627, USA
hlyu5@ur.rochester.edu, jluo@cs.rochester.edu

Abstract. Social media has emerged as a popular platform for sports fans to express their opinions regarding athletes' performance. The National Basketball Association (NBA) is widely recognized as one of the most popular sports leagues globally. However, an unfortunate aspect that has emerged in recent years is the presence of abusive fans within the league. Consequently, the focus of this research is to identify which NBA athletes experience abuse on Twitter and delve deeper into the underlying reasons behind such mistreatment. To address the research questions at hand, the study employs a curated set of keywords to query the Twitter API, gathering a comprehensive collection of tweets that potentially contain hate speech directed toward NBA players. A deep learning classification model is implemented, effectively identifying tweets that genuinely exhibit hate speech. We further use keyword search methods to detect the specific groups that are targeted by hate speech the most and identify topics of hate speech tweets. The findings of our research indicate that certain groups of athletes are particularly vulnerable to hate speech from fans. Notably, high-performing athletes, `Black` athletes, overweight athletes, short athletes, and athletes associated with the `LGBTQ` community are found to be highly susceptible to abusive remarks. Racism, physique shaming, play style, and anti-LGBTQ remarks are the major themes. These findings contribute to a broader understanding of the challenges faced by NBA athletes in the digital space and provide a foundation for developing strategies to combat hate speech and foster a more inclusive environment for all individuals involved in the NBA community.

Keywords: Hate speech · NBA · Social media · Natural language processing

1 Introduction

In recent years, professional athletes in the National Basketball Association (NBA) have increasingly expressed their concerns about being subjected to hatred and abuse from fans and media personnel on various social media platforms [11]. Among these platforms, Twitter has emerged as a prominent arena where fans can directly engage with players, making it a hotspot for hate speech

R. Thomson et al. (Eds.): SBP-BRiMS 2023, LNCS 14161, pp. 74–84, 2023.
https://doi.org/10.1007/978-3-031-43129-6_8

directed toward NBA athletes. Unfortunately, the prevalence of derogatory language and abusive behavior on Twitter persists despite efforts to combat it [11]. Consequently, basketball players in the NBA have become a vulnerable target group for hate speech abuse.

Within this context, it is essential to address two key questions: Who are the athletes experiencing abuse? And what are the underlying reasons behind this mistreatment? In this study, we curate a set of hate speech-related keywords to collect tweets that potentially contain hateful content against NBA players. We then employ a deep learning model to detect hate speech tweets. The keyword search methods are used to detect the specific groups of athletes that are targeted by hate speech. By analyzing the collected data, the study aims to uncover the major themes prevalent in these hate speech tweets. Next, we conduct correlation analysis on a series of players' performance statistics, their demographics, as well as their physical characteristics. Our study seeks to obtain insights into the underlying motivations behind hate speech abuse in the NBA.

2 Method

2.1 Hate Speech Detection

Detecting hateful content on Twitter is not a trivial task because users may use certain codes to avoid detection by automated systems [8,9]. Other challenges may include linguistic subtleties, varying definitions of hate speech, and limited access to data for training and testing such systems [7]. In our study, we first use keywords to collect tweets that may contain hateful content and then employ a transformer-based language model to perform the final classification.

Data Collection. We use Twitter's API - Tweepy, to gather tweets containing potential hate speech targeting NBA players. To collect such tweets, we first compile a list of hate speech-related keywords. Previous research has indicated that online hate speech can stem from various motivations, including but not limited to racial discrimination, gender-based targeting, and body shaming.[1] For instance, Powell *et al.* [10] found that transgender individuals experience higher rates of digital harassment and abuse overall, and higher rates of sexual, sexuality, and gender-based harassment and abuse, as compared with heterosexual cisgender individuals. By employing this methodology, we aim to gather a dataset that captures the diverse manifestations of hate speech directed at NBA players on social media. In particular, the keyword list is composed of *nigger, nigga, bitch, b*tch, n*gg*r, fuck, bum, motherfucker, bollock, wanker, dirty, lame, bozo, faggot, pussy, f*ck, piece of shit, sh*t, bastard, cock, gay, lesbian, fucker, fool, cunt, asshole, hate, stupid, useless, fraud, cost me, owe me, lost money, liar, trash, ass, overrated, flop, flopper, flopping, coward, choker, choke artist, loser,*

---

[1] https://www.news24.com/sport/tennis/commentator-dokic-hits-out-at-fat-shaming-trolls-at-australian-open-20230123.

choking, selfish, stat padder, ball hog, stat pad, soft, weak, retard, prick, dick, dickhead. The combinations of the names of current NBA players ($n = 461$, obtained from basketballreference.com) and hate speech-related keywords are used to query tweets through Tweepy. In the end, we identify a total of 503,424 tweets of potential hate speech targeting current NBA players.

Modeling. A tweet that contains hate speech-related lexicons might be an instance of *offensive language* instead of hate speech which is defined as "language that is used to express hatred toward a targeted group or is intended to be derogatory, to humiliate, or to insult the members of the group" [2]. Therefore, we further leverage a transformer-based language model to detect hate speech from the collected tweets. Transformer-based models have demonstrated exceptional performance in text classification tasks across various domains [1,6,13]. In particular, we first use an open-source hate speech dataset built by Davidson *et al.* [2] to train a BERT model [3]. We then use the trained model to detect hate speech from our data corpus.

The dataset of Davidson *et al.* [2] contains 24,783 tweets of three categories - *hate speech, offensive language,* and *neither.* To facilitate model training and evaluation, the dataset is split, allocating 90% for the training set and the remaining 10% for the testing set. We preprocess the dataset by removing stop words using the `wordcloud` package. We then use the `bert_en_uncased_preprocess` model to convert plain text inputs into tokens that are expected by BERT. The classifier is composed of a BERT encoder and an MLP prediction head. In particular, we choose the pre-trained `BERT-Small` model as the encoder, featuring four hidden layers composed of 512 nodes each. We opt for `BERT-Small` because of its capability in achieving *adequate* classification performance, while also being *efficient* in terms of computational requirements. The MLP module consists of three components: a dense layer, a dropout layer (dropout rate = 0.2) [12], and another dense layer for predicting labels. We use ReLU activations. The model undergoes a total of 80 epochs. The learning rate is 3×10^{-5}. To optimize the training process, we employ the AdamW optimizer [5] with a weight decay set to 0.

The model achieves an overall accuracy of 91.04 on the testing set of Davidson *et al.* [2], suggesting a good performance in hate speech detection. However, it is important to note that although the dataset of Davidson *et al.* [2] provides a valuable resource, the domain of our dataset *may not perfectly align* with theirs. Consequently, any potential domain shift between the two datasets may impact the model's performance when applied to our specific dataset. As a result, we further conduct an experiment to verify the robustness of the trained model on our dataset.

Robustness Verification. We sample another validation set of 150 tweets from our dataset. Three researchers read the tweets and independently label them into three categories (*i.e.*, hate speech, offensive language, and neither). The final label is assigned with the consensus votes from three annotators. The

Table 1. Top 10 hate speech keywords related to NBA athletes.

Rank	Word	Frequency
1	ass	1,786
2	hate	1,693
3	gay	801
4	stupid	781
5	people	627
6	white	617
7	man	570
8	nigger	541
9	dirty	463
10	racist	436

Fleiss' Kappa score of the three annotators is 0.35, indicating fair agreement. Subsequently, we evaluate the performance of our classifier using this manually labeled dataset. This three-class classifier achieves an accuracy of 79.33. Moreover, it exhibits a weighted F1 score of 79.59, a precision of 80.96, and a recall rate of 79.33. These results collectively demonstrate a commendable performance for a three-class classification problem. Finally, we apply our model to the entire collected tweets.

3   Results

From the dataset comprising 503,424 collected tweets, we find 3.33% ($n = 16,784$) of the tweets are classified as hate speech, and 60.11% ($n = 302,605$) are offensive language. The remaining 36.56% ($n = 184,033$) of the tweets are neither hate speech nor offensive language. We remove stopwords and apply lemmatization and tokenization to hate speech tweets. Table 1 shows the top 10 words that appear most frequently in hate speech on NBA athletes.

To identify the NBA athletes who were targeted by hate speech the most, we use the keyword search method. In particular, by leveraging the extracted player names and Twitter handles, we discover the top 50 NBA athletes who are subjected to the highest levels of hateful content. Table 2 shows the top 10 NBA athletes with the most associated hate speech tweets. Notably, the list of the 50 most hated athletes includes popular names such as Lebron James, Kevin Durant, Ja Morant, Steph Curry, Devin Booker, Anthony Davis, *etc.* Two primary reasons can contribute to the observed phenomenon. Firstly, popular players often attract more attention and discussions, thereby increasing the likelihood of encountering hateful content. The prominence of these players within the NBA creates a higher probability of hate speech directed toward them. Secondly, high-profile players and notable Twitter accounts tend to become targets for hate speech due to the potential for amplified online visibility [4].

To further characterize the targets of hate speech on NBA athletes, we use different sets of keywords to search for relevant tweets. The targeted groups mined are `Black`, `White`, `Jews`, `dirty players`, `LGBTQ`, `chokers`, `selfish players`, `fat players`, `racists`, and `short players`. Table 3 summarizes the keywords used for each group.

Table 2. Top 10 NBA athletes with the most associated hate speech tweets.

Rank	Player	# Tweets
1	Anthony Davis	3, 211
2	Ja Morant	2, 469
3	Anthony Edwards	2, 173
4	Mckinley Wright IV	1, 199
5	Lonnie Walker IV	1, 199
6	Alex Len	892
7	LeBron James	784
8	Russell Westbrook	596
9	Chris Paul	562
10	Kevin Durant	539

Table 3. Keywords of the targets of hate speech on NBA athletes.

Group	Keywords
`Black`	nigger, nigga, n*gg*r, black, niggers
`White`	white
`Jews`	jews
`Dirty player`	dirty, flop, flopper, flopping
`LGBTQ`	faggot, gay, lesbian
`Choker`	choker, choke artist
`Selfish player`	selfish, stat padder, ball hog, stat pad
`Fat player`	fat
`Racist`	racist
`Short player`	short, little, small

The group that experiences the highest degree of targeting is the `Black` community, with a significant count of 4,124 tweets specifically directed toward them. It is worth noting that out of the top 50 NBA athletes that are associated with the most hate speech tweets, 48 are of African descent, while 2 are of Caucasian descent. However, in 2022, approximately 71.8% of NBA players were African American.[2] These statistics raise important questions about the potential influence of racial bias in the criticism directed toward athletes.

Following closely, the LGBTQ community faces a substantial number of 2,938 tweets aimed at their community. The count of tweets targeting the White individuals ranks third, totaling 1,035 tweets. In fourth place, there are 698 tweets directed toward dirty players. Additionally, 470 tweets specifically target selfish players, while 468 tweets aim at individuals characterized as racists. Moreover, there are 212 tweets targeting fat players and 199 tweets focusing on short players. The Jewish community is the subject of 130 tweets, and 64 tweets are directed at individuals referred to as chokers. Figure 1 shows the tweet distribution of targeted groups.

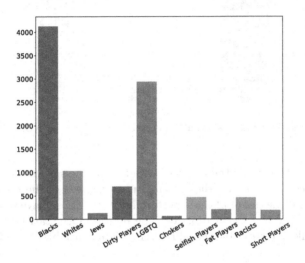

Fig. 1. Distribution of tweets related to the targeted groups of hate speech against NBA athletes.

Upon identifying the targeted groups within the hate tweets, these categories are subsequently organized into distinct topics, namely racism, physique shaming, play style, and anti-LGBTQ sentiments. More specifically, tweets about Black, White, and Jews are grouped into the racism topic. Tweets about fat players and short players are included in the physique shaming topic. The play style topic contains tweets about selfish players and chokers. Tweets about LGBTQ are included in the anti-LGBTQ topic. This classification enables a more comprehensive understanding of the underlying themes present within hate speech. The topic that emerges as the most prevalent is racism, with a support count of 5,289 instances. Following closely, the topic of anti-LGBTQ exhibits a

[2] https://43530132-36e9-4f52-811a-182c7a91933b.filesusr.com/ugd/403016_901e54ed015c44fb83df939d2070dc17.pdf.

support count of 2,940. Play style, on the other hand, garners a support count of 534, while physique shaming records a support count of 411. These figures highlight the relative prominence and occurrence of each topic within the analyzed hate speech tweets. Figure 2 shows the distribution of these topics.

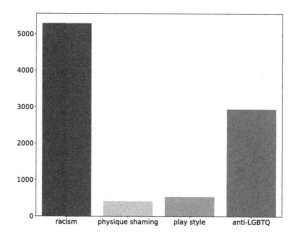

Fig. 2. Topic distributions of hate speech tweets related to NBA athletes.

To understand the potential correlation between hate speech tweets and players' performance, we compute the correlation coefficients of the number of hate speech tweets and a series of players' performance statistics, player demographics as well as their physical characteristics. The performance statistics of the NBA athletes are collected from basketballreference.com. Variables include:

- **Age**
- **G: Games Played.** The number of games in which a player has participated.
- **GS: Games Started.** The number of games in which a player was listed as a starter in the team's lineup.
- **MP: Minutes Played.** The number of minutes a player has been on the court during games.
- **TOV: Turnovers.** The number of times a player loses possession of the ball to the opposing team through errors such as bad passes, mishandling the ball, or offensive fouls.
- **Impact:** A player's influence or effect on the game. It encompasses various aspects of a player's performance that contribute to their team's success. The impact of a player can be evaluated through a combination of statistics, observations, and contextual analysis.
- **TS%: True Shooting Percentage.** It measures a player's shooting efficiency by taking into account their field goals, three-pointers, and free throws.
- **Usage:** It is a metric that quantifies the percentage of team plays or possessions that a player uses while they are on the court. Usage rate helps evaluate

the level of involvement and offensive responsibility a player has within their team's offensive system.
- **BMI: Body Mass Index.** It is a measure used to assess body composition and provide an indication of whether a person's weight is within a healthy range relative to their height.

The results revealed that the number of hate tweets demonstrated positive correlations with GS (Games Started), MP (Minutes Played), TOV (Turnovers), Impact, TS% (True Shooting Percentage), usage, and BMI (Body Mass Index). Conversely, hate tweet frequency showed negative correlations with age and G (Games Played) (Fig. 3).

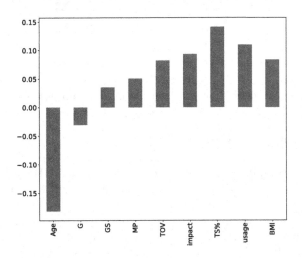

Fig. 3. Correlation coefficients between the number of hate speech tweets and variables including performance statistics, demographics, and physical characteristics of the top 50 most hated NBA athletes.

However, it is worth noting that MP (Minutes Played), TOV (Turnovers), Impact, GS (Games Started), TS% (True Shooting Percentage), and usage exhibit strong correlations with each other (Fig. 4). This suggests that their correlations with the number of hate tweets may be attributed to the fact that they are all performance metrics. Our analysis reveals that players who excel in their performance often become targets of hate speech, likely stemming from rival fans and individuals who may have financial stakes in outcomes, such as bettors.

Regarding the positive correlation observed between BMI and the number of hate speech tweets, we discover that a significant portion of the top 50 most hated NBA athletes consists of individuals categorized as overweight. Specifically, among these athletes, 17 individuals have a BMI exceeding 25. This suggests that their weight status might make them susceptible targets for fat shaming or height shaming through hate speech on social media platforms.

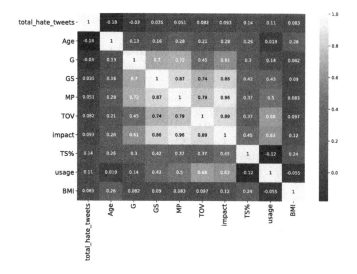

Fig. 4. Heatmap of correlations between the attributes of top 50 hated NBA athletes.

4   Discussions and Conclusions

In this study, we compile a list of hate speech-related and NBA athletes-related keywords to collect tweets that potentially contain hateful content toward NBA athletes. We then fine-tune a BERT model to classify collected tweets into hate speech, offensive language, and neither on an open hate speech dataset [2]. After examining the classifier performance on a manually labeled subset of our collected tweets, we find that out of the 503,424 tweets, 3.33% ($n = 16,784$) are classified as hate speech, and 60.11% ($n = 302,605$) are offensive language. Our model achieves an overall accuracy of 79.33, and a weighted F1 score of 79.59. These results demonstrate the effectiveness of our classification approach in discerning hate speech and offensive language within the collected dataset.

To gain a deeper understanding of the specific groups that are more susceptible to hate speech, we use the keyword search method. Through this process, we uncover notable patterns indicating that athletes belonging to the `Black` community and the `LGBTQ` community are disproportionately targeted with hate speech. Additionally, players who possess a distinct play style, as well as those who are shorter or overweight, emerge as prominent targets for such abuse. These findings shed light on the specific demographics and characteristics of athletes who are most likely to face hate speech within the NBA community. Racism, physical shaming, play styles, and anti-LGBTQ remarks are the major themes found in our collected dataset.

In conclusion, this study provides valuable insights into the prevalence of hate speech directed toward NBA athletes on social media platforms. By employing a combination of keyword searches and machine learning techniques, we have identified the targeted groups and major themes of hate speech within the NBA community.

Moving forward, further research can explore the impact of hate speech on the mental well-being of the targeted athletes and evaluate potential interventions to mitigate this issue. Additionally, analyzing the role of social media platforms and their policies in addressing hate speech toward athletes could contribute to fostering a safer online environment. It is essential to continue monitoring and addressing this ongoing problem to promote respect, inclusivity, and support for athletes across all platforms. By understanding hate speech in sports and taking proactive measures, we can work toward creating a positive and supportive environment for athletes to thrive both on and off the court.

Acknowledgments. This research was supported in part by the Goergen Institute for Data Science.

References

1. Chen, L., Lyu, H., Yang, T., Wang, Yu., Luo, J.: Fine-grained analysis of the use of neutral and controversial terms for COVID-19 on social media. In: Thomson, R., Hussain, M.N., Dancy, C., Pyke, A. (eds.) SBP-BRiMS 2021. LNCS, vol. 12720, pp. 57–67. Springer, Cham (2021). https://doi.org/10.1007/978-3-030-80387-2_6

2. Davidson, T., Warmsley, D., Macy, M., Weber, I.: Automated hate speech detection and the problem of offensive language. In: Proceedings of the International AAAI Conference on Web and Social Media, vol. 11, pp. 512–515 (2017)

3. Devlin, J., Chang, M.W., Lee, K., Toutanova, K.: BERT: pre-training of deep bidirectional transformers for language understanding. In: Proceedings of the 2019 Conference of the North American Chapter of the Association for Computational Linguistics: Human Language Technologies, Volume 1 (Long and Short Papers), pp. 4171–4186. Association for Computational Linguistics, Minneapolis, Minnesota, June 2019. https://doi.org/10.18653/v1/N19-1423, https://aclanthology.org/N19-1423

4. ElSherief, M., Nilizadeh, S., Nguyen, D., Vigna, G., Belding, E.: Peer to peer hate: hate speech instigators and their targets. In: Proceedings of the International AAAI Conference on Web and Social Media, vol. 12 (2018)

5. Loshchilov, I., Hutter, F.: Decoupled weight decay regularization. arXiv preprint arXiv:1711.05101 (2017)

6. Lyu, H., et al.: Social media study of public opinions on potential COVID-19 vaccines: informing dissent, disparities, and dissemination. Intell. Med. **2**(01), 1–12 (2022)

7. MacAvaney, S., Yao, H.R., Yang, E., Russell, K., Goharian, N., Frieder, O.: Hate speech detection: challenges and solutions. PLoS ONE **14**(8), e0221152 (2019)

8. Magu, R., Joshi, K., Luo, J.: Detecting the hate code on social media. In: Proceedings of the International AAAI Conference on Web and Social Media, vol. 11, pp. 608–611 (2017)

9. Magu, R., Luo, J.: Determining code words in euphemistic hate speech using word embedding networks. In: Proceedings of the 2nd Workshop on Abusive Language Online (ALW2), pp. 93–100 (2018)

10. Powell, A., Scott, A.J., Henry, N.: Digital harassment and abuse: experiences of sexuality and gender minority adults. Eur. J. Criminol. **17**(2), 199–223 (2020)
11. Reynolds, T.: NBA enacting zero-tolerance rules for abusive, hateful fan behavior (2019)
12. Srivastava, N., Hinton, G., Krizhevsky, A., Sutskever, I., Salakhutdinov, R.: Dropout: a simple way to prevent neural networks from overfitting. J. Mach. Learn. Res. **15**(1), 1929–1958 (2014)
13. Zhang, Y., Lyu, H., Liu, Y., Zhang, X., Wang, Y., Luo, J.: Monitoring depression trends on twitter during the COVID-19 pandemic: observational study. JMIR Infodemiol. **1**(1), e26769 (2021)

Feedback Loops and Complex Dynamics of Harmful Speech in Online Discussions

Rong-Ching Chang[1]([✉]), Jonathan May[2], and Kristina Lerman[2]

[1] University of California, Davis, CA, USA
`rocchang@ucdavis.edu`
[2] USC Information Sciences Institute, Marina del Rey, CA, USA
`{jonmay,lerman}@isi.edu`

Abstract. Harmful and toxic speech contribute to an unwelcoming online environment that suppresses participation and conversation. Efforts have focused on detecting and mitigating harmful speech; however, the mechanisms by which toxicity degrades online discussions are not well understood. This paper makes two contributions. First, to comprehensively model harmful comments, we introduce a multilingual misogyny and sexist speech detection model (https://huggingface.co/annahaz/xlm-roberta-base-misogyny-sexism-indomain-mix-bal). Second, we model the complex dynamics of online discussions as *feedback loops* in which harmful comments lead to negative emotions which prompt even more harmful comments. To quantify the feedback loops, we use a combination of mutual Granger causality and regression to analyze discussions on two political forums on Reddit: the moderated political forum *r/Politics* and the moderated neutral political forum *r/NeutralPolitics*. Our results suggest that harmful comments and negative emotions create self-reinforcing feedback loops in forums. Contrarily, moderation with neutral discussion appears to tip interactions into self-extinguishing feedback loops that reduce harmful speech and negative emotions. Our study sheds more light on the complex dynamics of harmful speech and the role of moderation and neutral discussion in mitigating these dynamics.

Keywords: Granger Causality · Moderation · Feedback Loop

1 Introduction

Despite efforts by social media platforms to mitigate harmful speech, it remains a widespread problem for millions of users who are routinely exposed to toxic speech and harassment online. Harmful speech in the form of cyberbullying has been suspected of contributing to population-level erosion of mental health of children and adolescents. Organized harassment and personal attacks contribute to the toxicity of many online communities. In more tragic cases, online hate speech directed at specific groups has been implicated in inciting real-world violence against ethnic or cultural minorities [18].

R. Thomson et al. (Eds.): SBP-BRiMS 2023, LNCS 14161, pp. 85–94, 2023.
https://doi.org/10.1007/978-3-031-43129-6_9

Given the breadth of the problem, researchers have focused on automatically detecting, monitoring, and mitigating harmful and toxic speech. Recent studies have shown that hate speech has echo chamber effects where users often flock together to spread hate speech and build up information cascades [10]. Moreover, hate speech tends to correlate with negative emotions [15,19], although the interactions between hate speech and negative emotions are not well understood.

In this paper, we study the dynamics of harmful speech in online discussions. We show that we can model the relationship between toxic comments and negative emotions as a *feedback loop* that amplifies both. We also examine the impacts of moderation on the dynamics of harmful speech. We analyze conversations on Reddit, a popular network of online discussion forums. Each forum, or subreddit, may have different community guidelines on what type of speech is allowed and what is moderated. Specifically, we investigate how the relationship between *harmful speech*, such as toxic or misogynistic comments, and *negative emotions*, such as anger, disappointment, fear, embarrassment, and sadness, affect the evolution of conversations on two subreddits dedicated to political discussions: *r/NeutralPolitics* and *r/Politics*. The first subreddit is moderated, which means that moderators remove comments deemed harmful. The *r/NeutralPolitics* forum also promotes neutral discussions without favoring or supporting any particular politicians and includes both pros and cons of a topic at the same time. The other subreddit forum, *r/Politics*. also has moderators; however, it follows broader rules such as limiting spam or hate speech only.

We hypothesize that harmful speech and negative emotions interact to create a *feedback loop*. The feedback loops create self-perpetuating "vicious cycles," where toxic comments lead to negative emotions, which encourage progressively more toxic comments. To establish the existence of feedback loops, we use (1) Granger causality demonstrating the mutually reinforcing nature of the interaction and (2) regression demonstrating the increase in harmful speech. We hypothesize that moderation with neutral discussion will decrease the effect of the feedback loop between harmful comments and negative emotions by curtailing harmful speech before it causes negative reactions. We explore this hypothesis through the following research questions:

– **RQ1:** *Do harmful comments and negative emotions (i.) reinforce each other, creating a feedback loop that (ii.) amplifies harmful speech and negative emotions?*
– **RQ2:** *Are there differences between feedback loops in moderated forums and forums that additionally promote neutral discussion?*

Our study suggests that the dynamics of online discussions are more complex than previously thought. Harmful speech interacts with negative emotions, which in the absence of neutral discussion may amplify harmful speech. Understanding the social dynamics of such speech will lead to more effective moderation strategies.

2   Related Work

Even though the causal relationship between harmful speech and negative emotions is not well understood, there is practical and theoretical evidence to suspect that there exists a self-reinforcing causal relationship between toxic comments and negative emotions. On the one hand, toxic and hate speech were found to co-occur with negative emotions. [15,19] On the other hand, the emotional contagion theory indicates that negative and harmful emotions can spread and influence others emotionally. Indeed, research has found a positive correlation between the number of negative posts by users to the number of negative posts users were exposed to [14]. Additionally, such vicious cycles have been found not only online but extending to offline behavior. Such *feedback loops* have been found between online hate speech and offline violence, which recursively leads to more online hate speech [16].

However, there are also reasons to suspect that there may not be such a causal relationship between harmful speech and negative emotions. For example, online hate speech and negative emotions might discourage some users from continuing to engage which could interrupt the formation of a vicious cycle [4,17].

3   Methods

We first describe the Reddit data, then present the methodology for detecting harmful comments and texts with negative emotions. After that, we outline the methodology for modeling the dynamics of harmful comments in discussions.

3.1   Data

We used Pushshift API [2] to collect data from *r/Politics* and *r/NeutralPolitics* subreddits. Pushshift is an archive of Reddit data. Each conversation or discussion consists of a post and subsequent comments. Both forums contain moderated political discussions; however, *r/NeutralPolitics* is an actively moderated forum, where moderators emphasize neutrality and cite sources in posts and comments. Comments that violate community guidelines and include insults, spam, or non-neutral content, are removed from the forum along with all subsequent comments in that thread. In total, we collected 3,819 and 9,397 conversations respectively from *r/Politics* and *r/NeutralPolitics*, consisting of 122,179 and 204,437 comments. On average, each discussion had 31 comments for *r/Politics* and 21 comments for *r/NeutralPolitics*. The conversations on *r/Politics* ranged from 2021-12-08 to 2022-06-27 while the conversations on *r/NeutralPolitics* ranged from 2017-02-26 to 2022-06-24.

3.2   Comment Classifications

We break down harmful speech into two main categories: toxic comments, and misogynistic comments. In this section, we describe our measures of toxicity, misogyny, and negative emotions.

Toxicity Detection. We used Detoxify,[1] a state-of-the-art API [12] for detecting toxic comments. This API recognizes a subset of speech that includes toxic or *obscene* comments, *threats, insults, identity attacks,* and *sexually explicit* language. This API is trained based on the toxicity dataset released by Jigsaw[2]. Each category represents joint labels identified by 10 annotators.

Emotion Detection. We used a fine-tuned model based on BERT and GoEmotions[3] to label emotions in comments. With more than 95,171 times of downloads, this is one of the most popular fine-tuned models for labeling emotions. GoEmotions [7] is a large dataset of Reddit comments that have been labeled with a range of emotions. We focus on *negative emotions*, specifically anger, disappointment, fear, embarrassment, and sadness.

Multilingual Misogyny/Sexism Detection. Misogynistic comments express contempt or dislike of women, while sexist comments promote harmful stereotypes about the role of men and women in society. There are several misogynistic and sexism datasets in different languages. However, there is no single general open-source sexism detection method. We fill this gap by training a multilingual classifier to recognize misogynistic and sexist comments.

We combined several multilingual ground truth datasets for misogyny and sexism (M/S) versus non-misogyny and non-sexism (non-M/S) [3,5,8,9,11,13, 20]. Specifically, the dataset expressing misogynistic or sexist speech (M/S) and the same number of texts expressing non-M/S speech in each language included 8, 582 English-language texts, 872 in French, 561 in Hindi, 2, 190 in Italian, and 612 in Bengali. The test data was a balanced set of 100 texts sampled randomly from both M/S and non-M/S groups in each language, for a total of 500 examples of M/S speech and 500 examples of non-M/S speech.

We used XLM-R [6], multilingual distill BERT and a Naive Bayes classifier with TF-IDF to create multilingual classifiers and compare their performance. We compared the performance of our classifier against Detoxify's [12] *sexually explicit* category.

We found the model using XLM-R[4] performs best across all languages ($F1 = 0.83$). The XLM-R model is fine-tuned with batch size of 16 over 4 training epochs, with a learning rate of $2e^{-5}$ and weight decay of 0.05. The DistillBert ($F1 = 0.77$) is multilingual and also fine-tuned with the same set of parameters as XLM-R, except for 3 epochs. The overall precision in DistillBERT was comparable to XLM-R (0.80) although it was better in some languages (Bengali). Naive Bayes had an F1 score of 0.76. Detoxify scores along the multilingual "sexually explicit" category provide a float between 0 to 1, which we binarize using the

---

[1] https://github.com/unitaryai/detoxify.

[2] https://www.kaggle.com/c/jigsaw-unintended-bias-in-toxicity-classification/data.

[3] https://huggingface.co/bhadresh-savani/bert-base-go-emotion.

[4] https://huggingface.co/annahaz/xlm-roberta-base-misogyny-sexism-indomain-mix-bal.

threshold of 0.5. Detoxify performed poorest on misogyny and sexism detection in comparison to other models ($F1 = 0.41$). The poor performance might be that Detoxify was trained on a single dataset, whereas in our case the dataset included comments from different social media platforms such as Twitter and Reddit.

3.3 Modeling Reddit Discussions

A discussion on Reddit is initiated by a post or a submission. Users can comment on a post or make another comment. We assume that users submit a new comment as a reaction or response to the previous comments [1]; therefore, we represent each discussion as a chronological sequence, i.e., a time series, of comments.

Because discussions may have different length, we standardized them by calculating each comment's relative position within the discussions. Specifically, given a post p that started a discussion of length n, we ordered the comments $c_1, \ldots, c_n$ in chronological order and calculated each comment's relative position in the discussion. For a comment i, its relative position $t = \frac{i}{n}$, which was a float between 0 and 1, where values near zero indicated comments near the start of the discussion and values near 1 indicated comments near the end of the discussion. In our experiment, we rounded each comment's relative position to two decimals, converting it to the percentile rank. This means that for discussions longer than 100 comments, there may have been more than one comment at the same relative position.

We calculated the toxicity, emotion, and misogyny scores (see Sect. 3.2) of each comment with their relative position in a discussion as time series. For binary features, we calculated the prevalence of the feature at a given position as a proportion. In other words, at a given position t across all discussions $p = 1, \ldots, N$, where N is the number of total posts in a subreddit, we calculated the fraction of all comments at that position with a binary feature $f = 1$. We could now measure how each time series evolved over the course of conversations and how these time series interacted with each other and across subreddits.

3.4 Measuring Feedback Loops with Mutual Granger Causality

Feedback loops create reinforcing interactions among toxicity and negative emotions. We represent the *feedback loop* as a bidirectional relationship where toxicity creates more negative emotions and vice versa. We use mutual Granger causality to test for this effect. Specifically, we test whether X Granger causes Y, and Y Granger causes X. Such mutual Granger causality exists when $\sigma^2(Y_t|Y_{t-1,\ldots 0}, X_{t-1,\ldots 0}) < \sigma^2(Y_t|Y_{t-1,\ldots 0})$ and $\sigma^2(X_t|Y_{t-1,\ldots 0}, X_{t-1,\ldots 0}) < \sigma^2(X_t|X_{t-1,\ldots 0})$. We use this approach to test for the existence of mutual Granger causality between harmful speech and negative emotions in each discussion with more than 10 comments. We then present the probability of such existence across each subreddit.

While mutual Granger causality provides a powerful tool to discover the underlying causal structure and the existence of feedback loops, it does not provide any information about the direction of the effect—whether it is reinforcing or extinguishing. To answer this question, we use regression to test whether the share of harmful comments and negative emotions increases or decreases over the course of a discussion.

4   Results

We process comments in Reddit conversations according to the above methodology. We identify *feedback loops* by using mutual Granger causality to test the self-reinforcing nature of the interactions between toxicity and emotions, and we use regression analysis to test whether these increase over the course of conversations.

4.1   Evolution of Harmful Speech

After standardizing the comment position within discussions in each subreddit (see Sect. 3.3), we create a time series for each category of harmful speech and negative emotion, which gives the probability that a comment at that position across all discussions in the subreddit expresses that emotion or toxic speech. We systematically measure the growth or decline of each category of harmful speech and negative emotion. Figure 1 shows OLS regression coefficients along with their statistical significance across all discussions in each subreddit and across *toxic discussions*, i.e., those with at least one harmful comment or negative emotion. A positive coefficient depicts an increase in hate speech or negative emotions as the conversation evolves.

Harmful speech and negative emotions demonstrate qualitatively different behaviors in the two different forums, the moderated political discussion in *r/Politics* and discussions that additionally promote neutral discussion in *r/NeutralPolitics*. All categories of harmful speech, except misogyny/sexism (M/S), increase over the course of a discussion in *r/Politics* subreddit, especially in *toxic discussions*. All coefficients are statistically significant (*p*-value < 0.001) in *toxic discussions*. In comparison, all categories of harmful speech show a small but statistically significant decrease over the course of discussions in *r/NeutralPolitics* subreddit. Among the negative emotions, anger, fear, and disappointment show a small but significant increase in the toxic discussions on the *r/Politics* subreddit, but except for anger, all other negative emotions decrease in the discussions on the *r/NeutralPolitics* subreddit, even in *toxic discussions*.

4.2   Feedback Loops

In this section, we present evidence for *feedback loops* in Reddit discussions by calculating mutual Granger causality between harmful comments and negative

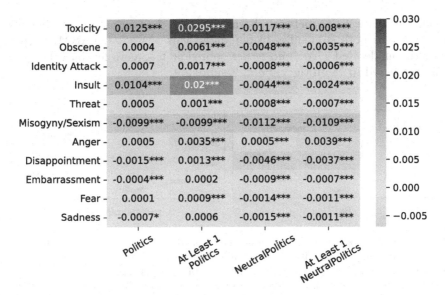

	Politics	At Least 1 Politics	NeutralPolitics	At Least 1 NeutralPolitics
Toxicity	0.0125***	0.0295***	-0.0117***	-0.008***
Obscene	0.0004	0.0061***	-0.0048***	-0.0035***
Identity Attack	0.0007	0.0017***	-0.0008***	-0.0006***
Insult	0.0104***	0.02***	-0.0044***	-0.0024***
Threat	0.0005	0.001***	-0.0008***	-0.0007***
Misogyny/Sexism	-0.0099***	-0.0099***	-0.0112***	-0.0109***
Anger	0.0005	0.0035***	0.0005***	0.0039***
Disappointment	-0.0015***	0.0013***	-0.0046***	-0.0037***
Embarrassment	-0.0004***	0.0002	-0.0009***	-0.0007***
Fear	0.0001	0.0009***	-0.0014***	-0.0011***
Sadness	-0.0007*	0.0006	-0.0015***	-0.0011***

Fig. 1. Regression slope of harmful speech and negative emotions aggregated over all discussions on the subreddits moderated political forum *r/Politics* and the moderated neutral discussion forum *r/NeutralPolitics*. Results with at least one harmful comment or negative emotion in both forums are also given. Asterisks represent p-values: (*) for p-value is less than 0.05; (**) if a p-value is less than 0.01; (***) if a p-value is less than 0.001.

emotions. Granger causality requires each time series to be stationary, meaning that its statistical properties do not change over time. We use Augmented Dickey-Fuller (ADF) to test for time series stationarity. The ADF tests had p-values less than 0.05, leading us to conclude that all the time series were stationary. We determine the optimal time lag for the Granger causality using the minimum Aikake Information Criterion (AIC) of the Vector Autoregressive (VAR) Model. The optimal lag was 11 for *r/Politics* and 16 for *r/NeutralPolitics*.

We found significant Granger causality between negative emotions and harmful speech, suggesting that they reinforce each other, creating feedback loops. The mutual Granger causality was tested in each discussion longer than 10 comments, and we present the probability of the existence across each subreddit. Figure 2 shows harmful and negative emotion pairs with a statistically significant mutual Granger-causality relationship. In the category of harmful speech, threats, obscenities, insults, identity attacks and toxic, misogynistic or sexist comments, all reinforce each other, creating feedback loops. Among the negative emotions, we found that anger, fear, embarrassment, sadness, and disappointment interact with categories of harmful speech. The color from blue to red of the edges represents the probability of mutual Granger causality across discussions in the given subreddit from low to high.

The feedback loops between harmful speech and negative emotions are present across both subreddits in our data. This answers our **RQ1**. Surprisingly, the moderated *r/NeutralPolitics* subreddit has more feedback loops, 90 total, compared to 78 on the *r/Politics* subreddit.

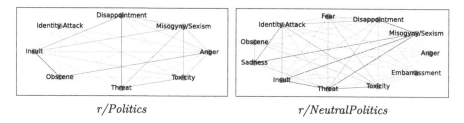

Fig. 2. Feedback Loops in online discussions. Each edge represents a statistically significant mutual Granger causality between categories of harmful speech (red) and negative emotions (green). The edge color changing from blue to red represents the probability from low to high of mutual Granger causality in each subreddit. (Color figure online)

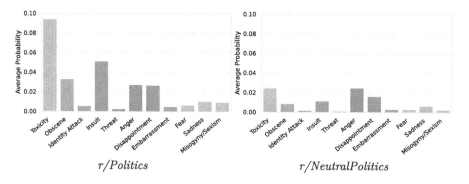

Fig. 3. Average probability of harmful comments and negative emotions in both subreddit

4.3   Feedback Loops and Harmful Speech

Do *feedback loops* increase toxicity of online discussions by amplifying harmful speech and negative emotions? This does not always appear to be the case. While there are more feedback loops in the *r/NeutralPolitics* subreddit (number of edges in Fig. 2(b)) than in the *r/Politics* subreddit, toxicity, and negative emotions (except for anger) generally decline over the course of the conversation in the moderated neutral discussion subreddit (Fig. 1). This suggests that even though feedback loops exist on this forum, neutral discussion helps to reduce them. Indeed, it appears that moderation creates self-extinguishing feedback loops that reduce negative emotions and harmful speech. As a result, there is less occurrence of harmful speech and negative emotions on *r/NeutralPolitics* (Fig. 3(b)). This answers our **RQ2**.

In summary, our results demonstrate that the moderation and neutral discussion in *r/NeutralPolitics* does not stop the feedback loops between the harmful speech and negative emotions; however, moderation and neutral discussion in *r/NeutralPolitics* has demonstrated an effect of decreasing the amplification of harmful speech and negative emotions as shown in Fig. 1.

5   Discussion and Conclusion

In this paper, we have depicted and introduced methodologies to model the dynamic relationship between harmful comments and negative emotions as a *feedback loop*. We also introduce a state-of-the-art multilingual misogyny and sexism detection language model. We demonstrate the existence of *feedback loops* in moderated political forum and political forum that additionally promote neutral discussions. Political discussions without the promotion of neutrality show stronger amplifying power in *feedback loops* between harmful comments and negative emotions than in *feedback loops* in moderated neutral political discussions. Moderated political forums that promote neutral discussion have shown self-reinforcing relationships to decrease harmful comments and negative emotions.

Even though our data is limited to two subreddits, both are closely tied to the same topic, US politics. Thus they are suitable for comparison. Our proposed method made an assumption that comment creation time is a good indicator of the ordering in participation in the conversations. These could potentially introduce bias as comments could be made without a discussant reading previous comments. A comment might also be deleted that would interrupt the ordering of a real conversation. Despite such potential drawbacks, our proposed methodology to model the *feedback loops* can be applied to any online discussion. We encourage future work to apply our methodology to further explore *feedback loops* besides political discussions.

We believe our work contributes to understanding the complex dynamics and relationships between harmful comments and negative emotions. To the best of our knowledge, we are the first paper that studies the causal relationship between toxic speech and negative emotions.

Acknowledgments. This material is based upon work supported in part by the Defense Advanced Research Projects Agency (DARPA) under Agreements No. HR00112290025 and HR001121C0168, and in part by the Air Force Office for Scientific Research (AFOSR) under contract FA9550-20-1-0224. Approved for public release; distribution is unlimited.

References

1. Alsagheer, D., Mansourifar, H., Shi, W.: Counter hate speech in social media: a survey. arXiv preprint arXiv:2203.03584 (2022)
2. Baumgartner, J., Zannettou, S., Keegan, B., Squire, M., Blackburn, J.: The pushshift reddit dataset. In: Proceedings of ICWSM, vol. 14, pp. 830–839 (2020)

3. Bhattacharya, S., et al.: Developing a multilingual annotated corpus of misogyny and aggression, pp. 158–168. ELRA, Marseille, France, May 2020. https://aclanthology.org/2020.trac-1.25
4. Carlson, C.R.: Misogynistic hate speech and its chilling effect on women's free expression during the 2016 us presidential campaign. J. Hate Stud. **14**, 97 (2017)
5. Chiril, P., Moriceau, V., Benamara, F., Mari, A., Origgi, G., Coulomb-Gully, M.: An annotated corpus for sexism detection in French tweets. In: Proceedings of LREC, pp. 1397–1403 (2020)
6. Conneau, A., et al.: Unsupervised cross-lingual representation learning at scale. In: Proceedings of ACL. ACL, July 2020. https://doi.org/10.18653/v1/2020.acl-main.747
7. Demszky, D., Movshovitz-Attias, D., Ko, J., Cowen, A., Nemade, G., Ravi, S.: Goemotions: a dataset of fine-grained emotions, July 2020. https://doi.org/10.18653/v1/2020.acl-main.372
8. Fersini, E., et al.: SemEval-2022 task 5: multimedia automatic misogyny identification. In: Proceedings of SemEval, pp. 533–549 (2022)
9. Fersini, E., Nozza, D., Rosso, P.: Overview of the Evalita 2018 task on automatic misogyny identification (AMI). EVALITA Eval. NLP Speech Tools Italian **12**, 59 (2018)
10. Goel, V., Sahnan, D., Dutta, S., Bandhakavi, A., Chakraborty, T.: Hatemongers ride on echo chambers to escalate hate speech diffusion. PNAS Nexus **2**(3), pgad041 (2023)
11. Guest, E., Vidgen, B., Mittos, A., Sastry, N., Tyson, G., Margetts, H.: An expert annotated dataset for the detection of online misogyny. In: Proceedings of EACL, pp. 1336–1350 (2021)
12. Hanu, L.: Unitary team. detoxify. github (2020). https://github.com/unitaryai/detoxify
13. Jha, A., Mamidi, R.: When does a compliment become sexist? Analysis and classification of ambivalent sexism using Twitter data. In: Proceedings of NLP+CSS, pp. 7–16 (2017)
14. Kramer, A.D., Guillory, J.E., Hancock, J.T.: Experimental evidence of massive-scale emotional contagion through social networks. Proc. NAS **111**(24), 8788–8790 (2014)
15. Min, C., et al.: Finding hate speech with auxiliary emotion detection from self-training multi-label learning perspective. Inf. Fusion **96**, 214–223 (2023)
16. Olteanu, A., Castillo, C., Boy, J., Varshney, K.: The effect of extremist violence on hateful speech online. In: Proceedings of ICWSM, vol. 12 (2018)
17. Salehabadi, N., Groggel, A., Singhal, M., Roy, S.S., Nilizadeh, S.: User engagement and the toxicity of tweets. arXiv preprint arXiv:2211.03856 (2022)
18. Stevenson, A.: Facebook admits it was used to incite violence in Myanmar. The NYT, 6 November 2018
19. Van Doorn, J.: Anger, feelings of revenge, and hate. Emot. Rev. **10**(4), 321–322 (2018)
20. Waseem, Z., Hovy, D.: Hateful symbols or hateful people? Predictive features for hate speech detection on Twitter. In: Proceedings of NAACL SRW, pp. 88–93 (2016)

Chirping Diplomacy: Analyzing Chinese State Social-Cyber Maneuvers on Twitter

Samantha C. Phillips(✉)(iD), Joshua Uyheng(iD), Charity S. Jacobs(iD),
and Kathleen M. Carley(iD)

CASOS Center, Software and Societal Systems Department,
Carnegie Mellon University, 5000 Forbes Avenue, Pittsburgh, PA 15213, USA
{samanthp,juyheng,csking,carley}@andrew.cmu.edu

Abstract. Governments around the world leverage social media to enact public diplomacy. In this article, we examine Chinese diplomatic communication on Twitter during two highly controversial events through a social cybersecurity lens: then-Speaker Pelosi's visit to Taiwan in early August 2022 and Taiwanese President Tsai's visit to the U.S. in early April 2023. We identify a small set of Chinese state-affiliated accounts that consistently tweet the most and are retweeted the most, demonstrating the highly centralized nature of China's external messaging. Using the BEND framework, we quantify social-cyber maneuvers used by the Chinese state to target U.S. and Taiwanese officials. We find they target individuals and ideas they perceive as direct challengers to the One China principle, neutralizing specific Taiwanese officials who support independence, while broadly dismissing and critiquing U.S. leaders, domestic affairs, and foreign policies. Our findings have implications for the study of online influence strategies and understanding China's broader diplomatic goals.

Keywords: public diplomacy · social-cyber maneuvers · social media · China

1 Introduction

In today's digital age, powerful actors use social media to participate in and influence public discourse. Social cybersecurity describes maneuvers through which such actors aim to change the collective narratives and networks of online conversation [2]. These maneuvers have profoundly affected societies and politics

This work was supported in part by the Knight Foundation and the Office of Naval Research grant Minerva-Multi-Level Models of Covert Online Information Campaigns, N000142112765, N000141812106, and N000141812108. Additional support was provided by the Center for Computational Analysis of Social and Organizational Systems (CASOS) at Carnegie Mellon University. The views and conclusions contained in this document are those of the authors and should not be interpreted as representing the official policies, either expressed or implied, of the Knight Foundation, Office of Naval Research, or the U.S. Government.

around the world, especially at the hands of nation-states [2]. Analyzing these online influence operations can unveil insights into their goals and strategies.

This article examines Chinese diplomatic communication through the lens of social cybersecurity. We analyze the actions of self-reported Chinese state-affiliated Twitter accounts in the context of Taiwan. The relationship between China and Taiwan is highly contentious, with the Chinese government advocating for the "unification" of Taiwan with the mainland based on the "One China"principle[1]. However, the current President of Taiwan, Tsai Ing-wen, firmly rejects these claims and maintains Taiwan's independence, placing the international community in a delicate position of having to choose between accepting China's stance or supporting Taiwan.

While the United States officially maintains diplomatic relations with China, it also maintains a robust unofficial relationship with Taiwan, employing a policy of strategic ambiguity[2]. Notably, however, two significant public meetings occurred between U.S. and Taiwanese government officials in the past year. The first involved then-U.S. House Speaker Nancy Pelosi's visit to Taiwan to meet with President Tsai, marking the first such visit by a speaker since Newt Gingrich in 1997. The second event was President Tsai's visit to the U.S., where she met with current House Speaker Kevin McCarthy, the highest-ranking U.S. official to meet with a Taiwanese president on American soil in almost three decades. Both encounters faced strong Chinese condemnation, leading to offline sanctions and warnings, alongside strategic online campaigns to sway public opinion.

Collectively, these events offer a valuable window into China's social-cyber maneuvers upon or against other state actors. This paper provides a comparative analysis of China's overt social media tactics employed in relation to different actors during these two distinct events. By leveraging quantitative psycholinguistics and network science, our research contributes to an empirical understanding of China's use of official communication channels to advance its diplomatic goals, with implications for understanding state actors on social media more broadly.

2   Related Work

Online information operations to shape narratives and influence public opinion are increasingly prevalent, especially during geopolitical conflicts. Inauthentic techniques to emulate human activity, such as automated bot accounts and puppet accounts using fake personas, have been effective at shaping public discourse and social networks [4,5]. These techniques are especially effective due to their low cost and social media users' difficulty with discerning humans from bots [3].

With the emergence of the social cybersecurity field, new analytical focus has expanded beyond *who* is performing online information operations, to *how* influence actions are conducted [2]. This captures the key insight that social media conversations are influenced not just by covert inauthentic accounts, but

---

[1] http://eu.china-mission.gov.cn/eng/more/20220812Taiwan/202208/t20220815_10743591.htm.

[2] https://www.cfr.org/backgrounder/china-taiwan-relations-tension-us-policy-biden.

also by overt actors such as nation-states. From this standpoint, understanding their actions to positively or negatively impact narratives and networks acquires greater importance. The BEND framework defines a collection of such social-cyber maneuvers divided into four categories [1]. The B category represents positive network maneuvers (back, boost, bridge, build) that increase the (perceived) importance of individuals and groups. The E category represents positive narrative maneuvers (engage, enhance, excite, explain) that promote ideas and messages. The N category contains negative network maneuvers (narrow, neglect, neutralize, nuke) that attempt to decrease the (perceived) importance of individuals and groups. Finally, the D category spans negative narrative maneuvers (dismay, dismiss, distort, distract) that undermine ideas and messages.

China's strategy towards using social media platforms embraces a *public diplomacy* approach, which extends regular diplomacy by shaping narratives to influence opinion in foreign audiences [9]. China's media outreach began as a way to "tell China's stories" through its own international telecommunications outlets to further Chinese interests [7]. Some studies have focused on the Chinese state's creation of self-referencing networks to centralize control over topics of conversation [12]. Others, meanwhile, explore China's aggressive "Wolf-Warrior" rhetoric and its complicated effects of simultaneously pushing Chinese nationalist sentiment while potentially polarizing views of China abroad [6].

However, despite broad interest in Chinese diplomatic activity, the current literature still lacks a systematic understanding of how the Chinese state utilizes social media in diplomatic contexts. By harnessing a social-cyber maneuvers approach, our work offers a novel quantitative mapping of China's online information tactics in response to contentious geopolitical developments.

3   Methods

3.1   Data Collection

Our data collection used a keyword search via the Twitter v1 API. We collected tweets containing the word "taiwan" between January 2022 and April 2023. We then binned our data around two events with approximately two weeks before and after each event to capture the effect of the visit on social cybersecurity maneuvers: a) Nancy Pelosi's visit to Taiwan from July 18, 2022 to August 17, 2022 and b) Tsai Ing-Wen's visit to the United States from March 14, 2023 to April 20, 2023. In total, we collected 1,493,722 tweets with 495,783 unique users.

3.2   State-Affiliated Labels

We use a previously curated list of over 350 Chinese Diplomatic, Government, and Media Twitter handles [8]. Previously, China's Twitter handles fell into these categories: China Government Official, China Government Organization, or China State-Affiliated Media. However, in April 2023, Twitter dropped all state-affiliated and government-official labels[3]. We use Twitter's past definition

---

[3] https://www.reuters.com/technology/twitter-removes-state-affiliated-media-tags-some-accounts-2023-04-21/.

of "state-affiliated" to mean that the "state exercises control over [an account's] editorial content through financial resources, direct or indirect political pressures, and/or control over production and distribution[4]. Government officials are any persons acting in an official function for the PRC. We use "Chinese state" and "Chinese state-affiliated" accounts interchangeably.

To label non-Chinese state officials, two authors independently labelled the accounts of interest as U.S. state-affiliated, Taiwanese state-affiliated, or neither. Because we are interested in U.S. and Taiwanese state officials in the context of China's public diplomacy efforts, we only labelled the 171 users that Chinese state-affiliated users mention during either event. The annotators had 100% agreement.

3.3   Network Analysis with ORA

ORA is a dynamic network analysis toolkit with specialized functionalities for high-dimensional social media networks[5]. We used ORA to convert nested Twitter data into meta-networks, or networks of networks comprised of agents or Twitter users, hashtags, URLs, and tweets, in addition to pairwise networks of these actors. We binned our data to capture the two weeks prior and two weeks after both events. To separate China's activity from the larger Twitter discussions, we then used the 209 Chinese state-affiliated accounts to extract their ego communities or all of the immediate connections of China's state-sponsored accounts. Our data was skewed towards the Pelosi visit, with 12,850 tweets and 21,710 users (5,014 tweets by 213 Chinese state accounts). The Tsai visit had 2,072 tweets and 5,377 users (923 tweets by 105 Chinese state accounts). For both events, over 80% of tweets by Chinese state accounts are in English and less than 5% are in Chinese according to Twitter's language tags.

3.4   BEND Maneuvers with ORA

To calculate the prevalence of BEND maneuvers, we also used the ORA software. We calculated the BEND maneuvers on the union of the ego networks associated with all the Chinese state accounts. This allowed us to focus on the BEND maneuvers that were relevant to the activity of Chinese state actors during the online conversations surrounding the Pelosi and Tsai visits. We specifically focused on the average level of BEND maneuvers associated with the identified Chinese state accounts throughout the visits of interest.

4   Results

4.1   Key Chinese State-Affiliated Actors

The first step in understanding the public diplomacy strategies of the Chinese government for Taiwan relations is to analyze the level of activity of state-

---

[4] https://help.twitter.com/en/rules-and-policies/government-media-labels.
[5] https://netanomics.com/ora-commercial-version/.

affiliated Twitter accounts. We consider the number of tweets and retweets of all Chinese state-affiliated accounts in each event as visualized in Fig. 1.

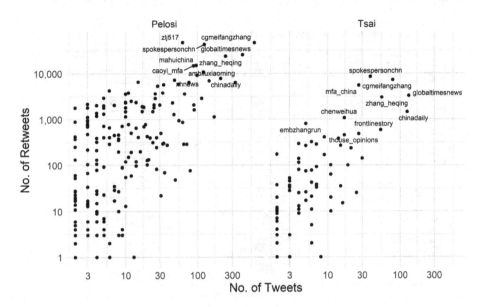

Fig. 1. Chinese state-affiliated accounts which produced the most tweets and were retweeted the most.

Our analysis shows that the same handful of Chinese state-affiliated users tweet the most and are likewise retweeted the most for both the Pelosi ($r = 0.692, p < .001$) and the Tsai visits ($r = 0.572, p < .001$). This suggests there is a set of government officials, such as Consul General of China in Belfast Zhang Meifang (cgmeigangzhang) and Assistant Minister of Foreign Affairs Hua Chunying (spokespersonchn), and affiliated media, like the Global Times (globaltimesnews), with the role of spreading China's influence and ideas on Twitter during contentious events such as these state visits.

Notably, Lijian Zhao, who was previously a Ministry of Foreign Affairs spokesperson, is absent from the Tsai visit dataset. He was removed from his position in January 2023 and has not tweeted since[6]. Overall, however, the same accounts dominate China's narrative about Taiwan during both events.

4.2   BEND Maneuvers Targeted at State Officials

To delve deeper into the social-cyber maneuvers used by the Chinese government on Twitter, we measure the degree to which each BEND maneuver is used in each tweet by Chinese state-affiliated accounts. In particular, as visualized in Fig. 2,

---

[6] https://foreignpolicy.com/2023/01/11/china-wolf-warrior-zhao-lijian-diplomacy/.

we examine the average value for each maneuver in tweets that mention other Chinese state actors, U.S. officials, and Taiwanese officials. Figure 3 provides examples of tweets containing BEND maneuvers for each target and event.

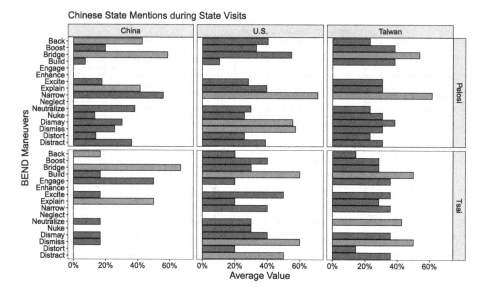

Fig. 2. Average BEND maneuvers Chinese state accounts in tweets mentioning other Chinese state officials, U.S. officials, or Taiwanese officials for the Pelosi and Tsai visits. Color is used for emphasis.

In mentioning fellow Chinese state accounts, positive maneuvers are generally used. As emphasized in Fig. 2, these include bridge (Pelosi: 0.586, Tsai: 0.667), back (Pelosi: 0.427, Tsai: 0.167), and explain (Pelosi: 0.414, Tsai: 0.500). Bridge and explain maneuvers notably hold for both Pelosi's visit to Taiwan and Tsai's visit to the U.S. These results imply that Chinese state accounts are often amplifying and enhancing the messages spread by other associated accounts, likely in an effort to maximize their spread and influence.

We also note the high use of the narrow maneuver (0.559) in tweets that mention other Chinese state officials during the Pelosi visit. This occurs when Chinese state-affiliated users mention multiple other Chinese state users and then critique, or distance themselves, from adversaries.

The influence strategies applied to U.S. and Taiwanese officials differ meaningfully between events. When Speaker Pelosi travelled to Taiwan, the Chinese government saw the U.S. taking an active role in Taiwan's affairs. They outwardly attack and reply to U.S. officials on Twitter, like Speaker Pelosi and Secretary of State Antony Blinken. This is demonstrated in their choice of maneuvers in the 135 tweets mentioning U.S. officials.

During the Pelosi visit, the dominant maneuvers in Chinese state tweets that mention U.S. officials are narrow (0.710), dismiss (0.573), and dismay (0.556).

Fig. 3. Examples of tweets demonstrating key BEND maneuvers. From the top left corner going clockwise: A) Bridge maneuver to connect and then threaten opposing Taiwanese leadership, B) Neutralize maneuver to diminish Tsai Ing-Wen's reputation, C) Build maneuver to strengthen the notion that there is a One-China coalition in Taiwan and D) Dismiss maneuver to dismiss the importance of Pelosi's visit to Taiwan. Note: all tweets replicated via TweetGen to respect Twitter's Terms of Service policies.

These negative network and narrative maneuvers challenge the U.S.'s position and authority by *dismissing* their credibility and impact, generating feelings of *dismay* around their actions, and *narrowing* U.S.-Taiwan connections. Figure 3 (D) contains an example of the dismiss maneuver targeted at Speaker Pelosi.

Their strategy towards Taiwanese officials during Pelosi's visit is to minimize their voice in the discussion, only mentioning Taiwanese officials in 13 of their tweets. When they do mention Taiwanese officials, they use high levels of bridge (0.538) and narrow (0.615) maneuvers. Ultimately, the Chinese government does not want Taiwan to act independently. They want to emphasize commonalities between China and Taiwan, *bridging* the two countries. Moreover, Chinese state accounts *bridge* Taiwanese officials who support Taiwan independence to effectively critique them, as shown in Fig. 3 (A). The narrow maneuver is used to sequester Taiwan from the U.S. and minimize their relationship.

On the other hand, when Taiwanese President Tsai travelled to the U.S. and met with now-Speaker McCarthy, the Chinese government focused online attacks on Tsai and other Taiwanese officials. In this case, they mentioned Taiwanese officials nearly two times as much as U.S. officials (19 and 10). In the few cases they mention U.S. officials, such as Speaker McCarthy, they use the build (0.600), dismiss (0.600), and distract (0.500) maneuvers. For example, they *build* up a protestor group during Tsai's visit and *dismiss* the authority of Congressman Raja Krishnamoorthi. We provide an example of the build manuever in Fig. 3 (C). Additionally, they *distract* from Tsai going to the U.S. by discussing China's military exercises in the Taiwan strait.

In tweets directed at Taiwanese officials, Chinese state accounts use build (0.500), dismiss (0.500), and neutralize (0.429). In particular, neutralize is used to target an individual or group to limit their importance and effectiveness. President Tsai and unofficial Taiwanese ambassador to the U.S. were the subject of many *neutralize* attempts. We provide an example of the neutralize maneuver in Fig. 3 (B). Additionally, they *build* the relationship between Tsai (and other supporters of Taiwanese independence) and the U.S. in an effort to discredit and *dismiss* both parties.

In sum, Chinese state-affiliated accounts targeted U.S. officials overwhelmingly when Pelosi went to Taiwan and Taiwanese officials when Tsai went to the U.S. They attack the U.S. more broadly, while focusing on Taiwanese leaders who support the independence movement.

5   Discussion

In this work, we examine China's public diplomacy efforts in the context of high-tension geopolitical events, investigating both their internal coordination and external targeting of U.S. and Taiwanese officials on Twitter. Using the BEND framework, we observe that throughout both the Pelosi and Tsai visits, Chinese state-affiliated accounts actively amplify messages posted by fellow Chinese state-affiliated users, especially through the back, bridge, and explain maneuvers. Nearly an identical set of accounts, combining government officials and media outlets, actively spread China's influence on Twitter through high numbers of tweets and retweets in both events. This demonstrates a highly centralized campaign for systematic messaging, echoing prior work on the Chinese state's digital diplomacy operations [10–12].

In juxtaposing the Pelosi visit to Taiwan with the Tsai visit to the U.S., we observe key insights into China's online influence strategies in managing diplomatic interests. During Pelosi's visit to Taiwan, China saw the U.S. taking an active role in developing relations with Taiwan. Through negative narrative and network maneuvers, Chinese state-affiliated accounts aim to delegitimize U.S. actions and reduce their significance for U.S.-Taiwan relations. At the same time, they strategically exclude Taiwanese voices in the online conversation. This seeks to deny Taiwan's agency as a participant in international diplomatic processes and thereby assert exclusive authority to speak on Taiwan's political status. This also aligns with previous analyses of public diplomacy strategies used by China, where they strategically ignore the opposition to limit their influence [10]. When Taiwanese officials are mentioned, it is through bridge and narrow maneuvers which increase the perceived connection between China and Taiwan, while decreasing the perceived connection between Taiwan and the U.S.

Conversely, during the Tsai visit to the U.S., it is Taiwan that disrupts China's desired political status quo. To counteract this, Chinese state-affiliated accounts reverse communication priorities, focusing more on Taiwanese officials than U.S. officials. Rather than Taiwan as a whole, Chinese social-cyber maneuvers target critiques and sanctions towards Taiwanese government leaders that support Taiwan's independence like President Tsai and unofficial Taiwanese ambassador to the U.S. Hsiao Bi-khim. At the same time, Chinese state-affiliated accounts use dismiss, distract, and build on U.S. officials, attempting to minimize their input on Taiwan's relationship with China.

Collectively, these findings demonstrate the utility of the BEND framework for systematically mapping a state actor's online actions to an international diplomatic context. Narrative and network maneuvers are significant not only in their direct impacts on online conversations, but also in the underlying political developments they reflect and reshape. The BEND framework emphasizes the variety of online actions that state actors can take to manage power shifts (and perceptions of these shifts) on the international stage. From this perspective, digital diplomacy cannot be seen one-dimensionally, but as potentially incorporating both positive and negative campaigns to manage diplomatic interests.

Based on these findings, forward-looking analyses can examine hybrid strategies of attack, seclusion, and selective amplification featured in the Chinese state's online information playbook in the region and beyond, especially with elections in Taiwan in January 2024. Our research was limited to two adversarial campaigns, but China remains the center of gravity in the Asia Pacific region. Furthermore, the present study focuses on overt state-affiliated accounts, but further research could investigate the use of covert networks and actors (e.g., bots) to further China's diplomatic aims. Finally, it is imperative to continue developing measures of impact of social-cyber maneuvers both online (e.g., diffusion of message) and offline (e.g., shift in public opinion) to assess the efficacy and design effective mitigation strategies.

References

1. Blane, J.T., Bellutta, D., Carley, K.M.: Social-cyber maneuvers during the COVID-19 vaccine initial rollout: content analysis of tweets. J. Med. Internet Res. **24**(3), e34040 (2022)
2. Carley, K.M.: Social cybersecurity: an emerging science. Comput. Math. Organ. Theory **26**(4), 365–381 (2020)
3. Everett, R.M., Nurse, J.R., Erola, A.: The anatomy of online deception: what makes automated text convincing? In: Proceedings of ACM SAC, pp. 1115–1120 (2016)
4. Ferrara, E., Chang, H., Chen, E., Muric, G., Patel, J.: Characterizing social media manipulation in the 2020 US presidential election. First Monday (2020)
5. Ferrara, E., Varol, O., Davis, C., Menczer, F., Flammini, A.: The rise of social bots. Commun. ACM **59**(7), 96–104 (2016)
6. Guo, J.: Crossing the "great fire wall": a study with grounded theory examining how China uses Twitter as a new battlefield for public diplomacy. J. Publ. Diplomacy **1**(2), 49–74 (2021)
7. Huang, Z.A., Wang, R.: Building a network to "tell China stories well": Chinese diplomatic communication strategies on Twitter. Int. J. Commun. **13**, 2984–3007 (2019)
8. Jacobs, C.S., Carley, K.M.: #WhoDefinesDemocracy: analysis on a 2021 Chinese messaging campaign. In: Thomson, R., Dancy, C., Pyke, A. (eds.) SBP-BRiMS 2022. LNCS, vol. 13558, pp. 90–100. Springer, Cham (2022). https://doi.org/10.1007/978-3-031-17114-7_9
9. Nye, J.S., Jr.: Public diplomacy and soft power. Ann. Am. Acad. Pol. Soc. Sci. **616**(1), 94–109 (2008)
10. Phillips, S.C., Uyheng, J., Carley, K.M.: Competing state and grassroots opposition influence in the 2021 Hong Kong Election. In: Thomson, R., Dancy, C., Pyke, A. (eds.) SBP-BRiMS 2022. LNCS, vol. 13558, pp. 111–120. Springer, Cham (2022). https://doi.org/10.1007/978-3-031-17114-7_11
11. Schliebs, M., Bailey, H., Bright, J., Howard, P.N.: China's Public Diplomacy Operations: Understanding Engagement and Inauthentic Amplifications of PRC Diplomats on Facebook and Twitter. Programme on Democracy and Technology, Oxford University, Oxford (2021)
12. Thunø, M., Nielbo, K.L.: The initial digitalization of Chinese diplomacy (2019–2021): establishing global communication networks on Twitter. J. Contemp. China, 1–23 (2023)

Vulnerability Dictionary: Language Use During Times of Crisis and Uncertainty

Wenjia Hu[1]([✉])[iD], Zhifei Jin[2][iD], and Kathleen M. Carley[2][iD]

[1] Human-Computer Interaction Institute, Pittsburgh, PA 15213, USA
wenjiah@andrew.cmu.edu
[2] Software and Societal Systems Department, Carnegie Mellon University,
Pittsburgh, PA 15213, USA
{zhifeij,carley}@andrew.cmu.edu

Abstract. Studying the expression of vulnerability – the psychological state that arises in moments of uncertainty, risk, and emotional exposure – is key to understanding how individuals cope during time of crisis. In this study, we synthesize past theoretical and qualitative work on vulnerability and present a psycholinguistic dictionary featuring seven different lexicons commonly associated with vulnerability language: accusation, incompetence, lack of trust, riot, out-group speech, agency, and helplessness. To validate our dictionary, we apply the identified lexicons to three different datasets and then generate samples for evaluation by two independent human labelers. The comparison between the human-labeled and machine-generated results demonstrates the dictionary's robustness, achieving a Cohen's Kappa of 0.66 in inter-rater reliability among two labelers and an overall accuracy of 0.74 when treating human label as the ground truth. With the dictionary, we aim to provide a nuanced understanding and detection of vulnerability language on social media content and provide a foundation for text feature extraction for traditional machine learning methods.

Keywords: Psycholinguistic dictionary · Vulnerability Language · LIWC · Natural Language Processing · Crisis

1 Introduction

Research on vulnerability plays a crucial role in understanding how individuals cope with adversity and uncertainty during times of crisis. Vulnerability refers to the emotional state that arises in moments of uncertainty, risk, and emotional exposure [3]. Individuals can experience vulnerability during personal crises, such as financial difficulties, family conflict, and health issues, as well as external crises like natural disasters, terrorist attacks, political riots, and public health crises [5]. For some, vulnerability allows for positive growth, such as increased psychological resilience and social connectedness [3]. For others, particularly those with limited resources, vulnerability is associated with chronic stress, poor mental

R. Thomson et al. (Eds.): SBP-BRiMS 2023, LNCS 14161, pp. 105–114, 2023.
https://doi.org/10.1007/978-3-031-43129-6_11

health, and compromised physical well-being [26]. The feeling of powerlessness and insecurity that accompanies vulnerability can lead individuals to seek certainty and control. Such feelings might promote cognitive biases and defense mechanisms such as blaming, scapegoating, and increased susceptibility to misinformation, conspiracy theories, and extremism [15].

As such, accurate measurement and representation are crucial for studying vulnerability. However, the current methods for measuring vulnerability are limited. The measurement of vulnerability mainly takes two forms: (1) Qualitative research methods, such as surveys, interviews, and manual coding of small data sets, and (2) natural language processing methods, such as sentiment analysis and topic modeling. While qualitative methods can provide accurate and nuanced evaluations of vulnerability [30], it becomes challenging to scale such evaluations to identify relevant posts from millions of social media posts generated during crisis events. On the other hand, natural language processing methods can be used to study the expression of vulnerability on a larger scale. However, most of the existing sentiment analyses only allow detection and measurement of crude proxies of vulnerability such as primary emotions like anger and fear [28]. Yet, the presence of emotion is not sufficient to indicate vulnerability, and vice versa. Vulnerability can manifest as a seemingly logical tweet accusing the public service of being incompetent during a natural disaster or a Facebook post sharing pseudoscientific health misinformation during a pandemic. Consequently, using crude proxies can oversimplify the concept and result in numerous methodological issues for subsequent analysis.

Our paper aims to fill the gap by offering a psycholinguistic dictionary that explores the nuanced language patterns related to the expression and description of vulnerability. Our contributions include: 1. synthesizing past theoretical and qualitative work to provide a nuanced overview of different aspects related to the expression of vulnerability. 2. developing and validating a set of lexicons related to vulnerability language. By presenting this psycholinguistic dictionary, we aim to enhance the understanding and analysis of vulnerability, offering insights into its expression in social media content during the time of crisis and how it impacts mental health and online extremism.

2   Background

2.1   Psycholinguistic Dictionaries

Psycholinguistic dictionaries are commonly used for sentiment analysis. These dictionaries operate on the assumption that the words one uses can reveal how one thinks, regardless of the topic discussed. Words such as pronouns (e.g., us, them) reflect our attentional preferences [9]. Popular dictionaries like Linguistic Inquiry and Word Count (LIWC) [2] and National Research Council (NRC) [20] present relevant lexicons for basic emotions, psychological processes, and personal concerns. These dictionaries are developed through either a closed-vocabulary approach, where experts curate and label a small set of vocabulary relevant to psychological processes [2], or an open-vocabulary approach, which expands a small set of words using word embeddings and employs crowd workers to annotate the words' relevance in each lexicon [20]. Despite the continuous advancements in the field of natural language processing, dictionaries remain popular in computational psychology and computational social science research due to their transparency and theoretical validity [9].

Specialized dictionaries are also developed to explore nuanced psychological processes that are not covered by lexicons in general dictionaries. These dictionaries take two approaches. The first approach is combining existing lexicons from general dictionaries. For example, the representation of moral emotion, "other-condemning", is created by combining the anger, disgust, and contempt lexicons from NRC [27]. The second approach is developing and validating new lexicons. For instance, the moral foundation dictionary presents five lexicons including care, fairness, loyalty, authority, and sanctity that are crucial to human moral reasoning processes [11].

2.2   Existing Vulnerability Measures

To the best of our knowledge, all existing studies measure vulnerability indirectly. Past work often analyzes vulnerablity-related language through basic emotions (e.g., anger, fear, and anxiety), thinking styles (e.g., certainty), and moral frameworks (e.g., fairness, harm, purity) from generic dictionaries like LIWC and NRC [1,24,25]. Machine learning algorithms that aim to provide a more comprehensive view of vulnerability language are often also built on features extracted through LIWC or NRC [17]. However, the limitation lies in the fact that basic emotions only encompass seven categories: happiness, surprise, contempt, sadness, fear, disgust, and anger. Consequently, the subtler nuances of vulnerability, such as feelings of powerlessness and lack of trust, may be overlooked when relying solely on these basic emotions for construction. Moreover, studying vulnerability solely through basic emotions like fear and anger can potentially lead to confusion between vulnerability and emotions. This confusion can have implications for subsequent analyses that rely on an accurate representation of the sentiment expressed in the content. Thus, in our dictionary, we present and validate new lexicons to provide a more nuanced representation of vulnerability.

3   Methods

We develop a vulnerability dictionary complementing existing general dictionaries. Similar to the approach used in the development of the grievance dictionary [29], we first identify relevant lexicons, taking out lexicons that are already covered by general and specialized dictionaries [2,20,29] and then iteratively improve the word selection within each lexicon. Finally, we assess the generalizability of the lexicons by evaluating them on previously unseen data.

3.1   Identify Relevant Lexicons

After examining the negative experiences of vulnerability in various contexts, we find that vulnerable individuals often underperform in work [22] and academic settings [8]. Following the existing framework develop for human evaluators [8,10], we identify common behavioral and language patterns including 1. experiencing grievance and discontent with workplace, school systems, and the authority; 2. complaining and blaming; 3. experiencing helplessness, seeing problems to be permanent, predicting defeat, giving up, absent from workplace or school; 4. and forming negative self-concepts including focusing on weaknesses, unfavorable self-comparisons. Based on these patterns, we decide to curate lexicons related to grievances, accusations, helplessness, and incompetence.

Additionally, we examine in-depth qualitative studies, including interviews [30], thematic analysis of written content [16], and large-scale analysis of social media content [24] to explore language patterns exhibited by vulnerable individuals facing personal or external crises. We find that individuals' language often includes complaints about specific events (e.g. personal health, financial and family issues) as well as general negative interpretations of these events, such as experiences of uncertainty, powerlessness, and a lack of trust towards authorities. Moreover, this language often reflects seemingly contradictory combinations of experiences, such as uncertainty, distrust of authority, and a yearning for certainty. Synthesizing findings from a wide range of mixed-method studies, we develop lexicons related to lack of trust, agency, and helplessness.

Finally, we explore the extreme end of vulnerability experiences, including susceptibility to radicalization and engagement with mis- and disinformation. Examining past research on the radicalization process, including terrorism, lone actors, and radicalists [13,23], we found that vulnerable individuals often perceive repression by out-groups, leading to the formation of victim-based identities [13]. They may seek alternative narrative frameworks through conspiracy theories and embrace radical ideologies [7]. The perceived victimhood and repression can give rise to discussions of protest, riots, and violence. However, since much of the violent talk and radicalization are already covered by the grievance dictionary [29], we only add two more lexicons of riot and out-group speech, to explore further.

3.2   Curate and Evaluate Word Selections for Each Lexicon

We follow a similar approach to the development of LIWC [2] by adopting a closed-vocabulary approach to curate the initial set of lexicons for our vulnerability dictionary. Based on prior theory, our dictionary encompasses seven lexicons including out-group speech, lack of trust, incompetence, agency, helplessness, accusation, and riot. We manually select 200–300 words and their variation of forms including nouns, verbs, adjectives, and plural forms for each lexicon. For example, for a word like "achieve," we include its other forms such as "achieved," "achievement," and "achieving." To test and improve the set of lexicons, we use NetMapper, a psycholinguistic analysis software [4]. NetMapper takes the lexicons and document lists as input and provides frequency counts of the words from each lexicon for every document as output. Here we use Twitter data for documents, where each tweet is treated as one document. For instance, if a tweet contains five words listed in the "agency" lexicon, with one word appearing twice, the agency count value would be six. Finally, we iteratively improve the word selection for each lexicon by analyzing the top-listed document for each lexicon.

During the development phase, we utilize a dataset of Twitter discussions around the ReOpen America campaign [18]. The original dataset consisted of 3.6 million unique users and 9.9 million tweets posted from April 1, 2020, to June 22, 2020. It was gathered from Twitter Academic API through keyword and hashtag-based searches using terms like "openup" and "reopen". We randomly select a subset of the dataset since data completeness was not crucial for evaluating the quality of the lexicon. We select the top- and bottom-listed documents for each lexicon based on count and manually label each tweet as either containing or not containing the relevant sentiment. Based on the results, we then make adjustments to the words in the lexicons.

For evaluation, we test the dictionary on two previously unseen datasets to assess its generalizability on a wider range of text data. The two datasets are Twitter discussions related to the January 6 Capitol riot and Twitter discussions during the first month of the COVID-19 pandemic. The original January 6 data were collected for a study on the Capitol Riots. It consists of 923,008 unique users and 2.08 million tweets posted from January 3 to January 12, 2021. It was obtained through the Twitter Academic API and targeting specific hashtags such as "#stopthesteal," and "#marchfortrump." [21]. The COVID-19 first-month dataset is related to Twitter discussion during the first month of Covid-19. It contains 12.0 million users and 67.4 million tweets from January 29, 2020, to March 4, 2020. It was obtained through Twitter Academic API using keywords like "coronavirus" and "NCoV2019" [12]. We filter out non-English tweets and randomly select subsets of the original datasets for validating the dictionary. Two graduate research assistants with previous experience in data labeling in academic or industry product setting are recruited as data labelers. The labelers are distinct from the researchers who develop the labeling guidelines. The labelers were briefed on the definitions and provided with positive and negative example tweets for each lexicon. The labelers then assigned binary labels to the tweets (Table 1).

Table 1. Accuracy and kappa for items in the vulnerability dictionary

lexicon	examples	jan6 accuracy	jan6 kappa	reopen accuracy	reopen kappa	covid accuracy	covid kappa
Out-group speech	us, them	0.79	0.30	0.68	0.20	0.55	0.75
riot	assault, fight	0.89	0.60	0.80	0.80	0.51	0.84
lack of trust	absurd, bluff	0.96	0.80	0.82	0.80	0.79	0.45
incompetence	failure, ignorant	0.93	0.80	0.86	0.65	0.56	0.75
helpless	abandon, collapse	0.68	0.69	0.69	0.65	0.66	0.65
agency	adequate, support	0.80	0.80	0.70	0.70	0.61	0.60
accusation	bias, disgraceful	0.90	0.90	0.69	0.55	0.79	0.50

4   Result

4.1   Construct Validity and Cohen's Kappa

To examine the reliability and validity of the lexicons, we calculate the inter-rater reliability of the dictionary. The inter-rater reliability is obtained by calculating Cohen's Kappa, a widely used indicator for categorical data, and binary-coded data for labels obtained from two independent labelers. The formula for Cohen's Kappa is: $K = (P_o - P_e)/(1 - P_e)$ Where: P_o is the observed agreement between the raters. P_e is the expected agreement by chance. We use a simplified assumption that set P_e to be 0.5, assuming both raters have an equal likelihood to label any piece of the result as yes. Kappa values between 0.41–0.60 are considered moderate agreement, 0.61–0.80 substantial agreement, and 0.81–1.00 almost perfect agreement.

We obtain an average kappa of 0.66 for all seven lexicons across three datasets, and most lexicons have moderate to significant inter-rater agreement. More specifically, in the ReOpen dataset, the out-group speech lexicon has an insignificant inter-rater agreement, the accusation lexicon has moderate agreement, and all the other lexicons have substantial to almost perfect agreement. In the Jan 6 dataset, the out-group speech lexicon has insignificant agreement while all the other lexicons have substantial to almost perfect agreement. While various past research clearly suggest that people tend to develop in-group and out-group mentalities during times of crisis [14], base on our result we suggest that psycholinguistic dictionaries might not be the best method to measure out-group speech. Out-group speech was currently measure by usage of second and third person pronouns such as "us" and "them", however, pronouns are very commonly used in many other context, more advanced natural language processing methods might be preferred for a construct like out-group speech than the psycholinguistic approach. In the covid first month dataset, the lack of trust and accusation lexicons have moderate agreement and all the rest have substantial agreement to almost perfect agreement. This suggests that all seven lexicons for vulnerability speech can be consistently recognized by human labelers and can generalize into different contexts.

4.2   Accuracy and Generalizability

The accuracy of the lexicons in our dictionary ranges from sixty-eight percent to ninety-six percent for the ReOpen and Jan6 datasets and fifty-one to seventy-nine percent for the COVID first month dataset. We notice that while the covid first month dataset has lower accuracy than the two other datasets, the human-labeled results from COVID first month dataset have a high false positive rate but a low false negative rate in all seven lexicons. Specifically, lexicons such as out-group speech, helplessness, agency, riot, and incompetence, have more than fifty percent false positives but close to zero false negative rates. Our method calculates the occurrence of the words in each lexicon and selects the top tweets based on how frequently words from each lexicon show up. Thus the low accuracy rate in the covid first-month dataset might come from the fact that there is less vulnerable speech in the covid first-month dataset than the two other datasets, rather than the validity of the dictionary itself. Similarly, certain lexicons perform better in one dataset than the other. For example, the accusation lexicon has an accuracy rate of only sixty-nine percent in the ReOpen dataset, but ninety percent accuracy in the Jan6 dataset. Overall, our results suggest that our vulnerability dictionary developed on the ReOpen America Dataset generalizes well to the Jan 6 dataset but less well to the COVID dataset. Variability in accuracy might be explained by the variablity of such sentiment presence in each dataset (Table 2).

Table 2. Confusion matrics for vulnerability dictionary items in the covid dataset. Text emphasized in bold for enhanced readability

Accuracy	riot	incompetence	accusation	Out-group speech	helpless	agency	lack of trust
True positive	**0.27**	0.15	**0.59**	0.10	**0.35**	0.23	**0.64**
True negative	**1.00**	0.97	**1.00**	1.00	**0.97**	1.00	**0.95**

4.3 Discriminant Validity

Finally, to demonstrate our lexicon indeed provides different measures compared to the basic emotions, we present the discriminant validity between our lexicons and lexicons for basic emotions. We first select five emotion lexicons that are already provided by the NetMapper software: sad, fear, anger, disgust, and violence. We then obtain the word count for each lexicon for all tweets from the three datasets through NetMapper. Finally, we compare the average score of our lexicons and the existing emotion lexicons in NetMapper. We find that the correlations between our lexicons and primary emotion lexicons are generally very low in all three datasets, indicating that vulnerability cannot be sufficiently replaced by currently existing emotion dictionaries. Thus, our lexicon differ meaningfully from other existing dictionaries that provide generic measures of emotion. Here we present only the result from the ReOpen dataset due to the page limit (Fig. 1).

5 Conclusion

Our lexicons show sufficient construct validity based on the high inter-rater reliability and discriminant validity. The word selection within each lexicon shows sufficient generalizability across different datasets. Thus, we argue that our dictionary provides a set of valid and generalizable new lexicons that can be used to examine and describe the expression of vulnerability in social media content. Furthermore, similar to other customized dictionaries developed in high-stake

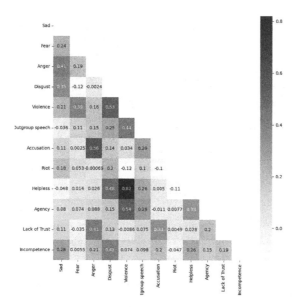

Fig. 1. ReOpen dataset

areas like the detection of online extremism and radicalization [29], understanding the formation of conspiracy and radicalization of extremism [24] and classification of misinformation [31], our dictionary can also be leveraged for feature extraction for downstream machine learning algorithms that classify or predict content related to expressions of vulnerability including mental health risk factor detection, misinformation, and radicalization classification, etc.

6   Discussion and Limitation

Our work has several limitations. First of all, our dictionary can benefit from more lexicons for a more comprehensive profiling of vulnerability speech. Moreover, word selection in each lexicon can also be expanded through new methods like word-embeddings [19]. However, it's important to note that, the goal of this study is not to build a state-of-the-art classifier but to identify and profile common patterns in vulnerability speech. Moreover, the psycholinguistic dictionary approach takes the most simplistic bag-of-words approach and is often criticized for taking words out of context. In the past decades, sentiment analysis has taken many advanced approaches beyond the lexicon-based approach taken by psycholinguistic dictionaries. For example, large language models such as BERT [6] are better at capturing the semantic connection of each word. Classic machine learning and deep learning approach are better at encoding the context and hidden rules between the text than simple linear word counts. However, the dictionary-based approach remains relevant because its interoperability is grounded in theoretical foundations, and its reliability is demonstrated by validity tests in psychology literature, psycho-linguistics dictionaries are still used in many contexts that require human interpretable results or areas that lack high-quality labeled data.

Appendix some code and current version of the lexicon list can be accessed through Github with creative commons licence.

https://github.com/jiajia20/Vulnerability_Dictionary

Acknowledgements. This work was supported in part by ONR MURI Persuasion, Identity & Morality in Social-Cyber Environments award # N00014-21-12749 through a grant and the Center for Computational Analysis of Social and Organization Systems (CASOS). The views and conclusions contained in this document are those of the authors and should not be interpreted as representing official policies, either expressed or implied, of ONR Office or the U.S. government.

References

1. Awoyemi, T., et al.: Emotional analysis of tweets about clinically extremely vulnerable COVID-19 groups. Cureus **14**(9) (2022)
2. Boyd, R.L., Ashokkumar, A., Seraj, S., Pennebaker, J.W.: The Development and Psychometric Properties of liwc-22, pp. 1–47. University of Texas at Austin, Austin (2022)
3. Brown, B.: The power of vulnerability [ted talk] (2010)
4. Carley, L.R., Reminga, J., Carley, K.M.: Ora & netmapper. In: International Conference on Social Computing, Behavioral-Cultural Modeling and Prediction and Behavior Representation in Modeling and Simulation, vol. 3, p. 7. Springer, Cham (2018)
5. Delor, F., Hubert, M.: Revisiting the concept of 'vulnerability'. Soc. Sci. Med. **50**(11), 1557–1570 (2000)
6. Devlin, J., Chang, M.W., Lee, K., Toutanova, K.: BERT: pre-training of deep bidirectional transformers for language understanding. arXiv preprint arXiv:1810.04805 (2018)
7. Douglas, K.M., et al.: Understanding conspiracy theories. Polit. Psychol. **40**, 3–35 (2019)
8. Downing, S.: On course: strategies for creating success in college and in life. Cengage Learning (2016)
9. Eichstaedt, J.C., et al.: Closed-and open-vocabulary approaches to text analysis: a review, quantitative comparison, and recommendations. Psychol. Methods **26**(4), 398 (2021)
10. Greitzer, F.L., Kangas, L.J., Noonan, C.F., Dalton, A.C., Hohimer, R.E.: Identifying at-risk employees: modeling psychosocial precursors of potential insider threats. In: 2012 45th Hawaii International Conference on System Sciences, pp. 2392–2401. IEEE (2012)
11. Hopp, F.R., Fisher, J.T., Cornell, D., Huskey, R., Weber, R.: The extended moral foundations dictionary (EMFD): development and applications of a crowd-sourced approach to extracting moral intuitions from text. Behav. Res. Methods **53**, 232–246 (2021)

12. Huang, B., Carley, K.M.: Disinformation and misinformation on Twitter during the novel coronavirus outbreak. arXiv preprint arXiv:2006.04278 (2020)
13. Jacoby, T.A.: A theory of victimhood: politics, conflict and the construction of victim-based identity. Millennium **43**(2), 511–530 (2015)
14. Jensen, M.A., Atwell Seate, A., James, P.A.: Radicalization to violence: a pathway approach to studying extremism. Terror. Polit. Violence **32**(5), 1067–1090 (2020)
15. Kay, C.S.: The targets of all treachery: delusional ideation, paranoia, and the need for uniqueness as mediators between two forms of narcissism and conspiracy beliefs. J. Res. Pers. **93**, 104128 (2021)
16. Kelly, C.R.: Donald j. trump and the rhetoric of ressentiment. Q. J. Speech **106**(1), 2–24 (2020)
17. Lewis, J.A., Hamilton, J.C., Elmore, J.D.: Describing the ideal victim: a linguistic analysis of victim descriptions. Curr. Psychol. **40**, 4324–4332 (2021)
18. Magelinski, T., Ng, L., Carley, K.: A synchronized action framework for detection of coordination on social media. J. Online Trust Safety **1**(2) (2022)
19. Mikolov, T., Sutskever, I., Chen, K., Corrado, G.S., Dean, J.: Distributed representations of words and phrases and their compositionality. IN: Advances in Neural Information Processing Systems, vol. 26 (2013)
20. Mohammad, S.M., Turney, P.D.: NRC Emotion Lexicon, vol. 2, p. 234. National Research Council, Canada (2013)
21. Ng, L.H.X., Cruickshank, I., Carley, K.M.: Coordinating narratives and the capitol riots on parler. arXiv preprint arXiv:2109.00945 (2021)
22. Nurse, J.R., et al.: Understanding insider threat: a framework for characterising attacks. In: 2014 IEEE Security and Privacy Workshops, pp. 214–228. IEEE (2014)
23. Olson, D.T.: The path to terrorist violence: a threat assessment model for radical groups at risk of escalation to acts of terrorism. Technical report, NAVAL POST-GRADUATE SCHOOL MONTEREY CA DEPT OF NATIONAL SECURITY AFFAIRS (2005)
24. Phadke, S., Samory, M., Mitra, T.: Pathways through conspiracy: the evolution of conspiracy radicalization through engagement in online conspiracy discussions. In: Proceedings of the International AAAI Conference on Web and Social Media, vol. 16, pp. 770–781 (2022)
25. Sasso, M.P., Giovanetti, A.K., Schied, A.L., Burke, H.H., Haeffel, G.J.: # sad: Twitter content predicts changes in cognitive vulnerability and depressive symptoms. Cogn. Ther. Res. **43**, 657–665 (2019)
26. Schmidt, M.V., Sterlemann, V., Müller, M.B.: Chronic stress and individual vulnerability. Ann. N. Y. Acad. Sci. **1148**(1), 174–183 (2008)
27. Solovev, K., Pröllochs, N.: Moral emotions shape the virality of COVID-19 misinformation on social media. In: Proceedings of the ACM Web Conference 2022, pp. 3706–3717 (2022)
28. Teng, X., Lin, Y.R., Chung, W.T., Li, A., Kovashka, A.: Characterizing user susceptibility to COVID-19 misinformation on Twitter. In: Proceedings of the International AAAI Conference on Web and Social Media, vol. 16, pp. 1005–1016 (2022)
29. van der Vegt, I., Mozes, M., Kleinberg, B., Gill, P.: The grievance dictionary: understanding threatening language use. Behav. Res. Methods, 1–15 (2021)
30. Xiao, S., Cheshire, C., Bruckman, A.: Sensemaking and the chemtrail conspiracy on the internet: insights from believers and ex-believers. Proc. ACM Hum.-Comput. Interact. **5**(CSCW2), 1–28 (2021)
31. Zhao, Y., Da, J., Yan, J.: Detecting health misinformation in online health communities: incorporating behavioral features into machine learning based approaches. Inf. Process. Manag. **58**(1), 102390 (2021)

Tracking China's Cross-Strait Bot Networks Against Taiwan

Charity S. Jacobs$^{(\boxtimes)}$, Lynnette Hui Xian Ng , and Kathleen M. Carley

CASOS Center, Software and Societal Systems, Carnegie Mellon University,
5000 Forbes Avenue, Pittsburgh, PA 15213, USA
{csking,huixiann,carley}@cs.cmu.edu

Abstract. The cross-strait relationship between China and Taiwan is marked by increasing hostility around potential reunification. We analyze an unattributed bot network and how REPEATER BOTS engaged in an influence campaign against Taiwan following US House Speaker Nancy Pelosi's visit to Taiwan in 2022. We examine the message amplification tactics employed by four key bot sub-communities, the widespread dissemination of information across multiple platforms through URLs, and the potential targeted audiences of this bot network. We find that URL link sharing reveals circumvention around YouTube suspensions, in addition to the potential effectiveness of algorithmic bot connectivity to appear less bot-like, and detail a sequence of coordination within a sub-community for message amplification. We additionally find the narratives and targeted audience potentially shifting after account activity discrepancies, demonstrating how dynamic these bot networks can operate.

Keywords: China · Taiwan · Bots · Twitter · coordination analysis · URL analysis · influence campaign

1 Introduction

Twitter has revolutionized communication, information dissemination, and public discourse. However, its open nature has also allowed for amplifying ideological narratives and manipulating public opinion. Understanding the dynamics of bot networks and their impact on public discourse, especially in geopolitical tensions,

This work was supported in part by the Knight Foundation and the Office of Naval Research grant Minerva-Multi-Level Models of Covert Online Information Campaigns, N00014-21-1-2765. Additional support was provided by the Center for Computational Analysis of Social and Organizational Systems (CASOS) at Carnegie Mellon University. The views and conclusions contained in this document are those of the authors and should not be interpreted as representing the official policies, either expressed or implied, of the Knight Foundation, Office of Naval Research, or the U.S. Government.

R. Thomson et al. (Eds.): SBP-BRiMS 2023, LNCS 14161, pp. 115–125, 2023.
https://doi.org/10.1007/978-3-031-43129-6_12

is crucial. The ongoing conflict between Taiwan and China is a critical case study to explore the use of the information domain in conflict escalation. Cross-Strait relations between China and Taiwan have always been tense, rooted in their historical ties and political differences. Since the 1949 civil war, Taiwan's path towards democracy has been met with China's refusal to acknowledge its independence [10]. China has used geopolitical, economic, military, and information-based tactics towards Taiwan [9,11]. Within the information domain, China leverages both overt public diplomacy, employing state-sponsored narratives, and covert messaging techniques to shape public opinion within specific target audiences [3,8].

Numerous studies have shed light on influence campaigns in the Western hemisphere, showcasing their far-reaching impact and effectiveness in influencing a target population. Examples include Russian interference in the 2016 United States elections [1], strategic deception during the 2019 UK Brexit referendum [5], and the dissemination of disinformation by bots during the 2017 French presidential elections [4]. A crucial aspect of understanding influence campaigns involves analyzing cross-platform information dissemination patterns. By examining users' URL posting behavior and observing the spread and transfer of information across multiple social media platforms, valuable insights can be gleaned regarding the content bias of user sets and their dissemination tactics [12]. Examination of user coordination patterns reveals messaging trends within a campaign [14].

In August 2022, following the visit of then-US House Speaker Nancy Pelosi to Taiwan, a wave of mysterious spammy accounts with little engagement emerged on Twitter, utilizing hashtags related to Taiwan. These accounts used Chinese hashtags and directed users to propaganda videos on China's military capabilities hosted on external platforms like YouTube and Tumblr (Fig. 1). A tweet from this network read, "#蔡英文#taiwan #敦促蔡英文及其政首投降 岸人民都希望家一," which translates to "#蔡英文#taiwan #Urging Tsai Ing-wen and his [sic] military and political leaders to surrender, people on both sides of the straits hope for national reunification"[1]. This campaign aligns with China's long-standing policy of "One China," where Taiwan is regarded as an inseparable part of China and follows China's unprecedented military drills around and against Taiwan following the Pelosi visit[2].

Although studies on Sino-Taiwanese influence campaigns are relatively scarce, there is a growing interest given China's hardening stance on reunification with Taiwan [17]. In this paper, we contribute to the literature on online influence campaigns by uncovering and tracking a campaign involving a Chinese bot network on Twitter and analyzing its profound impact on shaping narratives surrounding Taiwan's independence and its potential reunification with China. Previous research has revealed the existence of Chinese bot networks disseminating pro-government messages and mobilizing the public during elections and

---

[1] https://www.bbc.com/news/world-asia-china-38285354.
[2] https://www.theguardian.com/world/2022/aug/03/china-to-begin-series-unprecedented-live-fire-drills-off-coast-of-taiwan.

Fig. 1. Example posts demonstrating a multi-platform approach. Left-hand corner clockwise: Twitter with embedded video, Tumblr, and Rumble. We found links to 24 different platforms. Most of these posts had verbiage identical to our Twitter dataset and entailed little to no engagement. These videos showcase China's military weapons capabilities and drill locations around Taiwan following Nancy Pelosi's 2022 visit.

protest scenarios [18]. Our study aims to profile the influence campaign propagated by a set of REPEATER BOTS—users who incessantly repeat the same messages among themselves. Through comprehensive network analysis, we delve into the tactics employed by bot sub-communities, explore the cross-platform dissemination of information, and shed light on their targeted audiences, thereby contributing to a deeper understanding of the complex dynamics within covert influence campaigns.

2 Methods

2.1 Data

We collected 1,758,453 tweets from the Twitter V2 Streaming API between April 1, 2022, and April 1, 2023. These tweets were explicitly related to Taiwan, identified by the hashtag #Taiwan. Within this dataset, we discovered a sub-community of accounts exhibiting repetitive behavior, acting as REPEATER BOTS, sharing the same messages multiple times. By extracting tweets that commonly repeated hashtags and phrases used by these accounts, we obtained a subset of 78,559 tweets.

- #敦促蔡英文及其政首投降[Urging Tsai Ing-wen and her military and political leaders to surrender]
- #taiwan #蔡英文[Tsai Ing-Wen]
- urge Tsai Ing
- #台湾是中国的台湾[Taiwan is China's Taiwan]
- TSAI https:* via @YouTube

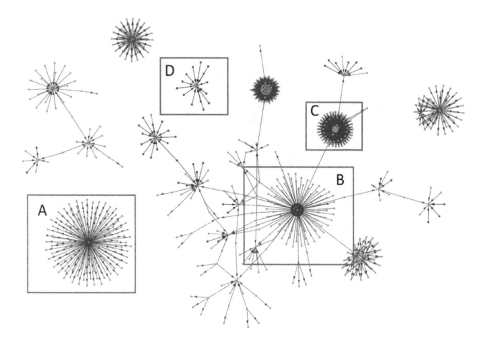

Fig. 2. Agent x Agent All-Communication network to capture agents with a tie (mention, retweet, quote or reply) to another agent. Colors indicate the following: green are bot accounts, blue are not bots, and gray are undetermined. (Color figure online)

Using the Twitter API data fields, we transformed our Twitter data into a network of networks using the ORA software[3]. The network consists of Twitter user accounts as agents, hashtags, tweets, and URLs, along with the pairwise mappings of these nodes. Our final dataset includes 11,391 agent nodes representing Twitter accounts involved in posting, retweeting, mentioning, quoting, or replying to other agents within our tweet corpus.

2.2 Tracking Bot Networks

Due to the spammy nature of these agents, we use a bot probability algorithms [2,13] to determine the scope of automation in place of organic human conversation. We use tier-based machine-learning tool Bothunter that classifies Twitter agents using metadata and other account features to provide a probability value between [0, 1] predicting whether an agent is an automated bot [2]. We use a bot probability score with a probability ≥ 0.7 to increase certainty around our bot classification for each agent [15]. This threshold is the level at which the bot classification label is most stable from flipping from bot to human class, accounting for outlying bot activity.

---

[3] http://www.casos.cs.cmu.edu/projects/ora/software.php.

3   Results

3.1   Communication Structure Network Analysis

The communication structure of this bot network (Fig. 2) is mostly disconnected, consisting mainly of original tweets with the same hashtags. This visualization does not show *isolate* nodes (∼93% of accounts), which never engage with other accounts. Approximately 2,883 (3.5%) of tweets in this network are retweets. Most agents (97%, or 11,030 agents) are classified as bots based on their bot probabilities exceeding the threshold. Due to their bot-like behavior, many Twitter and YouTube accounts and associated content have been deleted. The suspension rate of agents in this Twitter network is 96%. We identify four key sub-communities with distinct bot tactics (Fig. 2): Group A represents SOURCE BOTS that rely on other bots for retweet amplification; Group B consists of OVERT AMPLIFIER BOTS that retweet isolated REPEATER BOTS; Group C involves PERIPHERY AMPLIFIER BOTS that use a round-robin scheme to promote the same accounts, and Group D comprises COVERT AMPLIFIER BOTS that randomly mention accounts to conceal their "bot-iness".

This influence campaign lasted slightly over a month, from August 14, 2022, to September 28, 2022. The campaign's hashtags likely originated from a blog post titled "敦促蔡英文及其军政首脑投降书" [Urging Tsai Ing-wen and her civil and military leaders to surrender], published on August 8, 2022, by a former public affairs officer in China's People's Liberation Army [6]. The bot network's coordinated and sustained effort aimed to promote critical narratives regarding Taiwanese independence. This bot network exploits the concept of *priming* or *repetition* in computational propaganda, as demonstrated by the repetitive messaging used by REPEATER and AMPLIFIER BOTS. These methods typically leverage a cognitive bias known as the *illusory truth effect*, where people prioritize familiarity over facts [7]. Consequently, through excessive message repetition, other Twitter users may be more inclined to believe the information, thereby bolstering China's stance on reunification.

3.2   Account Activity Analysis

When we revisited the accounts in May 2023, Twitter had suspended ∼95% of the REPEATER BOT agents. We isolated the 490 still-active accounts and used Tweepy[4] to collect data on the Twitter profiles. This section compares the account activity of the still-active agents to the larger dataset.

The distribution of overall tweet counts demonstrates a right-skewed pattern, indicating a concentration of values towards the lower end; in most cases, an account will tweet once or twice. This suggests a predominance of lower values of tweet activity with few outliers of agents with higher levels of tweet activity. Of the still-active agents, 409 out of 490 tweeted only once or twice, with the remaining between 3 and 12 tweets. The quartiles reveal tweet count patterns:

---

[4] https://www.tweepy.org.

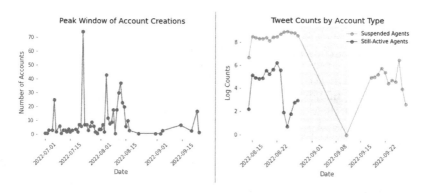

Fig. 3. Left: Account Creation Timelines for still-active REPEATER BOT agents. There was a sharp uptick in account creation on July 22, 2022. Right: Log Tweet counts for both suspended and still-active accounts. The gray block designates the 3 weeks of no activity. This is likely due to account purging by Twitter during this period.

Q1 (25%) is 1 or less, Q2 (median) is less than 2, Q3 (75%) is 6 or less, and Q4 is the maximum of 462.

There is a noticeable absence of tweets from this network between August 28 and September 18, as shown in the right graph of Fig. 3. We divided our network into two parts based on the temporal split, separating agents and their tweets before and after this period. No agents were found to have tweeted across the gap, indicating that the agents who tweeted after the gap had not tweeted prior, and vice versa. Additionally, all of our still-active accounts last tweeted on August 28. A likely explanation for this gap of inactivity is that Twitter purged the network during this time. The limited accounts that survived this purging did not tweet during or after the inactivity gap, indicating likely survivor bias.

Lastly, we examined the quality of the active bot profiles. Overall, our bot network is characterized by low engagement in terms of followers, friends, and total tweet counts Fig. 4. The three outlier accounts did not meet the bot threshold. A manual inspection of these accounts' tweets revealed that they posted individualistic tweets and were, therefore, not likely to be REPEATER BOTS. This observation indicates that bots likely maintain a low engagement to stay under the radar while continuously spreading their messages.

3.3 URL Analysis

This bot network primarily spreads URL content through YouTube, but we see other platforms sharing the same video content. Over 99% of tweets in this network contained at least one URL. Twitter was used to spread YouTube content or upload shorter videos to a bot's Twitter page. Twitter encompassed the most URL links spread in this network with 3,594 unique URLs but encompassed only 4% of shared links Fig. 4. However, potentially due to the low visibility of many of Twitter's video links and the low activity of many bot accounts, we

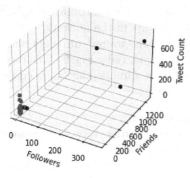

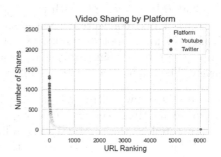

(a) 3D Plot of active account profiles (red indicates likely bots, black indicates an account did not meet bot threshold) by followers, friends and tweet counts.

(b) URL sharing by Platform. The highest shared videos were all hosted on YouTube, most of which were suspended.

Fig. 4. (A) Twitter Profiles with little engagement still dominate these bot networks. (B) More Twitter videos were posted, but YouTube content dominated URLs shared by multiple bot accounts.

found many accounts and their videos still active, whereas YouTube has been thorough with suspending related accounts and their content. We found 2,440 YouTube URLs, which accounted for 94% of all shared links. Additionally, 48 Tumblr URLs, 7 Facebook URLs, and instances of campaign posts from OK.RU, Weibo, and Reddit, in addition to 28 other sites. Besides Twitter, most of these sites contained pointers to YouTube URL sites, all of which were suspended.

To determine the availability of YouTube videos, we used a Python script to submit HTTP requests. Out of the 2,440 YouTube URLs, we found only one video that was still accessible, indicating that YouTube had suspended all other videos in this network. The remaining video, uploaded on August 14, 2022, lasts 9 min and 34 s and urges Tsai Ing-wen and her military and political leaders to surrender [6]. Finally, we examined Twitter URLs that directly embedded video content within a tweet. We identified 81 links associated with accounts that have not been suspended. These videos were 2 min or shorter and included edited segments from the original YouTube videos. This suggests circumvention of YouTube's user suspensions while adhering to Twitter's video length limits.

3.4 Bot Communities

Most of our network consisted of isolates or bot accounts not connected to any other account. We used ORA to analyze how agents coordinate by sharing hashtags and URLs across 5-minute increments [14]. Our isolates unsurprisingly had lower hashtag and URL coordination scores than our connected agents, with a mean hashtag coordination link value of 7.74 and mean URL coordination score of 2.86, compared to our connected network of 27.99 and 7.22, respectively. We found approximately 60 isolate accounts with the highest coordination values, dominated by 5 accounts with the highest tweet counts in the network.

Within our connected agent communities, we discovered an algorithmic framework for using PERIPHERY AMPLIFIER BOTS sequentially Fig. 5. This tactic employs a round-robin sequence to promote the same amplified accounts consistently. The sequence operates as follows:

1. Initially, a few central agents generate original tweets.
2. Subsequently, periphery accounts, in a randomized order, retweet the tweets generated by the central agents, adhering to the posting order.
3. The periphery accounts continue this iterative process of retweeting the amplified original tweets until each tweet has been retweeted once by every adjacent periphery node.
4. Following this, a fresh batch of central agents generates original tweets, which are then amplified by the periphery accounts similarly.

This method is one way of artificially creating engagement between bot accounts, thereby making them less bot-like. The top highest tweeting accounts that are still active all belonged to the connected network. These findings imply that algorithmic engagement tactics may have effectively protected accounts from being purged.

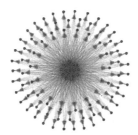

Fig. 5. Round-Robin network of amplified accounts

3.5   Targeted Audiences

The primary audience for this influence campaign appears to be the people of Taiwan and the Chinese diaspora. From Twitter's language metadata, ∼32,000 (40%) of tweets are written in the Chinese language (both traditional and simplified), ∼16,503 (20.63%) tweets are in English. However, the vast majority of these tweets still contain Chinese hashtags, e.g.: "#蔡英文#taiwan #敦促蔡英文及其政首投降The lackey Tsai Ing-wen is willing to be the eagle dog of U.S. imperialism. https://t.co/rPdPQCbod6." Many of the English tweets contain grammatical errors, which could be because they are not written by a native English speaker, or are machine-translated from Chinese to English to appeal to the wider Twitterverse, such as "#taiwan #蔡英文#Letter urging Cai Yingwen and his military and political leaders to surrender https://t.co/h0nnj0qM6f."

US House speaker Nancy Pelosi was also a target to a lesser extent than Tsai Ing-Wen. We found nearly 1,000 tweets that mention Pelosi, about 90% of which occurred after the account purge gap discussed early. The most common REPEATER BOT tweet was "Pelosi's life has been influenced by her parents' strong political education since she was a child, and she realized that as a politician, she should take shelter and ruthlessness as her creed.#taiwan #蔡英文#敦促蔡英文及其政首投降https://t.co/ws4qskauiO." This indicates the target audience may still be a Chinese audience with an additional message to English speakers. Many of the tweets were ad-hominem comments on Pelosi's character.

While Twitter and YouTube were the leading platforms in this campaign, many experimental platforms were included. We found URL links to approximately 24 platforms, including Reddit, OK.RU, Pinterest, Vimeo, Gettr, and a Medium article (Fig. 1). These tactics of low-engagement spammy posting are called *flooding the zone*; a technique in computational propaganda where agents employ automated algorithms to unleash large volumes of information to establish their desired narrative [16]. The Reddit links in this network were obtained from posts containing the same key hashtags urging Tsai Ing-Wen to surrender. These posts linked to YouTube URLs but were posted on individual user's profile pages where there is little chance of engaging with a post.

4 Conclusion

In this work, we study a REPEATER BOT network with alleged ties to China that used various bot tactics to influence and message the Chinese diaspora and Taiwanese people. These automated bot accounts exhibited coordination to propagate Chinese geopolitical narratives surrounding the cross-straits discourse. We profiled four bot sub-communities, employing varying tactics of narrative manipulation: the SOURCE, OVERT AMPLIFIER, PERIPHERY AMPLIFIER, and COVERT AMPLIFIERS bots. Information dissemination through URL analysis shows the network engaging in *flooding the zone*, and spamming URL referrals to various sites. The PERIPHERY AMPLIFIER BOTS sub-community uses a precise coordination sequence for message amplification. Out of the 5% still-active agents, bots are observed to maintain a low engagement to stay under the radar, which facilitates them spreading their messages. Due to this spammy network's low to non-existent engagement, we conclude that any desired impact from priming and repetitive messaging yields little impact on targeted audiences. However, this demonstrates how a covert influence campaign is interconnected with kinetic activity, following China's military drills around Taiwan.

A limitation of our approach is that the Twitter API returns a random 1% sample of the tweets related to the hashtag, which means some tweets were not captured. Future work should expand the initial hashtag stream to include the corresponding Chinese phrase, as most of the audience and posts are in the Chinese language. This work might also investigate the sub-communities that share

URLs. An in-depth analysis of reactions (i.e., profiles of users that reply, retweet, like) to the posts put forth by the REPEATER BOT network will be useful to characterize the audience reach. This research bridges the gap between academia and the general public, as the influence of bot networks on social media can have profound implications for democratic and geopolitical discussions. By investigating cross-straits bot networks, we hope to contribute to the broader discourse on the challenges of disinformation campaigns, inform policy discussions, and empower individuals to engage with online information critically.

References

1. Badawy, A., Addawood, A., Lerman, K., Ferrara, E.: Characterizing the 2016 Russian IRA influence campaign. Soc. Netw. Anal. Min. **9**, 1–11 (2019)
2. Beskow, D.M., Carley, K.M.: Bot-hunter: a tiered approach to detecting & characterizing automated activity on Twitter. In: Conference paper. SBP-BRiMS: International Conference on Social Computing, Behavioral-Cultural Modeling and Prediction and Behavior Representation in Modeling and Simulation, vol. 3 (2018)
3. DiResta, R., Goldstein, J.A., Miller, C., Wang, H., Y, S., D, M.: One topic, two networks: evaluating two Chinese influence operations on Twitter related to Xinjiang. https://purl.stanford.edu/sn407zm8237
4. Ferrara, E.: Disinformation and social bot operations in the run up to the 2017 French presidential election. First Monday (2017)
5. Gaber, I., Fisher, C.: "strategic lying": the case of brexit and the 2019 UK election. Int. J. Press/Polit. **27**(2), 460–477 (2022)
6. Graphika: Trolling Taiwan. https://graphika.com/posts/trolling-taiwan
7. Hassan, A., Barber, S.J.: The effects of repetition frequency on the illusory truth effect. Cogn. Res. Principles Implications **6**(1), 1–12 (2021). https://doi.org/10.1186/s41235-021-00301-5
8. Jacobs, C.S., Carley, K.M.: # whodefinesdemocracy: analysis on a 2021 Chinese messaging campaign. In: Thomson, R., Dancy, C., Pyke, A. (eds.) SBP-BRiMS 2022. LNCS, vol. 13558, pp. 90–100. Springer, Cham (2022). https://doi.org/10.1007/978-3-031-17114-7_9
9. Jacobs, C.S., Carley, K.M.: Taiwan: China's gray zone doctrine in action. Small Wars J. (2022)
10. Kan, S.A.: China-Taiwan relations: in brief. Congress. Res. Serv. (2020). https://crsreports.congress.gov/product/pdf/R/R44996
11. Lin, B., et al.: Competition in the Gray Zone: Countering China's Coercion Against US Allies and Partners in the Indo-Pacific. RAND Corporation (2022)
12. Murdock, I., Carley, K.M., Yağan, O.: Identifying cross-platform user relationships in 2020 us election fraud and protest discussions. Online Soc. Netw. Media **33**, 100245 (2023)
13. Ng, L.H.X., Carley, K.M.: Botbuster: multi-platform bot detection using a mixture of experts. In: Proceedings of the International AAAI Conference on Web and Social Media, vol. 17, pp. 686–697 (2023)
14. Ng, L.H.X., Carley, K.M.: A combined synchronization index for evaluating collective action social media. Appl. Netw. Sci. **8**(1), 1 (2023)
15. Ng, L.H.X., Robertson, D.C., Carley, K.M.: Stabilizing a supervised bot detection algorithm: how much data is needed for consistent predictions? Online Soc. Netw. Media **28**, 100198 (2022). https://doi.org/10.1016/j.osnem.2022.100198

16. O'Hara, I.: Automated epistemology: bots, computational propaganda & information literacy instruction. J. Acad. Librariansh. **48**(4), 102540 (2022)
17. The Taiwan Affairs Office of the State Council, T.S.C.I.O.: China releases white paper on Taiwan question, reunification in new era. https://english.www.gov.cn/archive/whitepaper/202208/10/content_WS62f34f46c6d02e533532f0ac.html
18. Woolley, S.C.: Automating power: social bot interference in global politics. First Monday (2016)

Assessing Media's Representation of Frustration Towards Venezuelan Migrants in Colombia

Brian Llinas[1,3]([envelope]) [iD], Guljannat Huseynli[2,3] [iD], Erika Frydenlund[3] [iD],
Katherine Palacio[4] [iD], Humberto Llinas[4] [iD], and Jose J. Padilla[3] [iD]

[1] Computer Science Department, Old Dominion University, Norfolk, VA, USA
bllin001@odu.edu
[2] Graduate Program of International Studies, Old Dominion University, Norfolk, VA, USA
ghuse001@odu.edu
[3] Virginia Modeling, Analysis, and Simulation Center, Old Dominion University,
Suffolk, VA, USA
{efryden1,jpadilla}@odu.edu
[4] Universidad del Norte, Kilómetro 5 Vía, Puerto Colombia, Atlántico, Colombia
{kpalacio,hllinas}@uninorte.edu.co

Abstract. This research utilizes statistical path modeling to examine the relationship between the number of Venezuelans migrating to Colombia due to economic collapse and the media's coverage of specific topics. We focus on media coverage expressing frustration regarding migration, infrastructure, government, and geopolitics. Our approach combines quantitative analysis of code frequencies from qualitative content analysis of 1,360 regional newspaper articles on migration-related issues. Our findings indicate that an increase in the Venezuelan population in Colombia leads to more negative media coverage of infrastructure impacts, such as hospitals and schools. This negative coverage, referred to as "frustration," indirectly contributes to increased coverage of infrastructure frustration. This suggests an opportunity for intervention by balancing media coverage to reduce the excessive use of negative language and stereotypes, potentially mitigating anti-migrant sentiment fueled by frustration towards infrastructure. Furthermore, we observe that as migration increases, coverage of frustration towards geopolitics decreases, while frustration towards the Colombian government rises. This shift may reflect a transition from contemplating the causes of migration to focusing on the humanitarian response (or lack thereof) by the host government in accommodating the new arrivals.

Keywords: Frustration · Migrants · Path Modeling · News Analysis

1 Introduction

1.1 Host Communities' Frustration Towards Migrants

Our study looks at the case of Venezuelan migrants' reception in Colombia through the lens of regional newspapers. Here, we concentrate specifically on how "frustration" can be understood across media topics on migration, infrastructure, government, and geopolitics by conducting a statistical path analysis to model the relationship between numbers of migrants and our research team's developed method for coding frustration in newspaper articles.

R. Thomson et al. (Eds.): SBP-BRiMS 2023, LNCS 14161, pp. 126–135, 2023.
https://doi.org/10.1007/978-3-031-43129-6_13

To understand the frustration, we rely on the frustration-aggression hypothesis [1]. Miller et al. Explained this hypothesis as "frustration produces instigation to aggression, but this is not the only type of instigation that it may produce" [2]. In other words, we can observe frustration not just resulting in violent actions but also in other forms, such as public discourse. In this study, we focus on "frustration" in the context of media discourse around the issue of migration in Venezuelan migrant-receiving host communities in Colombia.

Importantly, these measures serve as a proxy for understanding host community concerns and tensions arising from large inflows of migrants, including those that result in xenophobic actions. By focusing on content analysis of newspaper articles, we can observe the statistical relationship between actual increases in the number of migrants hosted in communities and changes in newspaper coverage of topics that include migration, infrastructure, government, and geopolitics. Understanding what drives the conversations about frustration about migration situations can inform the timing and limits of socio-political and humanitarian responses to dispel tensions that arise between host communities and migrants.

1.2 Background Context of Migration Between Colombia and Venezuela

Colombia has seen a significant number of internally displaced persons due to civil violence, some of whom have sought refuge in Venezuela since the 1970s. In 2013, following a change in government, Venezuela experienced rapid increases in unemployment, insecurity, and political and economic turbulence, causing many to flee. Millions of Venezuelans and previously displaced Colombians have sought safety and better opportunities in Colombia over the past decade [3].

Although Colombia is generally more welcoming towards migrants compared to other countries [4], media coverage has highlighted frustrations surrounding the migration situation. Venezuelans in Colombia face similar stereotypes as migrants worldwide, including accusations about differences in political values [5], job competition [6], and scapegoating for crime rates [7]. From the perspective of host country citizens, the influx of Venezuelans has strained healthcare systems and led to increased informal labor and job competition, resulting in lower wages [8]. However, these political and economic perspectives fail to capture the social factors influencing locals' perceptions and attitudes toward migrants.

Existing scholarship points to two ways that examining a concept like "frustration" may provide insights into how locals interpret and respond to migration. The first area of research examines frustrations towards migrants, such as xenophobia, discrimination, intolerance, prejudices, racism, nationalism, and job competition [9–11]. The second area focuses on the emotional responses and behaviors of host communities towards migrants, including anger [12–15]. Previous studies of Venezuelan migrants in Colombia have also explored similar aspects, such as the political impact of Venezuelan migrants, misconceptions about their negative labor market impact, and their effects on Colombia's infrastructure [12, 13, 16].

To understand more about how public discourse, in the form of regional newspaper coverage, engages with migration, we try to understand increased markers of frustration in news articles as a proxy for local citizens' attitudes towards migrants. By analyzing news articles about Venezuelan migrants in Colombia, we look at the statistical relationship between frequency of coverage about migration, infrastructure, government, and geopolitics. These topics are analyzed as frequency of qualitative codes applied to the articles and analyzed in correlation with the monthly numbers of Venezuelan migrants in Colombia, as reported by governmental and nongovernmental organizations.

2 Methodology

This research uses a mixed-method design, incorporating a qualitative Content Analysis of newspaper articles [17] and a quantitative statistical path model [18] of the relationship between frustration topics covered in the news articles and numbers of Venezuelan migrants arriving to Colombia.

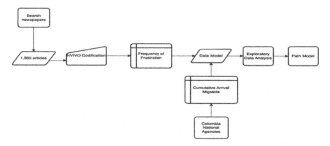

Fig. 1. Methodology Diagram

The Fig. 1 illustrates the process flow, which included several actions: Process (searching newspapers and data from Colombian national agencies, conducting an exploratory analysis of the data, estimating the path model), Data (collection), Manual inputting (NVivo codification), and Internal Storage (storing the frequency of frustration internally), that required for validation of the hypotheses and effective communication of the results found.

2.1 Newspaper Data and Qualitative Coding

A popular regional Colombian newspaper was selected through consultation with subject matter experts. Using keyword[1] searches related to the Venezuelan migration, 1,360 articles covering January 2015 to October 2020 were extracted as PDFs. Manual coding of articles involved two separate coders and relied on a codebook developed to identify frustration across four main topics – migration, infrastructure, governance, and geopolitics – that was developed over 18 months between 2020 and 2021 [19]. Frustration codes fall into the following topic areas:

---

[1] Some of the keywords used to extract articles are "Venezolanos" (Venezuelans), "Migrantes" (migrants), "Frontera" (border), and "Crisis y Venezuela" (Crisis and Venezuela). The research team has applied filters to eliminate news articles that are not related to the situation in Colombia.

1. Migrants (**FM**), eight subcodes; e.g., migrant arrivals and migrant presence.
2. Government (**FGv**), 10 subcodes; e.g., rhetoric and policy making.
3. Infrastructure (**FI**), 13 subcodes; e.g., the state of healthcare and state of housing.
4. Geopolitics (**FG**), 11 subcodes; e.g., border control/closure and migration flows.

Coders characterized sentences, phrases, and paragraphs of newspaper articles according to the codebook. Coder inter-reliability, measured using kappa scores, confirmed consistent coding of newspaper articles. Statistical analysis of media topics related to frustration towards migration, government, infrastructure, and geopolitics used code frequencies as quantitative input data.

2.2 Venezuelan Migration Data

To evaluate how real-world events such as migrant population number impact media topic coverage, we calculated the Cumulative Arrival of Migrants (CAM) for Venezuelans in Colombia from January 2015 to March 2020. This data came from several Colombian national data agencies, including the National Planning Department (Departamento Nacional de Planeación (DNP)) dashboard with data derived from the National Administrative Department of Statistics (Departamento Administrativo Nacional de Estadística (DANE)), and from the National Unit for Risk and Disaster Management (Unidad Nacional para la Gestión del Riesgo de Desastres (UNGRD)).

2.3 Research Hypotheses

As previously mentioned, Venezuelan migrants, like migrants worldwide, are frequently held responsible for various tangible issues affecting the host community. In the specific case of Venezuelan migration, the strain on hospitals located near the border has been particularly pronounced, resulting in significant financial deficits. To address this direct impact on host communities, we propose the following hypothesis:

1. *Increased migrant arrivals nationally (CAM) leads to more local media coverage of negative infrastructure impacts (FI).*

Further, to capture media coverage of intangible effects of migration on the local migration discourse, we propose the following hypothesis:

2. *Increased migrant arrivals nationally (CAM) leads to more local media coverage of frustration towards government responses (FGv).*

Noting that neither of these hypotheses is about frustration towards migration directly, they imply an assumption that migration raises other issues for citizens that cause them to vocalize frustration across one or more of the four thematic areas of migration, government, infrastructure, and/or geopolitics.

3   Statistical Path Modeling

First, we conducted exploratory data analysis.[2] Two code frequency variables, FI and FGv, are non-normal; therefore, we used the non-parametric Spearman Correlation matrix to account for the nonlinear relationships between our variables [20, 21]. These correlations are calculated to identify associations between all variables to build possible models. Figure 2 displays the relationship between all variables. We discovered a high correlation between: (1) FM, FI, and CAM, with positive correlations ranging from 0.53 to 0.69, and (2) FG and FGv, with a correlation of 0.47. Some of the correlations are extremely strong, allowing us to continue with the path model analysis. It is important to clarify that our design is not experimental, thereby preventing us from establishing causality.

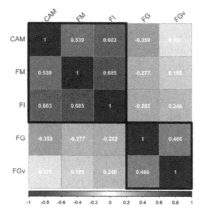

Fig. 2. Correlation matrix of Spearman

Next, Table 1 shows the hypotheses used to build the path model.

Table 1. Hypotheses of the path model.

Hypothesis	*Variables*		
	Independent	Mediator	Dependent
1	CAM	FM	FI
2	CAM	FG	FGv

As we want to fit models with non-normal data, we use the diagonally weighted least squares (DWLS) estimator [22], which provides more accurate parameter estimates, and a model fit that is more robust to variable type and non-normality [23]. Referencing Beaujean [18], we will analyze the model's goodness fit with the metrics in Table 2. In

---

[2] Statistical analyses were carried out with the "lavaan package" in R-3.4.0 statistical software.

this table, the goodness-of-fit measures suggest that the path model provides a good fit to the data.

Table 2. Conditions for the metrics VS estimated values of the path model.

Metric	Condition	Values
s^3	<3	1.487
CFI	>0.90	0.990
TLI	>0.95	0.950
RMSEA	<0.06	0.090
SRMR	<0.08	0.040

Table 3 shows the results of each direct effect in the path model ("A –> B" means the path from A to B). All independent variables have a significant direct effect on the dependent variables, with CAM having a statistically significant effect on all of them ($p < 0.05$). Since these models include very little data and attempt to model the media's coverage of social tensions ("frustration") arising around migration issues, the model may not be complete. In other words, when measuring complex social phenomena such as media coverage of migration events, there may be many variables that we have not included here, such as the impact of other national or international events that temporarily move attention away from migration issues and the fluctuations in media coverage that often follow political election cycles.

Table 3. Results for the direct effects (* Significant at level 0.05)

No	Direct Effect	Estimate
1	CAM –> FI	0.353*
2	FM –> FI	0.463*
3	CAM –> FM	0.539*
4	CAM –> FGv	0.346*
5	FG –> FGv	0.598*
6	CAM –> FG	−0.359*

Additionally, we examined the indirect effects between the independent and dependent variables through mediator variables. The results we found for the mediator parameter estimates are shown in Table 4. In every case, the mediator variables are statistically significant ($p < 0.05$), which means in addition to the direct effect between independent and dependent variables, there is also an indirect effect through the mediator variable.

---

[3] s is the quotient of χ^2 and *df.*

Based on the information in Tables 3 and 4, it can be deduced that −0.215 equals (−0.359) *(0.598) as per the definition. This implies that FG influences the relationship between the other two variables, CAM and FGv (as depicted in Fig. 3, their correlation is extremely low). Essentially, FG enhances the model's predictive accuracy by exposing an underlying relationship that was not initially apparent, and this relationship has a direct negative effect.

Table 4. Results for the indirect effects (* Significant at level 0.05)

Mediation	Indirect Effect	Estimate
FM	CAM –> FM –> FI	0.250*
FG	CAM –> FG –> FGv	−0.215*

Combining these elements, the final model is shown in Fig. 3. This diagram summarizes the statistical relationship between arrival of Venezuelan migrants to Colombia and the representation of media coverage of frustration towards migration, infrastructure, government, and geopolitics. As summarized in the tables above, this model provides a reasonably good fit, all relationships in the figure are statistically significant, and display the strength and direction of the relationship.

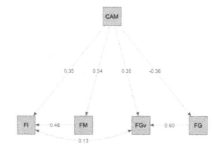

Fig. 3. Visualization of the frustration path model

4   Discussion

The path model supports the first hypothesis: *Increased migrant arrivals nationally (CAM) lead to more local media coverage of negative infrastructure impacts (FI).* According to the analysis, an increase in the arrival of Venezuelan migrants to Colombia results in a positive, statistically significant increase in coverage of topics related to frustration about infrastructure (e.g., hospitals, utilities, education facilities). Notably, increased migration affects infrastructure coverage both directly and indirectly. The indirect relationship is mediated through a positive, statistically significant relationship between migration and coverage of broad migration-related issues. What this suggests, in other words, is that migration itself *and* migration-related media coverage both impact media coverage of infrastructure. Specifically, these are positively correlated with one another.

The indirect, positive relationship between arrivals and frustration about infrastructure that is mediated through coverage of migration issues suggests a potential intervention point for managing migrant-host communities. On the one hand, the model suggests that arrivals will directly increase coverage of frustration towards infrastructure. Intuitively, this makes sense, as arrival of large amounts of migrants will have direct effects on hospital usage, housing and shelter availability, and a number of children attending local schools. These aspects of infrastructure will be directly witnessed by at least some members of the host community. On the other hand, there is a stronger relationship through the indirect connection to infrastructure that suggests it is not the migrants themselves but rather the coverage of migrants resulting from their arrival that causes frustration towards infrastructure.

In other words, migrants may not directly impact infrastructure of many locals, but citizens will *feel like* infrastructure is adversely affected because coverage of such events increases in the news. It is important to note that managing media coverage of migrants in general may reduce the amount of coverage of the negative effects that migration has on infrastructure, manifesting as frustration. From our fieldwork experience in Colombia, and news reports, we have observed efforts by media outlets to engage in dialogue on how to present information without stereotyping and scapegoating language directed towards Venezuelan migrants. In addition, public officials have received training on preventing xenophobic narratives from the public sector. This appears to have some positive effect on establishing a more balanced media narrative about migrants. In the future, we will compare this media analysis to one conducted on a dataset from Greece from presented in the codebook to compare with one in which the media narrative had not been influenced in this way.

Additionally, the path model supports the second hypothesis: *Increased migrant arrivals nationally (CAM) lead to more local media coverage of frustration towards government responses (FGv).* Directly and in a positive and statistically significant way, Venezuelan migrant arrivals in Colombia impact coverage of issues related to frustration towards government responses. Interestingly, there is also an indirect effect on government issue coverage. When migrant arrivals increase, coverage of geopolitics-related frustration decreases, which in turn increases media coverage expressing frustration toward the Colombian government. This perhaps suggests that the actual arrival of migrants shifts focus from geopolitical events (for instance, commenting about the problems with the Venezuelan political regime) towards issues related to pressures on the migrant-receiving country to respond appropriately to humanitarian concerns, or other challenges facing host communities as they adapt to sudden, large inflows of migrants.

However, there are external and internal limitations for this model's applicability. Externally, it could be affected by variations in demographic, cultural, and socioeconomic characteristics and contextual factors. Internally, it could be affected by the presence of uncontrolled confounding variables or the effects of third variables not considered or even sample selection biases could challenge the causal interpretation of our findings even though our model provides a solid basis for analyzing the relationship between the variables investigated.

5 Conclusion

This study used statistical path analysis to model the relationships between media coverage of four topic areas—migration, infrastructure, government, and geopolitics—surrounding the arrival of millions of Venezuelan migrants to Colombia. The statistical analysis used frequency of codes as quantitative input derived from qualitative content analysis of 1,360 regional newspaper articles over nearly six years of the Venezuelan migration situation. The findings provide direct statistical evidence of impacts on migrant arrival numbers and topical coverage in news media. The indirect relationship between migrant arrival numbers and frustration about infrastructure in the media is mediated by coverage of topics related to frustration towards migration. This presents an opportunity for the media to practice balanced approaches to reporting and reconsider capitalizing on tropes, stereotypes, and negative framing that may compound perceptions that the media convey about infrastructure (such as hospitals and schools) being negatively impacted by migrants. Given that the Colombian government has taken steps to train media professionals in managing xenophobic rhetoric, future research must compare the findings of this study to a case in which no such government intervention happened to moderate anti-migrant sentiments in the news.

Acknowledgment. This research is funded by grant number N000141912624 by the Office of Naval Research through the Minerva Research Initiative. This research is additionally funded by grant number P116S210003 by the US Department of Education. Liss Romero & Lia Castillo (Universidad del Norte, Colombia) and Lydia Cleveland & Madison Gonzalez (VMASC) created the codebooks and qualitatively coded the news articles.

References

1. Dollard, J., Miller, N.E., Doob, L.W., Mowrer, O.H., Sears, R.R., Ford, C.S., et al.: Frustration and aggression. Routledge (2013)
2. Miller, N.E., Mowrer, O., Doob, L.W., Dollard, J., Sears, R.R.: Frustration-Aggression Hypothesis (1958)
3. ICG: Hard Times in a Safe Haven: Protecting Venezuelan Migrants in Colombia. Latin America and Caribbean Report Series: International Crisis Group (2022)
4. Frydenlund, E., Padilla, J.J., Palacio, K.: Colombia gives nearly 1 million Venezuelan migrants legal status and right to work. The Conversation US (2021)
5. Holland, A.C., Peters, M., Zhou, Y.-Y.: Left out: how political ideology affects support for migrants in Colombia. Available at SSRN 3803052 (2021)
6. Santamaria, J.: When a stranger shall sojourn with thee: the impact of the venezuelan exodus on colombian labor markets. University of Minnesota, Department of Applied Economics, Working Paper, p. 51422 (2020)
7. Bahar, D., Dooley, M., Selee, A.: Venezuelan migration, crime, and misperceptions: a review of data from Colombia, Peru, and Chile (2020)
8. Borders, D.W.: Venezuelans in Colombia struggle to find health care: "This is a crisis" (2019)

9. Franceschelli, M.: Global migration, local communities and the absent state: resentment and resignation on the Italian island of Lampedusa. Sociology **54**(3), 591–608 (2020)
10. Graham, C.: Does migration cause unhappiness or does unhappiness cause migration? Some initial evidence from Latin America. Handbook of Happiness Research in Latin America, pp. 325–341 (2016)
11. Erlandsen Lorca, M.: New uses and new gratifications of digital diasporic media amongst same-language societies: a qualitative case study on the Venezuelan immigrant communities in Chile and in Colombia after the refugee crisis (2015-onwards) (2021)
12. Peñaloza-Pacheco, L.: Living with the neighbors: the effect of Venezuelan forced migration on the labor market in Colombia. J. Labour Market Res. **56**(1), 14 (2022). https://doi.org/10.1186/s12651-022-00318-3
13. Woldemikael, O.: Migration as Municipal Management: The Political Effects of Venezuelan Migration in Colombia. Harward University (2022)
14. Makuch, M.Y., Osis, M.J., Becerra, A., Brasil, C., de Amorim, H.S., Bahamondes, L.: Narratives of experiences of violence of Venezuelan migrant women sheltered at the northwestern Brazilian border. PLoS ONE **16**(11), e0260300 (2021)
15. Alrababa'h, A., Dillon, A., Williamson, S., Hainmueller, J., Hangartner, D., Weinstein, J.: Attitudes toward migrants in a highly impacted economy evidence from the Syrian refugee crisis in Jordan. Comp. Polit. Stud. **54**(1), 33–76 (2021)
16. Galindo, G., Navarro, J., Reales, J., Castro, J., Romero, D., Rodriguez, A.S., et al.: Immigrants resettlement in developing countries: a data-driven decision tool applied to the case of Venezuelan immigrants in Colombia. PLoS ONE **17**(1), e0262781 (2022)
17. Malici, A., Smith, E.S.: Political Science Research in Practice. Routledge (2018)
18. Beaujean, A.A.: Latent variable modeling using R: a step-by-step guide. Routledge (2014)
19. Cleveland, L., et al.: A Codebook for Analyzing Host Communities' Frustration Within Migration Situations (2023). https://doi.org/10.17605/OSF.IO/6MVGR
20. Puth, M.-T., Neuhäuser, M., Ruxton, G.D.: Effective use of Spearman's and Kendall's correlation coefficients for association between two measured traits. Anim. Behav. **102**, 77–84 (2015)
21. Schober, P., Boer, C., Schwarte, L.A.: Correlation coefficients: appropriate use and interpretation. Anesth. Analg. **126**, 1763–1768 (2018)
22. Yanuar, F., Nisa Uttaqi, F., Zetra, A., Rahmi, I., Devianto, D.: The comparison of WLS and DWLS estimation methods in SEM to construct health behavior model. Sci. Technol. Indonesia **7**(2), 164–169 (2022). https://doi.org/10.26554/sti.2022.7.2.164-169
23. Mindrila, D.: Maximum likelihood (ML) and diagonally weighted least squares (DWLS) estimation procedures: a comparison of estimation bias with ordinal and multivariate nonnormal data. Int. J. Digit. Soc. **1**(1), 60–66 (2010)

Human Behavior Modeling

Agent-Based Moral Interaction Simulations in Imbalanced Polarized Settings

Evan M. Williams[(✉)] [iD] and Kathleen M. Carley

Carnegie Mellon University, Pittsburgh, USA
{emwillia,carley}@andrew.cmu.edu

Abstract. Rising US polarization in recent years has negatively impacted many friend and family relationships. To determine the best moral strategies for facilitating cross-party communication, we create an agent-based simulation underpinned by Moral Foundations Theory to model small-group moral conversations where the majority of agents align with either liberal or conservative views. We find, contrary to what moral re-framing research has assumed, that loyalty may be the best moral foundation for facilitating cross-party communication. More research is needed to understand the depolarizing effects of moral arguments in group settings.

Keywords: Moral Foundations Theory · Simulation · Polarization · Agent-Based Modeling

1 Intro

Partisan conflict, polarization, and fractured hyper-partisan information streams present threats to the stability of American democracy. The recent rise in polarization has had widespread negative effects, ranging from shortening the length of family Thanksgiving dinners with politically-different family members to partisan violence [3,10]. Concerningly, there has also been a rise in affective polarization—out-party animosity—in recent years that is not clearly tied to ideological changes [20]. Depolarization research has made large strides in recent years, but studies often rely on text prompts and videos that may be difficult to apply in real-world scenarios. Online users in an inundated information environment might simply ignore depolarization content, and convincing holiday party guests to watch depolarization videos together may be challenging. This increasingly presents a challenge for families and friend-groups wherein simple cross-party communication can feel like navigating a mine-field.

In persuasion research, "moral reframing" has shown success in changing views and facilitating cross-party communication [8]. By re-articulating an argument in the moral frame of an audience, the argument is more likely to resonate. This effect has been found to hold across a diverse array of issues [7,8]. For

R. Thomson et al. (Eds.): SBP-BRiMS 2023, LNCS 14161, pp. 139–148, 2023.
https://doi.org/10.1007/978-3-031-43129-6_14

example, one study found that conservatives were more likely to support universal health care when exposed to rhetoric rooted in purity, i.e., "that sick people are disgusting" than when framed in terms of fairness, i.e. "access to health care is a right" [7]. However, to our knowledge, no work yet explored interactions between all moral-vice pairs on persuasion. Nor has this been done in the context of highly-polarized group settings.

In this work, we are interested in simulating how moral reframing strategies might play out in highly-polarized small-scale settings. For the set-up, one can imagine a small holiday party where agents begin with a position in a social network, and where the majority of guests have similar political viewpoints, i.e., 'liberal' or 'conservative'. We simulate 2 million small-group moral conversations held by 20 million simulated agents. We track group-level and individual-level responses to moral statements within the simulation. One random guest is in the minority party. Each agent at the initial time-step has a latent moral state, where the distributions of morals and vices are drawn from distributions dependent on the agent's political views. A moral statement is generated—perhaps someone makes a claim or highlights a piece of news—and each agent at the party takes turns reacting to the news. Agents update their moral state based on 1) their initial latent moral state, 2) their susceptibility to updating, which we link to their underlying social network, 3) the moral content of the original message, 4) The morality expressed by agents with whom the agent is close. With this simple simulation, we attempt to determine the best moral reframing strategies for facilitating positive and healthy exchanges in imbalanced and polarized small-group settings. More concretely, we examine the moral messages that yield the largest quantities of negative-to-positive sentiment (and positive-to-negative sentiment) transitions of minority-party and majority-party agents in simulated politically-polarized small-group conversations. We open source our model at https://github.com/EvanUp/PolarizedABM.

While Moral Foundations political research has primarily focused on reframing arguments in the moral frames associated with the intended audience, we find that loyalty, which is not disproportionately associated with liberal or conservative viewpoints, was the most effective rhetorical tool for generating positive conversations. This reinforces the social-identity origins theory of affective polarization and suggests that polarized agents benefit from evoking shared identities to blur in-group out-group divides [20]. We hope that this work encourages further research into the depolarizing effects of moral arguments and moral mimicry and hope that this can serve as a blueprint for future human subject depolarization research.

2   Related Works

In recent decades, Moral Foundation Theory (MFT) has been shown to have significant success in changing views and facilitating cross-party communication. MFT reduces morality to 5 separable virtue-vice pairs [14]. These pairs are care-harm, fairness-cheating, loyalty-betrayal, authority-subversion, and purity

- degradation. This framework is not culturally-dependent and provides a generalizable framework for discussing morality. Follow-on research has found that the five-factor model is stable across a diverse set of societies, even outside Western, Educated, Industrialized, Rich, Democratic (WEIRD) countries [5]. These proposed factors appear to capture universal innate moral characteristics and provide a language for discussing morality across fields. Conservatives and liberals have been found to rely on different sets of moral values [5,15]. Conservatives tended to place more emphasis on purity and authority, whereas liberals were found to place higher emphasis on fairness and care. Research has also found that moral arguments that don't resonate with the values of an audience can be ineffective or even backfire, entrenching people further in their beliefs [8,11]. Liberty/oppression have been proposed as a candidate for a 6th moral/vice category, but have not received widespread acceptance and have been excluded from major annotated corpora [17,24]. For these reasons we all choose to exclude liberty/oppression. Our modeling approach is inspired in part by Friedkin's Social Influence Model [9].

3   Methods

3.1   Assumptions

We make the strong assumption that people mimic moral arguments, particularly of those with whom they are close, and this assumption is built into the update function of agents. The complex relationship between morality, mimicry, and empathy is still not fully understood, but in experimental settings empathy can both guide and interfere with morality [4]. Previous research has shown that behavioral and emotional mimicry are extremely important in prosociality [6]. Morality-related information has been found to have primacy over sociability and confidence in forming impressions of others [1,19]. Mimicry also facilitates attitude convergence, leading the mimicker and mimicked to become more similar in attitudes, preferences, and opinions [6]. In a more recent study, participants interacted with a confederate presented as moral or as lacking morality. Researchers found that when the confederate lacked moral qualities, mimicry and postural openness were lower than for moral confederates [21]. While these studies do not explicitly consider moral mimicry, there appears to be clear interplay between morality and mimicry, and insofar as morality falls within attitudes, preferences, and opinions, there is a theoretical rationale for the assumption.

We also assume that messages with positive sentiment are more desirable than those with negative sentiment as we assume that positive sentiment is more conducive to healthy discussion than negative sentiment. Psychological research has long revealed that negative emotions can escalate into "Amygdala Hijack" where rational faculties are temporarily lost [13]. While we struggled to find average holiday party statistics, an Eventbrite poll found that the majority of Gen Z, Millenial, and Gen X respondents define a party as at least 10 people [12], so we set the number of agents in each conversation to 10.

3.2 Agents

We begin the simulation with a set of agents $U = \{u_1, \ldots u_n\}$ with an underlying
social network $\mathcal{G}$ where $\mathcal{G}$ is an unweighted and undirected Erdos-Renyi random
graph with an average degree of 0.3. We select this graph configuration because
it has been used in epidemiological social network simulation [23]. Edges between
users are homogeneous and represent kin or non-kin friendship as defined in the
General Social Science Survey [2]. Q users are assigned the majority political
party θ_1, and $Q - n$ users are randomly assigned the minority political party
θ_0 at the beginning of each iteration. Moral states $M \in \mathbb{R}^{10}$ are drawn from
a distribution based on Graham et al.'s findings that liberal and conservative
individuals rely on different sets of morals [15]. While the authors do not differ-
entiate between virtue and vices, they find that conservatives tended to place
more emphasis on purity and authority. We make the assumption that this is
true both for virtues and vices and so for conservative agents, Purity, Degradation, Authority, and Subversion are randomly drawn from $U(0.5, 1)$. All other
conservative moral-vice pairs are randomly drawn from $U(0, 1)$. In contrast, liberals were found to place higher emphasis on Fairness and Care. So similarly, for
liberal agents, Fairness, Cheating, Care, and Harm are drawn from $U(0, 1)$. All
remaining liberal moral-vice pairs are randomly drawn from $U(0, 1)$.

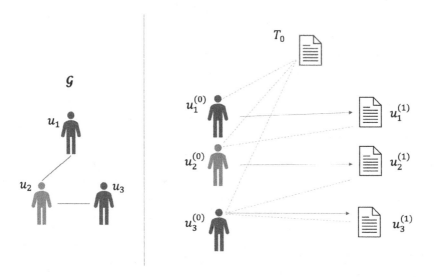

Fig. 1. Diagram of model set up with only 3 agents. On the left, we see the social
network $\mathcal{G}$ of agents u_1, u_2, u_3. On the right, the agents are exposed to an moral initial
message T_0 and respond in a random order, expressing their updated beliefs. Agents
all see (dashed-line) the initial message, but are only impacted by the messages of
users with whom 1) they are neighbors in $\mathcal{G}$ and 2) who have already spoken. After
observing, agents then produce a response $u_i^{(1)}$.

3.3   Emotion

We find that moral values are highly correlated with sentiment and that virtue-vice annotations can often be accurately mapped to sentiment with relatively simple models. To generate a mapping from morals to emotions, we train a model on the Moral Foundation Twitter Corpus (MFTC) [17]. The MFTC contains 35k tweets annotated with annotated virtues and vices by multiple annotators. We represent these annotations as a row-normalized matrix where rows are texts and columns reflect annotator virtue or vice labels. We use VaderSentiment to extract categorical sentiment labels from each Tweet [18]. We choose to drop neutral Tweets as 1) we are interested in messages that modify emotions 2) case studies in polarized environments tend to show extreme positive-negative sentiment polarization as well [16,22]. This results in a dataset with 13,370 negative sentiment tweets and 11,051 positive sentiment tweets. We find that values are strongly correlated with positive sentiment ($r = 0.44$) and that vices are strongly correlated with negative sentiment ($r = 0.45$)[1]. We apply a random 70-30 train-test split to the resulting corpus and train a logistic regression model to map the 10 moral states to binary sentiment labels. On negative sentiment, the model's F1 score is 0.81, on positive sentiment, the model's F1 is 0.76. More advanced models could likely increase accuracy—particularly those that consider text alongside moral values—but that is not the goal of this work. We elected to use logistic regression due to its speed and its interpretability. We use this model to map agents' initial moral states and final moral states to sentiment.

3.4   Model

Once agents and their networks are initialized, the group is exposed to an initial moral statement T_0, which is a 1-hot encoded vector representing one of the 10 virtues or vices. At time step 1, a random agent, u_i, with an initial moral state $u_i^{(0)}$ will observe T_0 and to produce a message $u_i^{(1)}$. At time step 2, another random agent u_j will observe the initial message T_0. If u_j a no edge in $\mathcal{G}$ with u_i, u_j's update function will be entirely based on the initial state $u_j^{(0)}$ and the content of the message T_0. However, if u_i and i_j have an edge in $\mathcal{G}$, u_j's moral state will still update based on $u_j^{(0)}$ and T_0, but will also include the message of u_i, $u_i^{(1)}$. The impact of T_0 and neighbors who have responded to T_0 is dampened by a susceptibility score, which for u_i, we define as the inverse of $(1+Degree(u_i))$. The intuition is that users with many strong social ties will feel like they can confidently express their views, whereas users with few social ties may feel more pressure to conform to the group. We note that a complete graph or a fully disconnected graph $\mathcal{G}$ would represent a special cases in which susceptibility

---

[1] As each observation can be assigned multiple virtues or vices from each annotator, we construct a table of 'virtue' and 'vice' probabilities using the annotators' label distribution, i.e. the percentage of annotations assigned to virtue or vice for each observation. We then calculate the pearson correlation coefficient between virtue/ vice probabilities and 2 binary sentiment variables (positive and negative).

weights are identical across all agents. The update function for any agent u_i can be written as:

$$u_i^{(1)} = \tanh(u_i^{(0)} + \frac{1}{S(u_i)}(T_0 + \frac{1}{n}\sum_{k=1}^{n} \mathcal{N}_{read}(u_i))) \tag{1}$$

where $S(u_i) = 1 + Degree(u_i)$. More intuitively, the user's previous moral state is updated based on the susceptibility of users to the original the original message T_0 and the average of agents with whom the agent shares an edge who have already spoken. We use tanh because it will help surface active moral states and decrease less-active moral states. We provide a diagram for the entire model set-up in Fig. 1.

After the model randomly iterates through all 10 users in the graph, the conversation is complete. We simulate 100,000 randomly-configured conversations and explore which moral arguments were most impactful in flipping the sentiment of the conversation. We are particularly interested in the moral values that correspond with the largest number of flips from moral states associated with negativity to moral states associated with positivity. Table 1 describes the set-up, controls, and variables of this virtual experiment.

Table 1. Virtual Experiment Description

	# Test Cases	Values
Control Variable		
Number of Majority Party Agents	1	9
Number of Minority Party Agents	1	1
Comments Per Agent	1	1
Number of Iterations	1	100,000
Average Degree	1	0.3
Initial moral State	10	0–1
Independent Variables		
Moral States	10	{0,1}
Majority Party	2	['liberal', 'conservative']
Num steps	10*100,000*10*2	20M
Dependent Variables		
Vice-Moral Transitions	1	0–20M
Moral-Vice Transitions	1	0–20M
Negative-Positive Minority Transitions	1	0–20M
Negative-Positive Majority Transitions	1	0–20M

4   Results

We are interested in the moral messages that cause negative → positive as well as positive → negative transitions for all agents as well as for agents in the minority party. For each majority party, for each moral virtue or vice, we simulate 100,000 conversations, each containing 10 agents. Therefore, for any given moral value and either majority party, 1 million agents have the opportunity to flip sentiment over the course of the simulation. We highlight the morals associated with the most overall (majority + minority) and minority-party transitions.

In our liberal-majority virtual experiment, the two moral states associated with the largest number of overall negative-to-positive sentiment flips were Purity (212K) and Loyalty (192K). The two values associated the largest number of conservative-agent (minority) negative-to-positive flips were loyalty (15.3k) and care (13.6k). While Purity resulted in the largest overall number of positive flips, its corresponding vice, degradation, resulted in the largest number of overall negative flips followed by harm. Harm and Cheating correspond with the largest number of negative flips for the conservative minority agents. The results of the liberal-majority virtual experiment are displayed in Table 2.

Table 2. Liberal Majority: flips from negative to positive and from positive to negative out for 1 Million agents in 100,000 conversations around each moral value. M→Pos and M→Neg are minority (conservative) flips to moral states corresponding positive or negative sentiment.

Morals	Agent→Pos	Agent→Neg	Minority→Pos	Minority→Neg
purity	**212743**	22193	12215	3102
authority	77801	62087	5061	6860
fairness	115120	50591	11501	4232
degradation	17789	**171317**	1445	14344
care	130247	46598	**13601**	3902
loyalty	**192011**	22837	**15331**	1806
subversion	22778	152616	1773	13037
cheating	28022	159890	1150	**20679**
harm	26534	**168757**	1073	**22308**
betrayal	25794	139378	1379	14462

In our conservative-majority virtual experiment, the two moral values corresponding to the largest number of negative-to-positive sentiment flips were care (170k) followed closely by loyalty (169k). For liberal minority agents, loyalty (21.5k) and purity (20.7k) were associated with the largest number of negative-to-positive flips. The vices corresponding the largest number of overall negative flips were cheating (202k) and degradation (117k). For liberal minorities, vices were harm (18k) and cheating (16.8k). Full results can be found in Table 3.

Table 3. Conservative Majority: flips from negative to positive and from positive to negative out for 1 Million agents in 100,000 conversations around each moral value. M→Pos and M →Neg are minority (liberal) flips to moral states corresponding positive or negative sentiment.

Morals	Agent→Pos	Agent→Neg	Minority→Pos	Minority→Neg
purity	129368	44429	**20758**	3424
authority	69837	65984	8685	6065
fairness	146652	32077	14190	3518
degradation	41693	**117116**	3570	14967
care	**170342**	26469	16338	2972
loyalty	**169803**	24987	**21551**	1853
subversion	43380	109259	3888	13601
cheating	17488	**202626**	2111	**16865**
harm	15284	217927	1843	**18189**
betrayal	34005	125420	3244	13538

5 Discussion and Limitations

Our simulation finds that loyalty—a foundation related to our tribal history—may be the strongest rhetorical approach for breaching highly imbalanced and polarized settings. Loyalty resulted in the most or second-most negative-to-positive flips for our overall conservative majority, overall liberal majority, conservative minority, and liberal minority simulations. These results make intuitive technical sense. In the majority-liberal simulation, liberals' fairness and care are already typically higher than conservatives' values for fairness and care, so introducing these arguments in the initial state does little in changing the overall group dynamic for the majority. The same can be said of Purity and Authority in the conservative-majority simulation.

While the result is explainable, it is surprising in some respects, as research has not strongly associated loyalty with either conservative or liberal beliefs. Unlike the virtue/vice pairs associated with conservatives and liberals respectively, loyalty was simply drawn from $U(0,1)$ for all users. Additionally, loyalty only had the third-highest coefficient in the logistic regression (after purity and care), so these results were not immediately anticipated. However, it does makes mathematical sense. As agents tend to have higher initial values for certain virtue-vice pairs, depending on their party, those values are less likely to flip. It also makes intuitive sense that loyalty should be the most powerful rhetorical tool for reaching across partisan divides. Loyalty evokes shared group membership and focuses on identities that unite us as humans. This result provides evidence in favor of theories that affective polarization originates in social identity [20]. Blurring in-group and out-group lines by evoking shared family, heritage, beliefs, national identity, etc. may help break down partisan barriers.

The largest limitation in this work is the lack of Empirical Validation. We employ input validation throughout the paper by highlighting the theoretical justifications of each of our decisions and linking the model to moral, persuasion, psychological, and polarization literature. Additionally, the format of our conversation, where each agent talks once and all agents take turns talking is unlikely. In reality, a subset of agents will probably talk the most and some agents may never speak at all. In future work, we would like to validate these results with a human-subject experiment to empirically explore the impact of loyalty rhetoric in bridging partisan divides.

6   Conclusion

We create an agent-based simulation to model small-scale moral conversations that one might encounter at holiday parties or in friend-groups. In contrast to literature on moral reframing, which attempts to frame arguments in the morality of the audience, our simulation finds that loyalty, a value not disproportionately associated with either conservatives and liberals—is the most effective moral value for facilitating positive cross-party communication. This result provides additional evidence in favor of viewing affective polarization as originating in social identity. Loyalty may be an effective rhetorical tool because it blurs the lines between in-group and out-group differences by focusing on shared identities.

Acknowledgements. The research for this paper was supported in part by the Office of Naval Research (ONR), MURI: Persuasion, Identity, & Morality in Social-Cyber Environments, and ONR Scalable Tools for Social Media Assessment under grants N000142112749 and N00014-21-1-2229. It was also supported by the center for Informed Democracy and Social-cybersecurity (IDeaS) and the center for Computational Analysis of Social and Organizational Systems (CASOS) at Carnegie Mellon University. The views and conclusions are those of the authors and should not be interpreted as representing the official policies, either expressed or implied, of the ONR or the US Government.

References

1. Brambilla, M., Leach, C.W.: On the importance of being moral: the distinctive role of morality in social judgment. Soc. Cogn. **32**(4), 397 (2014)
2. Burt, R.S.: Network items and the general social survey. Soc. Netw. **6**(4), 293–339 (1984)
3. Chen, M.K., Rohla, R.: The effect of partisanship and political advertising on close family ties. Science **360**(6392), 1020–1024 (2018)
4. Decety, J., Cowell, J.M.: The complex relation between morality and empathy. Trends Cogn. Sci. **18**(7), 337–339 (2014)
5. Doğruyol, B., Alper, S., Yilmaz, O.: The five-factor model of the moral foundations theory is stable across weird and non-weird cultures. Personality and Individ. Differ. **151**, 109547 (2019)
6. Duffy, K.A., Chartrand, T.L.: From mimicry to morality: the role of prosociality. (2017)

7. Feinberg, M., Willer, R.: From gulf to bridge: when do moral arguments facilitate political influence? Personality Soc. Psychol. Bull. **41**(12), 1665–1681 (2015)
8. Feinberg, M., Willer, R.: Moral reframing: a technique for effective and persuasive communication across political divides. Soc. Pers. Psychol. Compass **13**(12), e12501 (2019)
9. Friedkin, N.E., Johnsen, E.C.: Social influence and opinions. J. Math. Sociol. **15**(3–4), 193–206 (1990)
10. Frimer, J.A., Skitka, L.J.: Are politically diverse thanksgiving dinners shorter than politically uniform ones? PLoS ONE **15**(10), e0239988 (2020)
11. Gadarian, S.K., Van der Vort, E.: The gag reflex: disgust rhetoric and gay rights in American politics. Polit. Behav. **40**(2), 521–543 (2018)
12. Gervis, Z.: Research reveals how the average American defines a party (2021). https://swnsdigital.com/us/2019/08/research-reveals-how-the-average-american-defines-a-party/
13. Goleman, D.: Emotional Intelligence: Why It Can Matter More Than IQ. Bloomsbury Publishing (1996)
14. Graham, J., et al.: Moral foundations theory: the pragmatic validity of moral pluralism. In: Advances in Experimental Social Psychology, vol. 47, pp. 55–130. Elsevier (2013)
15. Graham, J., Haidt, J., Nosek, B.A.: Liberals and conservatives rely on different sets of moral foundations. J. Pers. Soc. Psychol. **96**(5), 1029 (2009)
16. Habibi, M.N., et al.: Analysis of Indonesia politics polarization before 2019 president election using sentiment analysis and social network analysis. Int. J. Modern Educ. Comput. Sci. **11**(11) (2019)
17. Hoover, J., et al.: Moral foundations twitter corpus: a collection of 35k tweets annotated for moral sentiment. Soc. Psychol. Pers. Sci. **11**(8), 1057–1071 (2020)
18. Hutto, C., Gilbert, E.: Vader: A parsimonious rule-based model for sentiment analysis of social media text. In: Proceedings of the International AAAI Conference on Web and Social Media, vol. 8, pp. 216–225 (2014)
19. Iachini, T., Pagliaro, S., Ruggiero, G.: Near or far? it depends on my impression: moral information and spatial behavior in virtual interactions. Acta Physiol. (Oxf) **161**, 131–136 (2015)
20. Iyengar, S., Lelkes, Y., Levendusky, M., Malhotra, N., Westwood, S.J.: The origins and consequences of affective polarization in the united states. Ann. Rev. Polit. Sci. **22**(1), 129–146 (2019)
21. Menegatti, M., Moscatelli, S., Brambilla, M., Sacchi, S.: The honest mirror: morality as a moderator of spontaneous behavioral mimicry. Eur. J. Soc. Psychol. **50**(7), 1394–1405 (2020)
22. Sanders, A.C., et al.: Unmasking the conversation on masks: natural language processing for topical sentiment analysis of COVID-19 twitter discourse. AMIA Summits Transl. Sci. Proc. **2021**, 555 (2021)
23. Stonedahl, F., Wilensky, U.: NetLogo Virus on a Network Model. Northwestern University, Evanston, IL, Center for Connected Learning and Computer-Based Modeling (2008)
24. Trager, J., et al.: The moral foundations reddit corpus. arXiv preprint arXiv:2208.05545 (2022)

Investigating the Use of Belief-Bias to Measure Acceptance of False Information

Robert Thomson[(⊠)][iD] and William Frangia

United States Military Academy, West Point, NY 10996, USA
robert.thomson@westpoint.edu, william.frangia@fulbrightmail.org

Abstract. Belief-bias occurs when individuals' prior beliefs impact their ability to judge the validity (i.e., structure) of an argument such that they are predisposed to accept conclusions consistent with their prior beliefs regardless of the argument's validity. The present study uses a minimal explanation paradigm to evaluate how United States Military Academy cadets assess the validity of arguments surrounding the pull-out from Afghanistan presented by different sources of authority. Participants exhibited a significantly greater likelihood of rejecting an invalid argument with true facts compared to accepting a valid argument with false facts, with overconfidence scores implying they were unaware of this difficulty in reasoning. We also found that participants were more critical of arguments about US capabilities coming from civilian sources. Results from the HEXACO personality assessment showed that task performance was positively correlated with perfectionism and inquisitiveness sub-scales, implying that those high in those measures were less likely to exhibit belief-bias. Even when factoring-in these traits, results revealed a small yet significant trend for participants to reject valid arguments from their peers compared with senior military and civilian counterparts. Overall, the present study shows a differential impact of belief-bias on true vs false facts, that this is influenced by the underlying source of the argument, and that personality traits mediate these effects.

Keywords: belief-bias · reasoning · personality

1 Background

There has been a renewed focus on critical thinking skills following the rapid rise of misinformation on the internet, especially surrounding major events such as COVID-19 [18]. With the advent of pre-trained large language models, it is easier than ever to create misleading information and populate it across the internet. Several recent studies have investigated the use of syllogistic reasoning studies to investigate how political leaning [2,5,13] and scientific explanation [21] impact critical reasoning skills. One failure of critical reasoning is known as the belief-bias effect [9]. The belief-bias effect occurs when individuals tend to ignore the validity (i.e., logical structure) of an argument and instead focus on the believability of its conclusions [9,10,14,15].

© The Author(s), under exclusive license to Springer Nature Switzerland AG 2023
R. Thomson et al. (Eds.): SBP-BRiMS 2023, LNCS 14161, pp. 149–158, 2023.
https://doi.org/10.1007/978-3-031-43129-6_15

1.1 Belief-Bias

For scientific and technical arguments, belief-bias is most prominent in novice individuals compared to experts. It is theorized that belief-bias occurs when individuals try to construct a mental model of the argument structure. Usually, simple mental models are created and evaluated until a good candidate explanation is reached, but in the case of belief-bias if a believable conclusions fits an early model, individuals are less likely to search for alternative models (or explanations) and are less likely to evaluate the validity of the argument's structure [15]. In other words, these prior beliefs might act as heuristics that interfere with critically reasoning about an argument's structure [24]. The coherence and simplicity of the argument's structure also enhance the believability of its conclusions [6,17,22] (for a review, see [11]).

Traditionally, belief-bias has been measured using syllogistic reasoning tasks. In [9] participants were presented with syllogisms that contained a conclusion that is contrary to social representations of smoking:

Major Premise: All things that are smoked are good for the health;
Minor Premise: Cigarettes are smoked.

Minor Cnclusn: Therefore, cigarettes are good for your health.   ∴

In this case, participants are likely aware that smoking is associated with health risks and cannot disregard their prior belief, focusing their response on the fact that the conclusion is false despite the fact that the argument has a valid logical structure (*modus ponens*: All A are B, A, therefore B). Participants exhibited lower accuracy judging the validity of syllogisms with unbelievable (usually false) conclusions. These results are not entirely limited to conclusions, as false premises (the major premise that all things smoked are good for you) may also impact belief-bias to a lesser extent.

This syllogistic reasoning paradigm has been shown to be effective at measuring belief-bias involving political ideology surrounding gun control [5], environmentalism [13], and immigration [2] among other topics. Furthermore, [2] found a differential impact of analytical thinking from participants with political ideologies, specifically that analytical thinking predicted overall performance on left-leaning participants but did not predict performance on right-leaning participants reasoning over right-leaning topics. The HEXACO personality inventory has previously correlated intellectual humility with susceptibility to consume misinformation as a form of belief-bias [3].

Overconfidence. Models of attitude change have suggested that subjective confidence can be used to determine the amount of processing that an individual will perform when provided with novel information [8]. A recent categorization study [20] found that exemplars associated with prior knowledge that was unrelated to the category structure interfered with correct categorization and also increased subjective confidence, leading to a strong overconfidence effect. This overconfidence effect was replicated in [21] showing that participants exhibited increased overconfidence when exhibiting belief-bias independent of changes in accuracy. That is, when accuracy drops, confidence did not drop to the same degree (or even increased). Interestingly, this research implies that individuals appear to be implicitly unaware of their bias.

1.2 Minimal Explanation Paradigm

A limitation in syllogistic reasoning studies is that the syllogisms are not ecologically valid, that is, they is not presented in a way that an individual would naturally read/hear in the media or online. [21] developed a minimal explanation paradigm which presents syllogisms in a more conversational manner, They also include an irrelevant explanation which does not impact that validity of the syllogism but adds an element to further promote belief-bias:

If [a Baje moves toward a Yulo then they will stick together] ^{P1}.
[A Baje moves toward a Yulo]^{P2} because [Bajes and Yulos are bound by a force]^{IE} that [attracts them]^{C*}
Brackets and superscripts highlight variables in the syllogisms and were not presented to the participants.

The premises are denoted by P1 and P2 and the conclusion is denoted by C. The irrelevant explanation is denoted by IE. While not technically a logical syllogism, it does have a structure whereby the conclusion is either consistent or inconsistent with the premises. To be compatible with prior literature, in this paper we refer to these arguments as having either valid or invalid form.

In [21], participants were predisposed to reject arguments that used intentional explanations (i.e., likes) compared to mechanistic (i.e., bound by a force), mediated by the source of the argument. Specifically, *scientists* were considered more reliable sources of information regarding inanimate phenomena while *people* were more reliable sources of information of animate phenomena. There was a tendency for participants to reject intentional explanations of inanimate phenomena (e.g., two physical items are attracted to each other because they *want* to be together) when the argument came from scientists, implying an expectation for scientists to talk in terms consistent with their expertise.

2 Present Study

A meta-analysis by [12] found that the importance of a belief is inversely related to attitude change. That is, it is hard to change the attitude of someone holding a strong belief on a topic. Supporting this theory, studies have demonstrated that words invoking vivid imagery are more persuasive [4,19]. This implies that participants holding stronger beliefs are more likely to exhibit belief-bias. An affectively-charged topic is thus more likely to exhibit belief-bias from those persons invested in the topic. The present study uses the Minimal Explanation

Paradigm [21] with the pullout of the US forces from Afghanistan in 2021 as an affectively-charged topic. The data was collected in early 2022 from cadets at the United States Military Academy. The goal of this study was to further validate the Minimal Explanation Paradigm methodology and evaluate whether there were any differences in performance between the acceptance of truthful information when compared against mis- and disinformation and whether the and whether this was impacted by the source of information. To ensure a balanced coverage, we also presented information that focused on US vs Adversary capabilities, as it may be the case that participants are predisposed to reject information about Adversary capabilities. This study also investigated whether any personality characteristics are correlated with degree of belief-bias.

We predict that cadet participants would be more likely to accept information about US capabilities coming from a military authority when compared against a civilian authority, and further that cadet participants would be critical of civilian sources talking about military capabilities. Similar to [21], we hypothesize that participants would be unaware of their own belief-bias, exhibiting a greater degree of overconfidence when belief-bias was present. We further predict that participants exhibiting high conscientiousness would be resistant to belief-bias.

2.1 Methods

Participants. A total of 90 participants completed this study from a pool of undergraduate cadets taking introductory psychology courses at the United States Military Academy. Cadets range in age from 17 to 27 years of age and roughly 78% are male. Participants received 1% course credit for participating.

Materials. Consistent with [21], sixteen training syllogisms consisted of eight examples of *modus ponens* (MP; If A then B. A, therefore B.) and *modus tollens*; (MT; If A then B. not-B, therefore not-A.). To avoid the effects of prior knowledge, practice syllogisms used nouns developed from CVCV (consonant-vowel-consonant-vowel; e.g., Baje, Yulo) non-words. For instance, the example below presents a valid modus ponens practice syllogism.

If there is a Sohi then there is a Loze
There is a Sohi.
_____
Therefore, there is a Loze   ∴

Thirty-two experimental syllogisms used the minimal explanation paradigm format. Task instructions explicitly stated that the irrelevant explanatory element did not affect the validity of the argument:

"An example of a valid *modus ponens* argument is as follows:

All things that are smoked are good for the health;
Cigarettes are smoked, *which may alter your sense of taste.*
_____
Therefore, cigarettes are good for your health.   ∴

The conclusion logically follows from the premises despite the irrelevant explanation ("which may alter your sense of taste"). It is still valid despite the initial premise being not true. If you see a statement of this form, you should report it as 'Valid'."

Each modified syllogism type (MP and MT) varied in terms of the validity (valid or invalid logical form), truth of the syllogism content (true or false fact), topic focus (US vs adversary capability/event), and source of the information (a peer leader, a civilian authority figure, a military authority figure). The truth of each syllogism was established by reviewing media reports validated by individuals at the United States Military Academy's Combating Terrorism Center. Two versions of each syllogism type were developed. An example of a valid MP syllogism based on a true fact that focuses on a US event is:

When [US forces leave a combat zone] they [render ineffective all equipment which was left behind].[P1]
The [US withdrew from Afghanistan in August][P2] because [the war was too expensive and there was no stomach for it][IE] and [left their equipment unusable for Taliban forces.][C*]
*Brackets and superscripts are for reader reference. The premises are denoted by P1 and P2 and the conclusion is denoted by C. The irrelevant explanation is denoted by IE.

Finally, each participant completed the HEXACO-60 personality assessment [1].

Procedure. Participants performed the study online via the Qualtrics survey system. Following Welcome and Instructions pages, participants completed 16 practice syllogisms with no feedback provided prior to the start of the experimental session. The experimental session consisted of three blocks of 32 modified syllogisms. The block consisted of participants being told that the source of the modified syllogisms was either one of the Brigade Tactical Officer (BTO; a high-ranking military leader at the United States Military Academy), a cadet First Sergeant (1SG; a cadet who advises company commanders on morale and discipline), and an Army civilian researcher/expert at the USMA Combating Terrorism Center (CTC). These three sources are well-known to the cadet population. The same 32 modified syllogisms were used in each block, but their order was randomized. The source order was counterbalanced between participants.

2.2 Results

The experimental design for each of the independent variables used a repeated-measures analysis of variance (ANOVA) consisted of 3 (Source: BTO, 1SG, CTC) x 2 (Validity: Valid or Invalid) x 2 (Fact Truth: True or False) x 2 (Topic: US or Adversary) variables. Syllogism type and the two replications were collapsed for the purposes of analyses consistent with [21]. We report Greenhouse-Geisser adjusted statistics where appropriate with the unadjusted degrees of freedom.

While a previous study [21] removed participants whose performance fell below chance (6 of 90), in the present study all participants who completed the experiment (89 of 90) were included in the analyses. While overall accuracy was slightly lower in the present study ($M = .61$ compared with $M = .66$ in [21], standard deviation was substantially higher ($SD = .40$ instead of $SD = .16$). This caused many participants to exhibit performance near chance levels (only 25 of 89 exhibiting significantly different than chance by two-tailed t-test, $<.05$). One difference which accounts for this effect is that the present population are cadets at a military academy who have substantially more time commitments (e.g., military training in addition to schooling) than counterparts at civilian universities. The additional fatigue and time-pressure may exacerbate the effect of any heuristics impacting their likelihood to exhibit belief-bias.

Response Accuracy. Supporting our hypothesis that the source of information influences participants' ability to accurately judge the validity of arguments, the ANOVA for proportion correct revealed a significant three-way interaction between Source, Fact Truth, and Topic, $F(2,178) = 4.33$, $MSE = .029$, $p = .015$, $\eta^2 = .046$. As seen in Fig. 1, participants exhibited a tendency to reject arguments about US capabilities from civilian researchers compared to the BTO (military leadership) reflected by a slight difficulty judging the validity of civilian arguments about US forces when the facts are truthful, while also exhibiting a slight improvement judging the validity of civilian arguments about US forces that are not truthful when compared against the same arguments coming from the BTO. Participants also exhibited a tendency to more often accept valid arguments about adversaries from civilian researchers when compared to their peer leadership (cadet 1SG).

The three-way interaction also qualifies a number of main effects and interactions. As seen in Fig. 2a an interaction between Fact Truth and Validity, $F(1,89) = 31.32$, $MSE = .068$, $p < .001$, $\eta^2 = .260$, qualified main effects of Fact Truth, $F(1,89) = 4.17$, $MSE = .047$, $p = .044$, $\eta^2 = .045$, and Validity, $F(1,89) = 15.24$, $MSE = .630$, $p < .001$, $\eta^2 = .146$. Participants exhibited belief-bias with a tendency accept true arguments regardless of whether they had a valid or invalid form. In fact, participants were no better than chance at judging true arguments with an invalid form. This may reflect a broader example of the availability heuristic, whereby rapid judgments are made based on information that comes to mind first, in our case information that was likely seen previously through media and military sources. Of particular note, Fact Truth and Validity did not interact with Source Authority as we initially predicted, implying no overall trend towards exhibiting more belief-bias solely on the basic of whom was saying the argument.

Overconfidence. Overconfidence was computed by subtracting response accuracy from participants' confidence. As we hypothesized (and similar to [20,21] participants exhibited greater overconfidence when they exhibited belief-bias, reflected by a significant interaction for Fact Truth and Validity, $F(1,89) =$

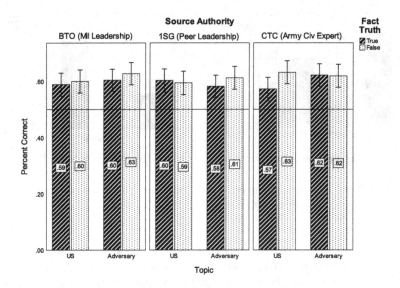

Fig. 1. Significant 3-way interaction between Source Authority, Fact Truth, and Topic. Error bars represent 95% CI. Horizontal bar represents chance (50%).

31.74, $MSE = .073$, $p < .001$, $\eta^2 = .263$. This interaction qualified a main effect for Validity, $F(1,89) = 13.65$, $MSE = .631$, $p < .001$, $\eta^2 = .133$, and a marginal effect for Fact Truth, $F(1,89) = 3.43$, $MSE = .052$, $p = .067$, $\eta^2 = .037$. As seen in Fig. 2b, participants were overall overconfident in their responses, with greater overconfidence for arguments with invalid form. We argue that this is primarily due to participants being unaware of their own difficulties with judging the validity of arguments with invalid form. This blindness to their own performance issues, especially when judging false information and malformed arguments, implies a lack of awareness of their own belief-bias. This may be exploited to increase the acceptance of mis- and disinformation or reject true information.

Impact of Personality Traits. To determine whether any particular personality traits were correlated with ability to judge the validity of arguments, participants completed the 60-question HEXACO personality inventory [1]. Prior research has found a correlation between intellectual humility and susceptibility to misinformation [3]. Unlike prior research, the present study conducted pairwise correlations between HEXACO measures and accuracy which resulted in significant effect of Inquisitiveness, $r(89) = .213$, $p = .045$ and Perfectionism, $r(89) = .211$, $p = .047$. High scores on these scales reflect high curiosity in the Openness to Experience domain (unexpected result), and high attention to detail in the Conscientiousness domain (as we hypothesized). Participants exhibiting greater curiosity and attention-to-detail exhibit overall greater task performance. To determine the degree to which any belief-bias still persists when correlated

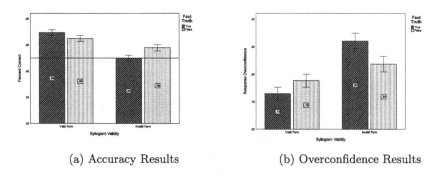

(a) Accuracy Results   (b) Overconfidence Results

Fig. 2. Significant interaction between Fact Truth and Validity. Error bars represent 95% CI. Horizontal bar represents chance (50%) performance.

personality traits are taken into account, a repeated-measures analysis of co-variance (ANCOVA) was performed which consisted of the same measures as above but with the addition of participants' Inquisitiveness score on the HEX-ACO as a co-variate. Interestingly there were no main effects of Fact Truth or Validity once participants' Inquisitiveness was accounted-for, nor did Topic interact with any other factors.

Unlike the principal analyses, this revealed a significant three-way interaction between Source, Fact Truth, and Validity, $F(2,178) = 3.14$, $MSE = .035$, $p = .047$, $\eta^2 = .035$, supporting our original hypothesis that participants would differentially consume information based on the source of the information. While only a small effect, participants exhibited a tendency to reject valid (but not invalid) arguments whose source was their peer leadership (1SG) compared to either civilian experts or military leadership, with a relatively stronger rejection of valid argument structure from false facts (a form a belief-bias). This qualified a Fact Truth and Validity interaction, $F(1,89) = 6.58$, $MSE = .066$, $p = .012$, $\eta^2 = .070$. Similar interactions occurred for overconfidence as well with a marginal interaction between Source, Fact Truth, and Validity $F(2,178) = 2.81$, $MSE = .039$, $p = .065$, $\eta^2 = .031$ and significant interaction between Fact Truth and Validity, $F(1,89) = 6.24$, $MSE = .071$, $p = .014$, $\eta^2 = .067$, showing the same lack of awareness of their own prior beliefs intruding on their decision-making. These results imply that participants exhibited belief-bias overall, and furthermore identify a concern that they were more likely to reject valid arguments from their peer leadership.

2.3   Discussion

The present study confirmed that participants exhibited belief-bias with predispositions to be more critical of civilian experts discussing US capabilities, while overall exhibiting a trend to be more critical of peer leadership when inquisitiveness is factored into analyses. Participants also appeared to be blind to their own belief-bias, exhibiting greater overconfidence when judging invalid arguments. In broader terms, we believe that stating the source of the argument is important when conducting belief-bias studies as otherwise participants may instead make an assumption about the nature of the argument's source and unduly introduce a confounder into the study (e.g., only a Democrat would ask about gun control in the first place). This concern was first brought up in [16] arguing that normative social influence is under-detected in reasoning studies.

Furthermore, while research predominantly discusses belief-bias in terms of judging the believability of the conclusion instead of the validity of the argument structure, there are possibly two distinct phenomena: 1) a tendency to reject valid arguments due to an unbelievable statement in the argument, and 2) a tendency to accept invalid arguments due to a believable statement in the argument. Both [21] and the present study have shown an asymmetry in belief-bias, in the present case where there were higher-order interactions with Validity. Participants exhibited a significantly greater likelihood of rejecting invalid argument with true facts compared to accepting a valid argument with false facts. The higher-order interaction with Topic (US vs Adversary) also implies that the belief-bias is topic-specific and was also seen in [23] where there was no correlation in degree of belief-bias between participants completing three different reasoning tasks. This asymmetry runs counter to the argument that belief-bias is just a response bias (i.e., a tendency to accept or reject arguments) [7]. We argue that future studies should further investigate this asymmetry to determine possible avenues to circumvent belief-bias in political or other polarizing topics. We also plan to investigate overconfidence effects to understand why individuals are implicitly blind to their bias, implying a failure in metacognitive monitoring. Further research is required to determine whether an intervention to raise metacognitive awareness of one's belief-bias would otherwise inoculate them against sources of believable mis- and disinformation.

Acknowledgements. This research was supported in part by the Office of Naval Research FY21 Multi-University Research Initiative Award N0001422MP00465. The views expressed in this work are those of the authors and do not necessarily reflect the official policy or position of the United States Military Academy, Department of the Army, Office of Naval Research, Department of Defense, or U.S. Government. The authors would like to thank Ms. Audrey Alexander and the Combating Terrorism Center at the United States Military Academy.

References

1. Ashton, M.C., Lee, K.: Empirical, theoretical, and practical advantages of the HEXACO model of personality structure. Pers. Soc. Psychol. Rev. **11**(2), 150–166 (2007)
2. Aspernäs, J., Erlandsson, A., Nilsson, A.: Motivated formal reasoning: ideological belief bias in syllogistic reasoning across diverse political issues. Thinking Reasoning **29**(1), 43–69 (2023)
3. Bowes, S.M., Tasimi, A.: Clarifying the relations between intellectual humility and pseudoscience beliefs, conspiratorial ideation, and susceptibility to fake news. J. Res. Pers. **98**, 104220 (2022)

4. Burns, A.C., Biswas, A., Babin, L.A.: The operation of visual imagery as a mediator of advertising effects. J. Advertising **22**(2), 71–85 (1993)
5. Calvillo, D.P., Swan, A.B., Rutchick, A.M.: Ideological belief bias with political syllogisms. Thinking Reasoning **26**(2), 291–310 (2020)
6. Douven, I., Schupbach, J.N.: The role of explanatory considerations in updating. Cognition **142**, 299–311 (2015)
7. Dube, C., Rotello, C.M., Heit, E.: Assessing the belief bias effect with rocs: it's a response bias effect. Psychol. Rev. **117**(3), 831 (2010)
8. Eagly, A., Chaiken, S.: Attitude structure. Handbook of social psychology **1**, 269–322 (1998)
9. Evans, J.S.B.T., Barston, J.L., Pollard, P.: On the conflict between logic and belief in syllogistic reasoning. Memory Cognition **11**(3), 295–306 (1983). https://doi.org/10.3758/BF03196976
10. Evans, J.S.B., Newstead, S., Allen, J., Pollard, P.: Debiasing by instruction: the case of belief bias. Eur. J. Cogn. Psychol. **6**(3), 263–285 (1994)
11. Gilovich, T., Griffin, D., Kahneman, D.: Heuristics and Biases: The Psychology of Intuitive Judgment. Cambridge University Press, Cambridge (2002)
12. Johnson, M.K., Raye, C.L.: False memories and confabulation. Trends Cogn. Sci. **2**(4), 137–145 (1998)
13. Keller, L., Hazelaar, F., Gollwitzer, P.M., Oettingen, G.: Political ideology and environmentalism impair logical reasoning. Thinking Reasoning, 1–30 (2023)
14. Markovits, H., Nantel, G.: The belief-bias effect in the production and evaluation of logical conclusions. Memory Cognition **17**(1), 11–17 (1989)
15. Newstead, S.E., Pollard, P., Evans, J.S.B., Allen, J.L.: The source of belief bias effects in syllogistic reasoning. Cognition **45**(3), 257–284 (1992)
16. Nolan, J.M., Schultz, P.W., Cialdini, R.B., Goldstein, N.J., Griskevicius, V.: Normative social influence is underdetected. Pers. Soc. Psychol. Bull. **34**(7), 913–923 (2008)
17. Pacer, M., Lombrozo, T.: Ockham's razor cuts to the root: simplicity in causal explanation. J. Exper. Psychol.: General **146**(12), 1761 (2017)
18. Roozenbeek, J., et al.: Susceptibility to misinformation about COVID-19 around the world. Royal Soc. Open Sci. **7**(10), 201199 (2020)
19. Schlosser, A.E.: Experiencing products in the virtual world: the role of goal and imagery in influencing attitudes versus purchase intentions. J. Cons. Res. **30**(2), 184–198 (2003)
20. Schoenherr, J.R., Lacroix, G.L.: Performance monitoring during categorization with and without prior knowledge: a comparison of confidence calibration indices. Can. J. Exper. Psychol. **74**(4), 302 (2020)
21. Schoenherr, J.R., Thomson, R.: Persuasive features of scientific explanations: explanatory schemata of physical and psychosocial phenomena. Front. Psychol., 12 (2021). https://doi.org/10.3389/fpsyg.2021.644809
22. Sloman, S.A.: When explanations compete: the role of explanatory coherence on judgements of likelihood. Cognition **52**(1), 1–21 (1994)
23. Thompson, V., Evans, J.S.B.: Belief bias in informal reasoning. Thinking Reasoning **18**(3), 278–310 (2012)
24. Trippas, D., Thompson, V., Handley, S.: When fast logic meets slow belief: evidence for a parallel-processing model of belief bias. Memory Cognition **45**, 539–552 (2017)

Simulation of Stance Perturbations

Peter Carragher[✉] [ID], Lynnette Hui Xian Ng [ID], and Kathleen M. Carley [ID]

Carnegie Mellon University, Pittsburgh, USA
{pcarragh,huixiann,kathleen.carley}@andrew.cmu.edu

Abstract. In this work, we analyze the circumstances under which social influence operations are likely to succeed. These circumstances include the selection of Confederate agents to execute intentional perturbations and the selection of Perturbation strategies. We use Agent-Based Modelling (ABM) as a simulation technique to observe the effect of intentional stance perturbations on scale-free networks. We develop a co-evolutionary social influence model to interrogate the tradeoff between perturbing stance and maintaining influence when these variables are linked through homophily. In our experiments, we observe that stances in a network will converge in sufficient simulation timesteps, influential agents are the best Confederates and the optimal Perturbation strategy involves the cascade of local ego networks. Finally, our experimental results support the theory of tipping points and are in line with empirical findings suggesting that 20–25% of agents need to be Confederates before a change in consensus can be achieved.

Keywords: Agent Based Modelling · Social Influence · Simulation

1 Background

ABM has a rich history in social network analysis [5,8,9,11–13]. Will et al. [13] categorized such studies into those that investigate endogenously emerging networks, exogenously imposed networks and co-evolutionary networks. Endogenous studies look at how social network structure evolves as a function of the set of agent states. Conversely, exogenous studies keep the network structure constant and model changes in agent states based on this structure. The co-evolutionary approach is a hybrid of both and models the interplay between agent states and network structure. Despite being a closer fit to genuine SI processes, this approach is relatively understudied [13]. Differing timescales between endogenous & exogenous effects complicate the matter.

Ng et al. [10] demonstrated that endogenous and exogenous features are equally important in predicting pro/anti-vaccine stance flips on Twitter. The SI model used to make this prediction is an exogenous one based on Friedkin's foundational social influence (SI) model [5], where the influence weight matrix W is static and precalculated from self-reports. In contrast Macy et al. [9] develop the Hopfield model, a co-evolutionary approach for the simulation of stance change. It has been shown that both exogenous [11] and co-evolutionary models can converge to a hyper-polarized state [9].

ABM has also been used to highlight the vulnerability of the SI model to manipulative actors Ross et al. [12] find that as little as 2–4% well-positioned bots are capable

R. Thomson et al. (Eds.): SBP-BRiMS 2023, LNCS 14161, pp. 159–168, 2023.
https://doi.org/10.1007/978-3-031-43129-6_16

of tipping the majority opinion. Of note is the variance in estimations of tipping points, with Centola et al. [4] showing that 25% of users must commit to certain language use before it gains traction. This motivates the investigation of a wide variety of environmental conditions in the simulation of intentional stance perturbation.

2   Simulation Objectives

In our simulation, we wish to observe the effect of intentional stance perturbations on the overall stances in the network. The process of a group of Confederate agents perturbing a social network simulates the real-world scenario of an influence operation, wherein disingenuous operatives strategically promote subversive and provocative content with the intention of catalyzing a shift in or preventing the formation of consensus.

2.1   Research Questions

In this study, our objective is to explore deliberate perturbations' impact on influence networks and their effect on individual stances. Using a co-evolutionary SI model that simulates how stance changes with respect to interpersonal influence, we seek to understand strategies that maintain influence while perturbing the network.

R1 How can Confederates manipulate stances effectively without losing influence?

R2 Considering factors such as susceptibility, network structure, and stance, which agents make for effective Confederates?

Finally, we investigate the potential for intentional perturbations to disrupt the prevailing consensus, shedding light on opinion dynamics within the influence network.

R2 To what extent can intentional perturbations change an established consensus?

By addressing these research questions, we will gain a deeper understanding of intentional perturbations in influence networks, their impact on stances, and strategies to optimize their effectiveness.

2.2   Contributions

By adopting a co-evolutionary SI model, we explore the tradeoff between influence and stance within social networks. Our simulations capture the intricate relationship between the emergence of the influence network and its impact on individual stances.

Secondly, we propose an evaluation criterion based on consensus, which serves as a measure to assess intentional stance perturbation. By analyzing changes in consensus over time, we can gauge the success of perturbations by the magnitude of stance change. Furthermore, our findings reveal the efficacy of targeted nudging strategies in perturbing stances. Well-targeted nudges yield the largest stance perturbation, underscoring the importance of utilizing the structure of the influence network.

Lastly, we find support for empirical findings on the theory of tipping points that suggest 20–25% of agents need to be Confederates to bring about a new consensus [4]. These contributions significantly enhance our understanding of influence dynamics, intentional stance perturbation, and consensus formation within social networks.

Table 1. Definitions of Terminology used in this study

Terminology	Definition
Stance	Opinion on a topic using a scale of two extremes, −1.0 and 1.0
Susceptibility	How vulnerable an agent is to the changes of stance/influence
Influence	How much impact/sway an agent has towards its neighbors
Confederates	The group of agents that are perturbing the network
Perturbation	The "nudging" of stances of other agents in the network
Tipping point	The timestep of the simulation where majority stance changes

Table 2. Definitions of Symbols used in this study

Symbol	Definition
i, j	identification of agents
$y(t)$	stance at timestep t
W	influence weight matrix, row-normalized
A	diagonal matrix of actor susceptibilities to influence
α	stance update rate
λ	influence update rate
μ_y	average stance at time t over N non-Confederate agents
θ	threshold of Confederate influence
k	node at timestep t with maximum global influence
w_i^g	Confederate i's total ('global') network influence
w_i^l	i's influence over the top M agents in i's ego-network

3   Model Definition

This simulation study performs an examination of the effect of Confederate agent selection and stance perturbation strategies on the overall stance in the network. In Table 1 and Table 2, we define some of the terminology and symbols used in this study, including the scale-free influence network. At each timestep t, the stance vector is updated, followed by the influence matrix, using two interdependent recurrence relations.

Following Friedkin's model [5], the stance update equation represents the exogenous effects of the influence network; an agent's stance is incrementally nudged towards the average stance of those who have influence over them. The model relates influence matrix W to stances y at time t as per Eq. 1. Note, the diagonal suseeptibility matrix A is scaled by a stance learning rate $\alpha = 0.001$.

$$y(t) = AW(t)y(t-1) + (I-A)y(1) \qquad (1)$$

Equation 2 describes the endogenous update rule for influence matrix W. Based on the Hopfield model [9], the influence update equation makes this a co-evolutionary model. We introduce an additional reverse relation between W & y based on the homophily process, to enable influence to co-evolve with stance. We also introduces the influence update rate, $\lambda = 0.01$, also known as the rate of structural learning [9].

$$W(t+1) = \lambda y_t y_t^{\mathsf{T}} + (1-\lambda)W(t) \tag{2}$$

The relation between the influence and stance update rates, λ and α, is crucial. Since the model dynamics are extremely sensitive to these update rates, it is also important to choose λ and α such that performance comparisons between various perturbation strategies can be made. That is, when the simulation converges, it has reached a stable polarised state where there is not an absolute majority. We find that the proposed values achieve this, as demonstrated in Fig. 2. In a stable polarised state we can compare strategies based on how many agents change stances, as approximated by the average stance of non-Confederate agents.

Concretely, the case for $\alpha > \lambda$ is trivial, as the simulation converges quickly as defined by the influence network topology. Here agents adjust their stance to match their ego networks faster than they can update their influence weights based on the homophily principle of the influence update rule. Our experiments use $\lambda >> \alpha$ so that agents adjust their influence based on homophily first and foremost. The result is a changing network structure that introduces a tradeoff for Confederates between maintaining influence and perturbing stance.

3.1 Influence-Stance Tradeoff

Adopting a radical position risks triggering resistance and skepticism, potentially leading to a loss of trust and influence. Operatives may establish trust by posing as like-minded individuals, sharing relatable experiences, and providing seemingly credible information. Therein, they face a dilemma; they may adopt a less extreme stance that aligns with users' existing beliefs to potentially increase their influence, or they attempt to change the narrative with extremist content and risk triggering resistance in return.

The homophily principle in our model encapsulates this tradeoff ($\lambda y_t y_t^{\mathsf{T}}$ in Eq. 2); as Confederate stances diverge from the average stance of the network, their influence decreases. Equation 1 illustrates this point with a single Confederate trying to perturb an 80-node network using the conversion perturbation strategy.

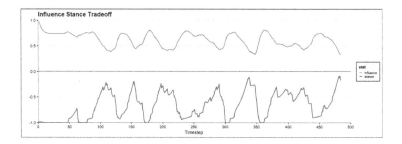

Fig. 1. In an 80-node network, a single Confederate struggles to perturb consensus while maintaining influence using the conversion perturbation strategy. The Confederate maintains a −1 stance until its influence begins to drop at timestep 90. It then raises its stance until its influence rebuilds. This repeats at timestep 130, the beginning of a distinctly cyclic pattern.

3.2   Confederate Selection

With this trade-off in mind, we propose three strategies for determining Confederates.

Maximum influence: we select the most influential agents according to weighted out-degree in the influence matrix at time $t = 0$; $W_{max} = argmax_j \Sigma_i^N W(0)_{ij}$.

Minimum susceptibility: we select the least susceptible agents, skewing the remaining agents towards being more susceptible; $A_{min} = argmin_j A_{jj}$

Random selection (control): Confederates are selected uniformly at random.

3.3   Perturbation Strategy

Confederates perturb the social network, causing the average stance of the agents to change. We define three strategies for the perturbation of a scale-free network: the conservative strategy, the conversion strategy, and the cascade strategy.

The conservative approach (Eq. 3) is to perturb stance if the Confederate's influence is above some threshold θ. Whenever influence drops below θ, we set the stance to μ_y; the average over N non-Confederate stances at time t: $\mu_y = \Sigma_i^N y(i,t)/N$.

$$y(i,t) = \begin{cases} \mu_y & \Sigma_j^N w(j,i) \leq \theta \\ -1 & \Sigma_j^N w(j,i) > \theta \end{cases} \qquad (3)$$

The conversion strategy (Eq. 4) is based on "nudging" the network towards the desired stance (–1). It is therefore a continuous function where we scale the magnitude of the perturbation by the current level of influence as in Fig. 1. When influence is low, the stance perturbation is less aggressive but when influence is high, the perturbation results in a more extreme stance (closer to –1). Here, w_i^g is the global influence factor of Confederate i on all N non-Confederate agents (Eq. 5). We normalize using the influence of k, the agent at timestep t with the maximum global influence.

$$y(i,t) = \mu_y^g + w_i^g * (-1 - \mu_y^g) \qquad (4)$$

$$w_i^g = \Sigma_j^N w(j,i) / \max_{k'} \Sigma_j^N w(j,k') \qquad (5)$$

The cascade strategy (Eq. 6) is derived from the conversion strategy (Eq. 4). The key difference for cascade is that we take a local approach when calculating Confederate influence, looking only at each Confederate's ego network. We rank all agents by the influence the Confederate has over them and sum the influence weights from the top M most influenced agents (Eq. 7) where $M = N/10$. This more targeted approach is named after the well known cascade effect which highlights the role of network structure in information diffusion.

$$y(i,t) = \mu_y^l + w_i^l * (-1 - \mu_y^l) \qquad (6)$$

$$w_i^l = \Sigma_j^M w(j,i) / \max_{k'} \Sigma_j^M w(j,k') \qquad (7)$$

4   Methodology

Our simulation experiment involves perturbations by a group of Confederates across scale-free networks. The ABM is implemented in the Construct framework [2,3,8]. This framework reads in a scale-free network and simulates the SI model as defined in Eqs. 1 and 2 until convergence. Results are averaged over five replicates.

To begin our experiments, we construct a series of generalized scale-free networks. A scale-free network is a network where the node degrees follow a power law distribution [1]. This leads to a characteristic hub structure that mimics a social network setting, where there are some nodes that have many connections, while others are more isolated. These networks are constructed using the preferential attachment model, with five replicates per network size. Experiments on alternative network constructions such as Small World, Erdos-Renyi, and Core Periphery, are beyond the scope of this study.

We construct networks with N agents, ranging from 10 to 150. We provide two attributes to each agent: stance and susceptibility. Stance, y, is initially 1, reflecting a state of consensus. Susceptibility, s is a random variable drawn from a Normal distribution; $s \sim N(0.1, 0.1)$. Next we choose a Confederate selection strategy, the percentage of agents to choose as Confederates, and a Perturbation strategy which defines the evolution of each Confederates stance. Confederate agent susceptibility is set to zero.

Finally, we run the simulation until convergence; that is, the point where the mean stance change of all N non-Confederate agents over the previous 30 timesteps is less than 0.001. At that point, the simulation is terminated.

To measure the optimality of Confederate strategies, we calculate $\hat{\mu}_y$, the mean stance of non-Confederate agents at convergence. As such lower is better, indicating the success of Confederate perturbations in driving network stance towards -1.

4.1   Virtual Experiments Setup

Our virtual experiment setup, detailing the independent, dependent and control variables are summarized in Table 3.

Table 3. Definitions of Terminology and Symbols used in this study

Variable	Type	Range	Value	Number
Number of agents, N	Independent	[10, 150]	10, 20, 30... 150	14
Percent Confederates	Independent	[0, 100]	5, 10, 15,..., 40	8
Perturbation strategy	Independent		Eqs. 3, 4, and 6	3
Agent selection strategy	Independent		W_{max}, A_{min}, Random	3
Mean stance at convergence, $\hat{\mu}_y$	Dependent	$[-1.0, 1.0]$	$[-1.0, 1.0]$	$\mathbb{R}$
Convergence timestep	Dependent	$\mathbb{N}$	[50, 200]	$\mathbb{N}$
Initial stance, y	Control	$[-1.0, 1.0]$	$-1, 1$	2
Susceptibility, s	Control	[0, 1.0]	$s \sim N(0.1, 0.1)$	$\mathbb{N}$
Influence update rate, λ	Control	[0, 1.0]	0.01	1
Stance update rate, α	Control	[0, 1.0]	0.001	1

5   Results and Discussion

We detail four key results; convergence to a polarized state, optimal Perturbation strategies, optimal Confederate selection strategies, and the observation of tipping points.

Convergence to a state of stable polarization. We first observe that stances in a network do eventually come to convergence. Figure 2 illustrates the change in agent stances across time for an 80-node network, where each line represents an agent. The stances in the simulation eventually converge into one of the two extremes, 1.0 and −1.0.

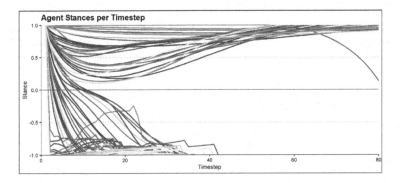

Fig. 2. Simulation of the change in agent stances for an 80-node network. Each line represents an agent's stance over time. Stances eventually converge into one of two extremes, 1 and –1.

Cascade is optimal. In examining the perturbation strategies against network size and means stances, we observe that the optimal Perturbation strategy is the cascade strategy. That is, an optimal strategy for the changing of stances involves the nudging of agents that are in a Confederate agent's direct neighborhood. This is illustrated in Fig. 3, where the line of mean stance for the cascade strategy is the lowest (best). We also note that the conversion strategy fares the worst out of the three strategies, indicating that global nudging strategies are not optimal in general and require precise targeting.

Influential Confederates are best. Out of the three Confederate selection strategies, influential agents are the best Confederates. This is observed in Fig. 4, where Confederate agent selection strategies are plotted against each other. We also note a clear trend of diminishing returns where Confederate strategies are less successful as the network size increases. This hints at the underlying resilience of scale free networks.

Minority stance tipping points exist. We observe that Confederate agents can create a tipping point wherein the network converges to the stance of the minority Confederate agents. Figure 5 shows tipping point ranges between 20–25% for the three different Perturbation strategies applied, mirroring empirical evidence from Centola et al. [4]. The cascade strategy requires the least Confederates to cause the tip, while the conversion strategy requires the most Confederates to tip the overall stance. This mirrors the findings for the stance perturbation experiment (Fig. 3).

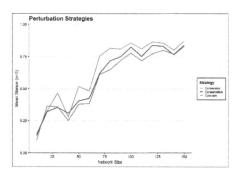

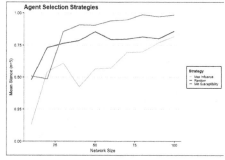

Fig. 3. Comparison of the perturbation strategies, lower is better. The cascade strategy is optimal.

Fig. 4. Comparison of the agent selection strategies, lower is better. Influential Confederates are optimal.

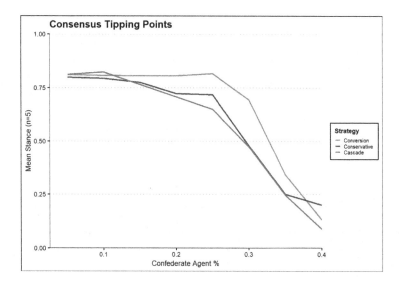

Fig. 5. With >20–25% of network agent as Confederates, mean stance shifts rapidly. Results are averaged over five 80-node networks, using the maximum influence selection strategy.

6 Validation

Our results match stylized facts that can be used to determine known phenomena. States of consensus are reached when users achieve social conformity with each other [2, 3, 11], which is observed through stance convergence (Fig. 2) and tipping points (Fig. 5) within our work. This is also observed within the principles of homophily, giving rise to the adage "birds of a feather flock together".

We find that influential people are best positioned to affect other people's stances (Fig. 4). Studies on targeted attacks on scale-free networks have shown that higher degree nodes have larger effects on the overall network and our experimental results provide the same observation [6, 12]. Similarly, the optimal perturbation strategy in our model is to begin by effecting a change towards people that are close to you, i.e. your ego-network (Fig. 3). This is in-line with the theory of information cascades [6].

Finally, the tipping points observed in our experimental runs are similar to previous research in the literature [4]; 20–25% of the network are required as Confederates to change the stance of the network. These results lend our model theoretic validity and face validity.

7   Limitations and Future Work

Validation is challenging; it is clearly unethical to precipitate consensus change in a real social network. A lack of real-world data further complicates the matter, as observations of stance changes in real-world networks are extremely rare. Ng et al. [10] find that only 1% of Twitter users changed their stances towards the Coronavirus vaccine over a one-year period. Most Twitter users maintained the same stance. An area of future research is to identify and process real event data with stance and opinion annotation change across time and adjust our formulation to the observed data.

Despite these limitations, it is important to understand the circumstances under which agents in a social situation will change their opinion, which has implications for persuasion and disinformation research. Differences in social networking platforms may play a role; where account histories are easily accessible, for instance, modeling suspicion as a function of stance volatility would counteract certain Confederate strategies. To account for this, Eq. 2 could be modified to incorporate a regularization parameter that penalizes highly volatile agents. Additionally, experimenting with other models for the influence network such as Small World, Erdos-Renyi, and Core Periphery, may explain how network structure affects the success of the Cascade strategy.

Given the emergence of stable polarized states within this model, perhaps the most natural and pertinent avenue for future work is the recovery from such a state. A body of theory, and in particular Krackhardt's notion of simmelian ties, supports such an effort [7]. The core building block of our model is that of the dyadic tie. Extending this model to capture triadic relationships would enable the interrogation of the role non-prejudiced third parties play in recovering from the polarized state.

8   Conclusion

In this work, we formulate and test a co-evolutionary social influence model to simulate intentional stance perturbation in scale-free networks. We define three strategies for selecting a set of Confederates that will perturb the network: maximum influence, minimum susceptibility, and a random selection for a control experiment. We design three Perturbation strategies that Confederates use to probe the network: conservative, conversion, and cascade strategies. Our results show that influential agents are the best choice of Confederate, that the optimal Perturbation strategy involves the targeted nudging of local ego networks and that there exists a range of tipping points for group consensus. We hope that this simulation sheds light on the effectiveness of intentional stance change within a social network and the manner in which successful social influence operations are run.

Acknowledgements. This work was supported in part by the Office of Naval Research grant (N000141812106) and the Knight Foundation. Additional support was provided by the Center for Computational Analysis of Social and Organizational Systems (CASOS) at Carnegie Mellon University. The views and conclusions in this document are those of the authors and should not be interpreted as representing the official policies, either expressed or implied, of the Knight Foundation, Office of Naval Research, or the U.S. Government.

References

1. Broido, A.D., Clauset, A.: Scale-free networks are rare. Nat. Commun. **10**(1), 1–10 (2019)
2. Carley, K.M.: Group stability: a socio-cognitive approach. Adv. Group Process. **7**(1), 44 (1990)
3. Carley, K.M., Martin, M.K., Hirshman, B.R.: The etiology of social change. Top. Cognitive Sci. **1**(4), 621–650 (2009)
4. Centola, D., Becker, J., Brackbill, D., Baronchelli, A.: Experimental evidence for tipping points in social convention. Science **360**(6393), 1116–1119 (2018). https://doi.org/10.1126/science.aas8827
5. Friedkin, N.E., Johnsen, E.C.: Social influence and opinions. J. Math. Soc. **15**(3–4), 193–206 (1990), publisher: Taylor & Francis
6. Jalili, M., Perc, M.: Information cascades in complex networks. J. Complex Netw. **5**(5), 665–693 (2017)
7. Krackhardt, D.: The ties that torture: Simmelian tie analysis in organizations. Res. Sociol. Organ. **16**(1), 183–210 (1999)
8. Lanham, M.J., Morgan, G.P., Carley, K.M.: Social network modeling and agent-based simulation in support of crisis de-escalation. IEEE Trans. Syst., Man Cybern.: Syst. **44**(1), 103–110 (2013)
9. Macy, M.W., Kitts, J.A., Flache, A., Benard, S.: Polarization in dynamic networks: a hopfield model of emergent structure. Dynamic Social Network Modeling and Analysis, pp. 162–173 (2003). Washington DC: National Academies Press
10. Ng, L.H.X., Carley, K.M.: Pro or Anti? a social influence model of online stance flipping. IEEE Trans. Network Sci. Eng., pp. 1–18 (2022). https://doi.org/10.1109/TNSE.2022.3185785
11. Rajabi, A., Gunaratne, C., Mantzaris, A.V., Garibay, I.: On countering disinformation with caution: effective inoculation strategies and others that backfire into community hyperpolarization. In: Thomson, R., Bisgin, H., Dancy, C., Hyder, A., Hussain, M. (eds.) SBP-BRiMS 2020. LNCS, vol. 12268, pp. 130–139. Springer, Cham (2020). https://doi.org/10.1007/978-3-030-61255-9_13
12. Ross, B., Pilz, L., Cabrera, B., Brachten, F., Neubaum, G., Stieglitz, S.: Are social bots a real threat? An agent-based model of the spiral of silence to analyse the impact of manipulative actors in social networks. Eur. J. Inf. Syst. **28**(4), 394–412 (2019). https://doi.org/10.1080/0960085X.2018.1560920
13. Will, M., Groeneveld, J., Frank, K., Müller, B.: Combining social network analysis and agent-based modelling to explore dynamics of human interaction: a review. Socio-Environ. Syst. Model. **2**(16325), 10 (2020) https://doi.org/10.18174/sesmo.2020a16325, https://sesmo.org/article/view/16325

Integrating Human Factors into Agent-Based Simulation for Dynamic Phishing Susceptibility

Jeongkeun Shin[✉], Kathleen M. Carley, and L. Richard Carley

Center for Computational Analysis of Social and Organizational Systems,
Carnegie Mellon University,
5000 Forbes Avenue, Pittsburgh, PA 15213, USA
{jeongkes,carley,lrc}@andrew.cmu.edu

Abstract. Many researchers focus on developing virtual testbeds to assess the magnitude of cyberattack damage and evaluate the effectiveness of cyber defense strategies in different cyber attack scenarios. These testbeds provide a controlled and cost-effective environment for simulating attacks and studying their impact on organizational security. One of the major challenges in developing such testbeds is accurately capturing the human factors in cybersecurity. Phishing attacks, in particular, exploit human vulnerabilities and can lead to significant security breaches. However, modeling and simulating human susceptibility to phishing in virtual environments is complex due to the dynamic nature of human behavior and the interplay of various factors. This paper addresses this challenge by proposing an agent-based modeling framework that incorporates human factors to simulate dynamic phishing susceptibility. The framework allows modeler to specify and assign weights to various human factors, such as personality traits and training history, which influence individuals' susceptibility to phishing attacks. By leveraging this framework, modelers can run virtual cyber attack simulations with dynamically changing phishing susceptibility among the simulated end user agents. This enables the testing and evaluation of different cyber defense strategies in the context of realistic human behavior.

Keywords: Agent-based Simulation · Phishing Susceptibility · Human Factors

1 Introduction

Virtual testbeds have become a vital tool for assessing the impact of cyberattacks in diverse scenarios for various reasons. First, virtual testbeds provide a controlled and safe environment to simulate cyberattacks and their consequences without risking harm to actual computing systems [1]. Second, virtual testbed is more cost-effective compared to real-world experiments since they allow researchers to explore various attack scenarios and evaluate different defense strategies without the need for additional costs [1,2]. Lastly, virtual

© The Author(s), under exclusive license to Springer Nature Switzerland AG 2023
R. Thomson et al. (Eds.): SBP-BRiMS 2023, LNCS 14161, pp. 169–178, 2023.
https://doi.org/10.1007/978-3-031-43129-6_17

testbeds facilitate sensitivity analysis, empowering researchers to easily conduct various experiments and compare the outcomes of cyber attack damage across diverse settings [3]. These settings encompass organizations of varying sizes, different human networks among end user agents within the organization, and various cyberattack frequencies. By manipulating these parameters, researchers can gain valuable insights into the impact of these factors on the extent and consequences of cyber damage. By leveraging the benefits of virtual testbeds, researchers can design and customize experiments according to their specific research objectives, adjust parameters, analyze the impact of various factors on cyberattack outcomes, evaluate and optimize cyber defense strategies in different contexts, thereby revealing the most effective approaches for mitigating cyber threats.

Building a virtual testbed presents a significant challenge: accurately modeling human factors in cybersecurity. Phishing attacks, a prevalent technique used by cybercriminals, exploit human errors, making it crucial to incorporate human factors for an accurate simulation of potential cyberattack damage. According to IBM, more than 95% of cyber incidents originate from human errors, with common mistakes including double-clicking suspicious URLs or attachments [4]. However, achieving accurate modeling is complex due to the influence of various human factors on an individual's phishing susceptibility, such as demographic information, personality traits, and past experiences. Moreover, phishing susceptibility dynamically changes over time as individuals naturally forget organizational cybersecurity policies, encounter new zero-day attacks, and undergo cybersecurity training. Addressing these challenges is essential to enhance the realism and effectiveness of virtual testbeds for studying cyberattacks and evaluating defense strategies.

In this paper, we propose a user-centric approach in the OSIRIS framework [8,9], allowing modelers to specify and assign weights to different human factors to determine the phishing susceptibility of individual end user agents. By incorporating these human factors, we capture the personalized nature of phishing susceptibility and provide a more realistic simulation environment. Additionally, we incorporate the dynamics of phishing susceptibility through the use of a forgetting curve model and cybersecurity education, a common cyber defense strategy. Lastly, we show the results of cyberattack simulations with various cybersecurity education intervals.

2   Related Works

In recent agent-based models of virtual testbeds, the vulnerabilities of different agent types are commonly predefined using empirical data. However, a limitation of these models is that the vulnerability values remain static throughout the simulation. For example, in the Cyber-FIT [6,7] model, agents' vulnerabilities are determined based on terrain type, environment, and the cyber situational awareness. Similarly, in the OSIRIS framework [8], end user agents' phishing susceptibility is derived from real-world survey data, adjusted for their predetermined levels of cybersecurity expertise and motivation [9]. However, these models do not consider many important human factors, such as demographic information and personality traits, that influence the phishing susceptibility of individual end user agents. As a result, the phishing susceptibility across agents tends to be relatively uniform. Furthermore, in both frameworks, the susceptibility values do not dynamically change during the simulation in response to various human factors. Factors like stress levels, fatigue levels, and natural forgetfulness can significantly impact individuals' susceptibility to phishing attacks. However, the existing models do not account for these dynamic changes in phishing susceptibility.

Several researchers have explored the relationship between various human factors and phishing susceptibility. Uebelacker and Quiel, for instance, introduced the Social Engineering Personality Framework (SEPF) and provided a theoretical understanding of how each of the Big Five personality traits [5] influences an individual's vulnerability to social engineering attacks [10]. Parrish et al. developed a phishing susceptibility framework that comprehensively captures the influence of personal factors, Big Five personality factors, experimental factors, and attack factors on phishing susceptibility [11]. Tornblad et al. summarized 32 predictors encompassing various human factors including personality traits, demographics, educational background, cybersecurity experience and beliefs, platform experience, email behaviors, and work commitment style that are potentially relevant in predicting the susceptibility of individuals to phishing attacks [12].

Furthermore, there have been many empirical studies that analyze the relationship between human factors and phishing susceptibility. Lin et al. conducted a study examining individuals' susceptibility to spear-phishing emails, considering factors such as demographics, education level, health status, and internet usage patterns [13]. They developed multiple multilevel logistic regression models, exploring different combinations of human factors to better understand the dynamics of susceptibility to phishing attacks. Lawson et al. conducted a study utilizing multiple linear regression models to predict phishing susceptibility based on personality traits [14]. Purkait et al. conducted an investigation on various human factors to develop multiple regression models aimed at predicting users' ability to identify phishing websites [15].

When constructing a virtual testbed for simulating cyberattack scenarios, determining which phishing susceptibility regression model to import from empirical research can be challenging. Rather than aiming to find the optimal model, we propose allowing modelers to select the phishing susceptibility model of their choice to import into the OSIRIS framework [8]. By offering the flexibility to try multiple models, modelers can observe the simulation results and make informed decisions based on their specific research objectives.

3   Dynamic Phishing Susceptibility Model

This section presents the implementation of a dynamic phishing susceptibility model within the OSIRIS framework. We demonstrate how regression models can be imported to assign unique phishing susceptibility values to individual end user agents based on their specific human factors. Additionally, we explore the dynamic nature of phishing susceptibility, taking into account people's natural forgetfulness and the impact of cybersecurity education during the simulation.

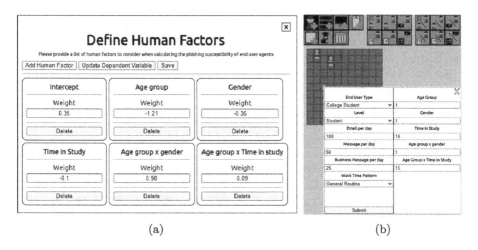

(a)   (b)

Fig. 1. User Interface of OSIRIS (a) User Interface for Specifying Human Factors and Assigning Weights (b) User Interface for Defining End User Agents and Entering Human Factor Values.

In Fig. 1, the user interface of the OSIRIS framework [8,9] is depicted. As illustrated in Fig. 1 (a), the modeler can specify the human factors to consider for calculating the phishing susceptibility of the end user agent. The initial intercept is fixed, and the modeler can add new human factors by clicking the 'Add Human Factor' button and assigning weights to each factor. As shown in Fig. 1 (a), we have entered the variables and weights from Lin et al.'s first multilevel logistic regression model [13]. The Fig. 1 (b) illustrates the process of deploying an end user agent in the gridworld. The user interface prompts the modeler to enter specific values for the assigned human factors. By inputting different values for each human factor, the end user agent is assigned varying levels of phishing susceptibility corresponding to the entered values.

When importing regression models, modelers need to be cautious about the dependent variable, as different researchers may calculate the output differently. For instance, Lin et al.'s logistic regression model determines the probability of the subject clicking at least one suspicious link out of the 21 spear-phishing emails delivered during the experiments [13]. On the other hand, Purkait's model calculates the number of phishing websites correctly identified by the subjects [15]. In such cases, modelers should modify the model's output to represent the probability of clicking the suspicious link in each phishing email by utilizing the 'Update Dependent Variables' option demonstrated in Fig. 1 (a). They can achieve this by typing in the equation that transforms the model's outcome into phishing susceptibility for each email.

3.1 Cybersecurity Education and Forgetting Curve

In addition to assigning unique phishing susceptibility values to end user agents based on their human factors, the OSIRIS framework [8,9] allows modelers to apply dynamic changes in the phishing susceptibility of individual end user agents while the simulation is running. Two key components for achieving these dynamic changes are cybersecurity education and the implementation of a forgetting curve model [16].

Cybersecurity education plays a vital role in improving individuals' awareness and knowledge of phishing attacks [17,18,21]. Modelers can specify the frequency of cybersecurity education intervals, such as once a month, once a quarter, or once a year. The OSIRIS framework incorporates these education periods and simulates their impact on end user agents' phishing susceptibility. Through regular education sessions, end user agents' awareness is reinforced, reducing their vulnerability to phishing attacks. The effectiveness of phishing education can vary due to factors such as the quality of the education program and the expertise of the lecturers. Consequently, establishing a fixed measure of its effectiveness is challenging. Instead, our approach involves giving the modeler a flexibility to determine the range of effectiveness for phishing education. Modelers can define the extent to which the education reduces phishing susceptibility. For instance, if the modeler sets the effectiveness range between 30% to 60%, an end user agent with an initial phishing susceptibility of 3% would experience a reduction in susceptibility ranging from 0.9% to 1.8%. This modeler-defined range allows for greater customization and accommodates the varying impact of phishing education in different contexts.

Furthermore, the forgetting curve model [16] has been incorporated into the OSIRIS framework to account for the natural decay of knowledge and awareness over time. Considering the extensive research on the retention of cyber education effects [19,20], our framework simulates the gradual loss of knowledge and awareness as end user agents move further away from their cybersecurity education sessions. This gradual fading of knowledge may result in an increased susceptibility to phishing attacks. Initially assuming a 100% retention rate of cybersecurity education, we acknowledge that over time, memory naturally deteriorates, leading to a decline in the education's effectiveness. To capture this phenomenon, modelers are prompted to enter the retention rate of the education at a specific time point. For example, if the modeler determines that the retention rate after 1 week is 80%, they can input $P(10080\,\text{min}) = 80$ (the OSIRIS framework's minimum unit of time is a minute). By combining the initial retention rate, $P(0) = 100$, right after the education and the modeler-provided data, the OSIRIS framework automatically constructs an exponential decaying curve to simulate the forgetting process of end user agents. This approach allows for a more realistic representation of how memory retention rate impacts phishing susceptibility over time. In summary, the end user agent's phishing susceptibility will be calculated based on the formula below.

$$PhishingSusceptibility(T) = OPS \cdot (1 - (ELCE \cdot e^{RT}))$$

In the formula, the phishing susceptibility of the end user during the simulation is updated every tick by multiple factors. OPS represents the original phishing susceptibility of the End User Agent, which is calculated based on various human factors that the modeler specified. ELCE represents the effectiveness of the last cybersecurity education, which is randomly selected from a modeler-defined range of the effectiveness of the cybersecurity education. The decay rate (R) is automatically calculated using the retention rate information at a specific moment provided by the modeler and assumption that the retention rate right after the education is 100%. T represents the total time passed (in ticks or minutes) since the last cybersecurity education.

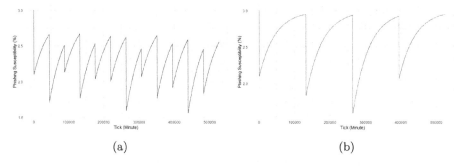

(a)   (b)

Fig. 2. Phishing Susceptibility of the End User Agent with Different Cybersecurity Education Intervals (a) Cybersecurity Education Interval: 1 month (b) Cybersecurity Education Interval: 3 months.

In Fig. 2, the dynamic phishing susceptibility of an individual end user agent is shown for different cybersecurity education intervals: 1 month and 3 months. The simulation was conducted over a 1-year period with an initial phishing susceptibility (OPS) set to 3%. The effectiveness of the last cybersecurity education (ELEC) was randomly selected from a range of 25% to 50% for each education session. The graph clearly demonstrates that more frequent phishing education intervals result in a lower level of phishing susceptibility. This finding highlights the importance of regular education sessions in reducing vulnerability to phishing attacks.

4   Virtual Experiment

In this section, we conduct a virtual experiment using the OSIRIS framework's phishing campaign in 'normal mode' [9]. In this mode, the cybercriminal agent regularly sends spearphishing emails to end user agents, and the end user agents rely solely on their phishing knowledge to determine whether an email is phishing or not. Similar to the previous OSIRIS paper [9], 40 employees will be deployed in the virtual organization. We will utilize Lin et al.'s first logistic regression model to predict the probability of an individual clicking at least one malicious phishing link among 21 phishing emails [13]. This prediction will be based on three human factors and two interactions between human factors: age group (0 = young, 1 = older), gender (0 = male, 1 = female), time in study (1–21), age group x gender, and age group x time in study. The Beta values for each variable, as shown in Fig. 1(a), are imported. Random values will be assigned to each human factor variable for the end user agent, and the logistic regression model will automatically calculate the corresponding value. The resulting logit value will then be translated into the probability that each end user agent will click at least one phishing link in the 21 phishing emails. Finally, this probability will be further translated into the phishing susceptibility for each phishing email.

To determine the values for effectiveness of the cybersecurity training memory retention rate, we will utilize the empirical research conducted by Kumaraguru et al. [19]. According to their findings, the phishing susceptibility before the embedded training was 82%. Immediately after the training, the susceptibility decreased to 32%. After a one-week delay, the susceptibility slightly increased to 36%. Based on these values, we can calculate that the effectiveness of the cybersecurity training (ELCE) was approximately 60.975%, indicating a reduction in susceptibility. To apply some variation, the effectiveness of each cybersecurity training will be randomly selected from a range of 57.5% to 62.5%. This range allows for slight variations in the training effectiveness, adding diversity to the simulations. Furthermore, the retention rate after one week was measured at 92%. Using this information, OSIRIS will automatically calculate the memory decay rate (R) by analyzing the changes in phishing susceptibility over time.

With these initializations, the end user agents in the simulation will have dynamic phishing susceptibility as the simulation progresses. We will manipulate the intervals of cybersecurity education and observe the overall number of phishing emails clicked within the organization over a one-year simulation period. The simulation plan, including the varied education intervals, is summarized in Table 1. The results of the simulation, including the proportion of phishing emails clicked by end user agents during the simulation, are summarized in Fig. 3.

The virtual experiment results revealed that reducing the cybersecurity education interval resulted in a decrease in the proportion of phishing emails clicked within the organization. This finding highlights the importance of regular and frequent cybersecurity education. To find the optimal cybersecurity education interval, organization leaders should consider the trade-off between reducing phishing incidents and minimizing cybersecurity education costs and worktime disruption. Our framework provides insights for decision-makers to establish an effective cybersecurity strategy by identifying the optimal education interval.

Table 1. Simulation Summary.

Type	Name	Implication
Input	Organization	Virtual 40 end user agents and computing device agents constructed with OSIRIS framework [8,9]
	End User Agent	Human-like agents, each with a unique phishing susceptibility determined by three human factors (Age group, Gender, and Time in Study), and two interactions between these factors (Age group x Gender, Age group x Time in Study)
	Phishing Campaign	Phishing campaign in OSIRIS framework [9]
Output	Phishing Damage	The proportion of phishing emails clicked by end user agents during the simulation
Parameter	Simulation Time	1 year (525,960 ticks)
	Training Intervals	Time interval at which cybersecurity education and training sessions are conducted for end user agents within an organization (None, 365 d, 180 d, 90 d, 60 d, 30 d, 15 d, 7 d)
	Number of Simulations	100

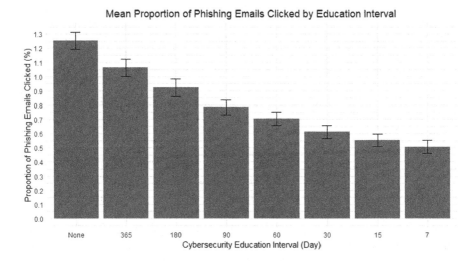

Fig. 3. Virtual Experiment Result.

5   Discussion and Conclusion

In this study, we recognize the limitation of assuming complete delivery of cybersecurity knowledge during education to end user agents. In reality, individual concentration levels during education can vary due to factors like personality traits, conditions on the education day, and loyalty to the organization. These factors influence learning effectiveness and phishing susceptibility. In future work, we will address this limitation by incorporating these human factors and investigating their impact on learning and susceptibility to phishing attacks.

Our main contribution lies in the development of a customizable simulation framework that integrates empirical studies evaluating initial phishing susceptibility based on various human factors with studies examining dynamic changes in susceptibility through factors like phishing education and memory retention. This framework enables the implementation of realistic and dynamic phishing susceptibility of human agents during the simulation runtime. Decision-makers can leverage this testbed to replicate their organization, conduct virtual phishing campaigns, assess the impact of phishing with different education periods, and determine the optimal and cost-effective duration for cybersecurity education.

Acknowledgement. The author(s) disclosed receipt of the following financial support for the research, authorship, and/or publication of this article: This research was supported in part by the Minerva Research Initiative under Grant #N00014-21-1-4012, and by the center for Computational Analysis of Social and Organizational Systems (CASOS) at Carnegie Mellon University. The views and conclusions are those of the authors and should not be interpreted as representing the official policies, either expressed or implied, of the Office of Naval Research or the US Government.

References

1. Zografopoulos, I., Ospina, J., Liu, X., Konstantinou, C.: Cyber-physical energy systems security: threat modeling, risk assessment, resources, metrics, and case studies. IEEE Access **9**, 29775–29818 (2021)
2. Crussell, J., Kroeger, T.M., Brown, A., Phillips, C.: Virtually the same: comparing physical and virtual testbeds. In: 2019 International Conference on Computing, Networking and Communications (ICNC), pp. 847–853. IEEE, 2019
3. Carley, K.M.: Computational organization science: a new frontier. In: Proceedings of the National Academy of Sciences 99, no. suppl_3, 7257–7262 (2002)
4. IBM: IBM security services 2014 cyber security intelligence index. (2014)
5. Gosling, S.D., Rentfrow, P.J., Swann, W.B.: A very brief measure of the big-five personality domains. J. Res. Pers. **37**(6), 504–528 (2003)
6. Dobson, G.B., Carley, K.M.: Cyber-FIT: an agent-based modelling approach to simulating cyber warfare. In: Lee, D., Lin, Y.-R., Osgood, N., Thomson, R. (eds.) SBP-BRiMS 2017. LNCS, vol. 10354, pp. 139–148. Springer, Cham (2017). https://doi.org/10.1007/978-3-319-60240-0_18
7. Dobson, G.B., Carley, K.M.: A computational model of cyber situational awareness. In: Thomson, R., Dancy, C., Hyder, A., Bisgin, H. (eds.) SBP-BRiMS 2018.

LNCS, vol. 10899, pp. 395–400. Springer, Cham (2018). https://doi.org/10.1007/978-3-319-93372-6_43

8. Shin, J., Dobson, G.B., Carley, K.M., Richard Carley, L.: OSIRIS: organization simulation in response to intrusion strategies. In: Social, Cultural, and Behavioral Modeling: 15th International Conference, SBP-BRiMS 2022, Pittsburgh, PA, USA, September 20–23, 2022, Proceedings, pp. 134–143. Springer International Publishing, Cham (2022). https://doi.org/10.1007/978-3-031-17114-7_13

9. Shin, J., Richard Carley, L., Dobson, G.B., Carley, K.M.: Modeling and simulation of the human firewall against phishing attacks in small and medium-sized businesses. In: 2023 Annual Modeling and Simulation Conference (ANNSIM), pp. 369–380. IEEE (2023)

10. Uebelacker, S., Quiel, S.: The social engineering personality framework. In: 2014 Workshop on Socio-Technical Aspects in Security and Trust, pp. 24–30. IEEE (2014)

11. Parrish, J.L., Bailey, J.L., Courtney, J.F.: A personality based model for determining susceptibility to phishing attacks. Little Rock: University of Arkansas, 285–296 (2009)

12. Tornblad, M.K., Jones, K.S., Siami Namin, A., Choi, J.: Characteristics that predict phishing susceptibility: a review. In: Proceedings of the Human Factors and Ergonomics Society Annual Meeting, vol. 65, no. 1, pp. 938–942. Sage CA: Los Angeles, CA: SAGE Publications (2021)

13. Lin, T., et al.: Susceptibility to spear-phishing emails: effects of internet user demographics and email content. ACM Trans. Comput.-Hum. Interact. (TOCHI) **26**(5), 1–28 (2019)

14. Lawson, P., Zielinska, O., Pearson, C., Mayhorn, C.B.: Interaction of personality and persuasion tactics in email phishing attacks. In: Proceedings of the Human Factors and Ergonomics Society Annual Meeting, vol. 61, no. 1, pp. 1331–1333. Sage CA: Los Angeles, CA: SAGE Publications (2017)

15. Purkait, S., De Kumar, S., Suar, D.: An empirical investigation of the factors that influence Internet user's ability to correctly identify a phishing website. Inf. Manage. Comput. Secur. **22**(3), 194–234 (2014)

16. Ebbinghaus, H.: Memory: a contribution to experimental psychology. Ann. Neurosci. **20**(4), 155 (2013)

17. Kumaraguru, P., Rhee, Y., Acquisti, A., Faith Cranor, L., Hong, J., Nunge, E.: Protecting people from phishing: the design and evaluation of an embedded training email system. In: Proceedings of the SIGCHI Conference on Human Factors in Computing Systems, pp. 905–914 (2007)

18. Kumaraguru, P., Sheng, S., Acquisti, A., Faith Cranor, L., Hong, J.: Lessons from a real world evaluation of anti-phishing training. In: 2008 eCrime Researchers Summit, pp. 1–12. IEEE (2008)

19. Kumaraguru, P., et al.: Getting users to pay attention to anti-phishing education: evaluation of retention and transfer. In: Proceedings of the Anti-phishing Working Groups 2nd Annual eCrime Researchers Summit, pp. 70–81 (2007)

20. Kumaraguru, P., et al.: School of phish: a real-world evaluation of anti-phishing training. In: Proceedings of the 5th Symposium on Usable Privacy and Security, pp. 1–12 (2009)

21. Sheng, S., Holbrook, M., Kumaraguru, P., Faith Cranor, L., Downs, J.: Who falls for phish? A demographic analysis of phishing susceptibility and effectiveness of interventions. In: Proceedings of the SIGCHI Conference on Human Factors in Computing Systems, pp. 373–382 (2010)

Designing Organizations of Human and Non-Human Knowledge Workers

David Mortimore[1,2](✉) ⓘ, Raymond R. Buettner Jr.[1] ⓘ, and Eugene Chabot[3] ⓘ

[1] Naval Postgraduate School, Monterey, CA 93943, USA
{dbmortim1,rrbuettn}@nps.edu
[2] Naval Undersea Warfare Center Division, Keyport, Keyport, WA 98345, USA
[3] Naval Undersea Warfare Center Division, Newport, Newport, RI 02841, USA
eugene.j.chabot.civ@us.navy.mil

Abstract. A robust body of research demonstrates that intentionally designing organizations generally has positive impacts on their performance and, by extension, goal attainment. Furthermore, a predominant view exists of such organizations as human-centric systems. However, the increasing ubiquity of artificial intelligent agents means it might be time to reimagine organizations as ecosystems composed of both human and non-human knowledge workers—and to purposefully design them as such. In the context of the development and production of a key component for a new electronic device, this paper provides two computational scenarios that describe how organizations might employ artificial intelligent agents and compares the impacts on performance. Based upon these results, this paper recommends that future studies investigate organizational missions most likely to benefit from non-human knowledge worker employment, the assignment of expertise and role negotiation between human and non-human knowledge workers, and when operational performance targets make employment of non-human knowledge workers worthwhile.

Keywords: Non-Human Knowledge Workers · Organizations · Organizational Design · Human-Machine Ecosystems · Organizational Performance

1 Introduction

Over the last 70 years, scholars and practitioners alike have found that well-designed organizations generate more robust organizational performance, which enhances goal attainment. Therefore, the purposeful design and engineering of organizations conditions goal attainment, such as value generation for stakeholders [1, 2]. The considerable body of work describing the effects organizational designs have on performance has generally considered organizations as human-centric systems [1–7]. However, organizations increasingly employ artificial intelligent agents in their organizational technologies. Therefore, the view of organizations as human-centric systems likely limits the relevancy of earlier organization theory (OT) and computational organization theory (COT) studies to modern and emerging organizations. Furthermore, the general treatment of artificial intelligent agents as parts of an organization's set of technical systems likely places unnecessary limits on organizational performance, thereby raising the question—*can a more relevant framework for designing organizations be developed?*

This paper reimagines organizations as complex systems in which a technical core composed of human knowledge workers (HKWs) and non-human knowledge workers (NHKWs) collaboratively perform tasks to attain goals. Departing from a general view of organizations as human-centric systems, this recharacterization of organizations and organizational technologies is likely to result in more optimal performance [7] and more relevant OT and COT studies. This paper begins with an overview of knowledge work, HKWs, and NHKWs. Next, organizational technologies are described, along with impacts NHKWs might have on them. An elementary example is then provided, using information from an empirical study regarding the launch of a personal electronic device (PED). Lastly, recommendations for future studies are offered.

1.1 Knowledge Work

Knowledge work drives the economies of the United States and other countries, making the performance of organizations characterized by knowledge work important to scholars and practitioners, alike.[1] *Knowledge work* refers to tasks that apply knowledge to knowledge and generate knowledge [8], such as analyzing military intelligence and scientific research, respectively. In comparison, *service work* applies knowledge to generally routinized tasks [8], such as troubleshooting hardware systems. Knowledge work and the processing of relevant information are inextricably linked because knowledge and information, themselves, are inextricably linked.[2] Therefore, organizations provide a particularly germane framework for investigating the performance of knowledge work and the impacts that NHKWs might have on organizational performance and, ultimately, goal attainment.

1.2 Human Knowledge Workers

HKWs are those individuals for whom knowledge work is the primary attribute of the tasks they perform. Importantly, it is the nature of a task that determines if it is knowledge work, not the nature of the individual performing the task [8]. Decision-making exemplifies knowledge work; decision-making intrinsically involves applying knowledge to knowledge in the gathering and processing of information to choose between options [9–12]. In contrast, presenting the results of choosing between options represents service work because it involves applying knowledge to performing a generally routinized set of activities (i.e., developing and delivering a presentation). Further, an individual can perform both knowledge and service work. The same individual who decided between investment options (knowledge work) can also develop and deliver a presentation on the results (service work). Ultimately, the volumetric proportion between knowledge and service work performed characterizes an individual as a knowledge or service worker.

---

[1] *Organizations* are fundamentally information processing and communication systems [2–4].

[2] *Knowledge* is information structured with perspectives, intuition, and experience [12–13]. *Information* is data with relevant context and *data* are numerals and symbols (logical and mathematical) with minimal context.

1.3 Non-Human Knowledge Workers

NHKWs co-exist with HKWs to perform tasks contributing more directly to organizational performance and, by extension, goal attainment. Despite their foundational algorithmic nature, NHKWs are more than advanced artificial intelligent agents—they are synthetic knowledge workers purposefully designed and encoded into organizations to relieve HKWs of cognitive tasks [14]. What distinguishes NHKWs from other forms of artificial intelligent systems is the conjunction of knowledge work, more comprehensive incorporation into organizational structures, algorithmic power, and task-level assignments commensurate with HKWs. The more comprehensive integration into organizational structures and technologies enables NHKWs to contribute more impactfully because NHKW performance is not limited unnecessarily by sub-optimal employment. Although adequate algorithmic power is a necessity, it is not a sufficient condition for an artificial intelligent agent to perform as a NHKW. This means that, although an artificial intelligent agent might have comparable or greater algorithmic power than a NHKW, the more limited organizational integration limits the impact they have on performance. Meanwhile, expert, robotic process automation, and similar technical systems generally lack the organizational encoding, algorithmic power, and operational employment characteristics necessary to perform knowledge work and, therefore, as NHKWs.

2 Organizational Technologies

An organization's technology describes how it accomplishes its mission and attains goals. In other words, an organizational technology refers to how an organization transforms raw inputs, including information and knowledge, into outputs, such as products and services [14, 15]. The technical core is composed of individuals, like knowledge workers, involved in the technological transformation process. Technical core personnel use techniques (i.e., methods performed via processes and mechanisms) to perform tasks, which are individual activities or sets of activities. To accomplish tasks, technical core personnel make use of technical systems, which are physical and non-physical resources, such as intellectual, hardware, software, and facility capabilities.

Divisions of labor between technical core members generate interdependencies that can impact organizational performance. *Interdependence* describes the degree to which outcomes rely upon inputs from others, including knowledge, information, and materials [16]. Three forms of interdependencies generally characterize organizations: pooled, sequential, and reciprocal [17]. *Pooled interdependence* describes cases in which resources needed by multiple tasks are centrally located and there is no workflow between the tasks themselves, such as providing heating and cooling independently to multiple work areas. *Sequential interdependence* describes cases in which the outputs of one task are inputs to a subsequent task, such as on an assembly line. *Reciprocal interdependence* describes cases in which the outputs of one task (task A) are used subsequently in other tasks (tasks B and C) and the outputs of the subsequently performed tasks are returned to the original task (task A) for use, such as in research and development. Concurrent task performance generally results in a greater volume of reciprocal interdependencies, which necessitates more robust task coordination because of the increased likelihood that changes in tasks B or C generate changes in task A [18]. This added collaboration workload can significantly affect organizational performance.

3   An Example

The launch of a new PED in 1998 necessitated the accelerated development of a then-new application specific integrated circuit (ASIC) and production of several prototypes in less than a year. The aggressive project schedule resulted in additional task inter-dependencies and coordination workload, which added to the total workload volume, thereby impacting task and organizational performance [18]. By using *POW-ER*, a validated engineering COT software application, and the validated computational model of the ASIC project [18–21], potential impacts of NHKWs on task and organizational performance are explored.[3]

3.1 POW-ER

POW-ER is a computational modeling and simulation software application that provides scholars and practitioners a means of engineering organizations. *POW-ER* enables users to investigate the impacts that organizational designs, task assignments and interdependencies, and hidden work have on performance. In many cases, the hidden workload associated with task coordination, rework, and resolving ambiguity between team members is not included in organizational models [18–20]. Consequently, the impacts that such activities have on task and organizational performance receives inadequate attention until too late—when the project is in, or about to be in, extremis necessitating herculean efforts to attain organizational goals. To address hidden workload, *POW-ER* uses the construct of total workload volume to capture the sum of task work, rework, and coordination work volumes and their impacts on performance.[4]

3.2 An Accelerated Launch

The 1998 launch of a new PED and development of a new ASIC, in particular, provides a robust backdrop for investigating impacts NHKWs might have on organizational performance. Meeting the deadline for the tradeshow at which the company planned to announce its new PED meant the project team had to accelerate the design and manufacture of an ASIC to roughly five months—a little more than half the normally allotted time [18]. Such an aggressive schedule resulted in team members performing tasks concurrently that they might have otherwise performed sequentially, which introduced additional reciprocal interdependencies. This, in turn, increased the coordination workload on team members [18–20]. Because ASIC development would consume nearly

---

[3] *POW-ER* is the enhanced version of the *Virtual Design Team* (*VDT*) developed by the Center for Integrated Facility Engineering [18–20]. *SimVision* is the commercialized version.

[4] *Coordination work volume* is the sum of communication, decision-making, and waiting time volumes [18–20]. *POW-ER* measures work volumes in full-time equivalent (FTE)-days, which represents the equivalent working time of an individual in a 24-h period. For the ASIC project, an FTE-day was eight hours [18, 21].

half the time the company had before the trade show, significant design and production delays could have proved disastrous for the company. Therefore, rapidly identifying and addressing emergent issues added to the existing supervisory and coordination workload.

Figure 1 provides a simplified organizational model of the ASIC design and production effort. The project starts with team members developing the needed specifications (i.e., the *Develop Spec* task) before next performing three tasks—*Implement Data Model*, *Implement User Interface (UI)*, and *Implement Analysis System*—concurrently with reciprocal dependencies between them [21]. Each of these three tasks provide inputs to the next task, *Integrate Systems*, which subsequently feeds the *Systems Integration Test* task. Completion of the *Systems Integration Test* task results in attaining the *Ready for Systems Test* milestone, meaning ASIC development is ready for its *UI Stress Test* and *Analysis Stress Tests*, which are performed concurrently. The *Software (SW) Design Coordination* task provides the *Data Architect*, *SW Project Manager*, *UI Team*, *Analysis Team*, *Integration Team*, and *Customer Representative* team positions a means to coordinate task performance and resolve issues in a *Group Status Meeting*. Individuals and sets of individuals filling team positions compose the technical core. A more comprehensive discussion regarding the operationalization of OT and COT theoretical constructs in *POW-ER* is available in [18–22].

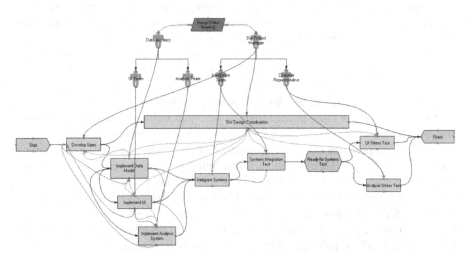

Fig. 1. The baseline organizational design for ASIC development and production depicts tasks, milestones, interdependencies, team positions, supervisory relationships, task assignments, communication linkages, and rework. Tasks and milestones are represented with yellow and blue trapezoids, respectively. Task interdependencies are denoted with black arrows. Team positions are represented with green silhouettes of humans; supervisory relationships are denoted with black arrows between positions. Task assignments are identified by blue arrows. Communication linkages are indicated by green arrows and team meetings are represented with magenta parallelograms. Potential rework between interdependent tasks is depicted with red arrows. Source: [21].

3.3 Comparative Trials

Two toy-problem [19] experiments provide for preliminary investigation into the conjecture that NHKWs more optimally impact organizational performance when performing tasks versus activities, *ceteris paribus* [14].[5] To take advantage of prior empirical research [18, 21] and for simplicity, all tasks are considered knowledge work, making all team members knowledge workers. Using the baseline scenario [21], two experiments provide likely lower and upper bounds on NHKW impacts on organizational performance, based upon activity- versus task-level work assignments.

The baseline scenario includes probabilistic estimates for four project-level properties—communication, noise, functional exceptions, and project exceptions—strengthening the realism of computational results. *Communication* represents the likelihood that team members need to exchange information with others based upon the degree of reciprocal interdependencies and task ambiguity [18–22]. *Noise* estimates the probability that communications unrelated to the ASIC project disturb personnel. *Functional exceptions* represent the probability a team member must rework part of their own tasks without impacting others. *Project exceptions* represent the chance problems arise that impact interdependent tasks and generate rework. The probabilities assigned to the four properties are 0.2, 0.1, 0.1, and 0.1, respectively [21].

The first experiment represents an artificial intelligent agent performing activity- versus task-level work, as an artificial intelligence (AI)-enabled task monitoring system (TMS) that continuously monitors project status and alerts team members of issues that could affect performance. To model this scenario, the parameters of the *Group Status Meeting* were modified such that, instead of a single 90-min meeting each week [21], team members receive updates every workhour. To minimize the impacts of such frequent updates on task performance, the duration of updates is limited to one minute. This scenario should result in more limited impacts on organizational performance because of the sub-optimal employment of artificial intelligent agent capabilities.

The second experiment scenario represents NHKW employment by substituting a NHKW for a HKW. In this case, the artificial intelligent agent performs knowledge work, possesses adequate algorithmic power, is organizationally encoded, and operates at the task level. Substituting a NHKW for a HKW on the *Analysis Team*, which consists of a single HKW in the baseline scenario [21], generates a more relevant comparison. This scenario should result in more significant impacts on performance because it more fully employs NHKW capabilities. Table 1 summarizes the three scenarios.

For the purposes of this paper, it is assumed that there is no variance in HKW productivity throughout the day, there is no change in rework volume because of task

---

[5] Empirical investigation of how NHKWs might affect organizational performance is limited to toy problems regarding task-level work prior to a more comprehensive investigation.

Table 1. Scenario comparison.

	Baseline	Experiment 1	Experiment 2
Role of the artificial intelligent agent	None	AI-enabled TMS	*Analysis Team* NHKW
Performs knowledge work		X	X
Possesses adequate algorithmic power		X	X
Is organizationally encoded		X	X
Performs task-level work			X
Anticipated impacts on organizational performance		Limited	More significant

fatigue, and the productivity of one NHKW is equivalent to four HKWs.[6] For each scenario, 1,000 simulations were run using a seed value of 1.0 for comparability.

3.4 Results

Figure 2 displays simulation results from the three scenarios with project tasks along the vertical axis and calendar dates along the horizontal axis. Red-colored bars indicate critical path tasks, blue-colored bars identify tasks not on the critical path, and grey-colored bars identify the float, or slack, available before the task impacts project completion. Diamonds represent project milestones and simulated task durations are displayed for the baseline, AI-enabled TMS, and *Analysis Team* NHKW scenarios from top-to-bottom.

In the baseline scenario simulation, the ASIC project team is projected to complete its work in 230 workdays. The task that most significantly impacts project completion and, therefore, organizational performance is *Implement Analysis System* with a simulated duration of 98 workdays. An AI-enabled TMS, the second scenario, has minimal impact; simulated results indicate the project will finish one day sooner. In contrast, the substitution of a NHKW for a HKW, the third scenario, generates more significant impacts with simulated project completion 59 workdays (26%) sooner than the baseline scenario. This same substitution further results in a 73-workday (75%) reduction in the duration of the task that most significantly impacts organizational performance in the baseline scenario. Impacts on other task and organizational performance parameters, such as rework volume, are generally negligible for the three scenarios.

---

[6] During an eight-hour workday, it is assumed a HKW takes a 30-min meal break, two 15-min breaks, and loses an hour to non-project disruptions, meaning a HKW performs six hours of task-related work during one FTE-day. In contrast, the nature of a NHKW means such breaks are likely unnecessary and that a NHKW does not experience the same performance limitations as HKWs [23]. Thus, a single NHKW can work a total of 24 h in a 24-h period, the equivalent of four HKWs.

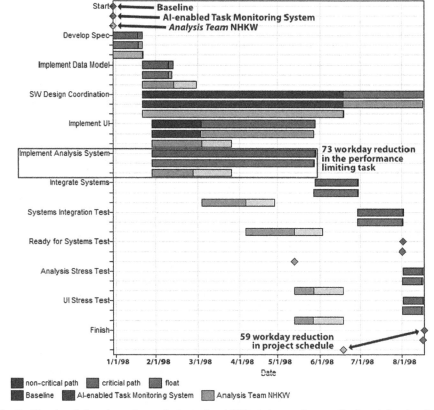

Fig. 2. Simulated durations, in work-days, for ASIC project tasks are displayed for the three scenarios with diamonds representing project milestones Adapted from: [21].

4   Discussion

Earlier studies that demonstrate intentionally designing organizations result in more optimal performance also appear to apply to ecosystems composed of HKWs and NHKWs. Designing such organizations means the reconception of a popular view of organizations as human-centric systems [1–7]. This paper juxtaposes two scenarios—an organization that incorporates artificial intelligent agent capabilities at the activity level with an organization that more optimally employs a NHKW at the task level. The former scenario also represents an organization that incorporates an artificial intelligent agent as a technical system, while the latter scenario represents an organization that employs the same capabilities as a member of its technical core.

Preliminary results, while limited, are telling and consistent with expectations. First, when the ASIC project team more optimally employs a NHKW, the impacts to simulated organizational performance are significant—an estimated 26% reduction in project duration. Second, the same results eclipse the simulated results from the less optimal approach of using the same artificial intelligent agent capabilities as a technical system. The results are likely because the NHKW is not limited by limitations on human performance [14, 23]. Notably, only the NHKW *Analysis Team* scenario resulted in a simulated project duration (171 days) close to the allotted five-month window.

This paper has several limitations. First is the relatively narrow discussion of the effects NHKWs might have on organizational design and performance. Only two cases, both rudimentary, are considered, thus meriting a more thorough assessment. Second, the exploration of NHKW characteristics is generally limited to the overarching construct and their operational employment. A more comprehensive investigation of relevant characteristics is needed. Third, assumptions regarding knowledge worker productivity warrant refinement. Notwithstanding these limitations, the use of a validated COT tool and information from a prior empirical study make the results more compelling by eliminating effects from using a non-validated tool and uncorroborated information.

Future studies should further explore the extent to which NHKWs impact organizational design, performance, and, by extension, goal attainment. Four recommended focus areas are: (a) the organizational missions most likely to benefit from NHKW employment; (b) the identification and assignment of expertise within HKW and NHKW organizations; (c) role negotiation between HKWs and NHKWs; and (d) when operational performance targets make NHKW employment worthwhile.

5 Conclusion

It is not a matter of *if*, but *how* organizations will more optimally employ artificial intelligent agents. A rich body of scholarly work, corroborated by practice, demonstrates that purposefully designing organizations positively impacts performance. The increasing presence of artificial intelligent agents means it is likely time to reimagine and redesign organizations as ecosystems composed of HKWs and NHKWs. In the context of an ASIC development and production project, this paper presented two scenarios and computationally demonstrated that intentionally designing and incorporating NHKWs into organizations can significantly impact organizational performance. The accelerating use of such artificial intelligent agents makes the rapid formalization and operationalization of NHKWs an imperative—for scholars and practitioners, alike.

Acknowledgements. This research is supported by the Office of Naval Research Cooperative Autonomous Swarm Technology, and Cognitive Science and Human & Machine Teaming programs.

References

1. Burton, R., Obel, B.: Strategic Organizational Diagnosis and Design: The Dynamics of Fit, 3rd edn. Springer Science Business Media, LLC, New York (2004)
2. Galbraith, J.: Organization Design. Addison-Wesley Publishing Co., Menlo Park (1977)
3. March, J., Simon, H.A.: Organizations. John Wiley & Sons Inc., New York (1958)
4. Cyert, R., March, J.: A Behavioral Theory of the Firm. Cambridge University Press, Cambridge (1963)

5. Mayo, E.: The Social Problems of an Industrial Civilization. Andover Press, Andover (1945)
6. Daft, R., Lengel, R.: Organizational information requirements, media richness and structural design. Manage. Sci. **32**(5), 554–571 (1986)
7. National Academies of Sciences: Engineering, and Medicine: A Decadal Survey of the Social and Behavioral Sciences: A Research Agenda for Advancing Intelligence Analysis. National Academies Press, Washington D.C (2019)
8. Drucker, P.: The rise of the knowledge society. Wilson Q. **17**(2), 52–71 (1993)
9. Arrow, K.: The Limits of Organizations. W. W. Norton & Company Inc., New York (1974)
10. Cohen, M., March, J., Olsen, J.: A garbage can model of organizational choice. Adm. Sci. Q. **17**(1), 1–25 (1972). https://doi.org/10.2307/2392088
11. Mortimore, D., Canan, M., Buettner, R., Jr.: Two probability theories and a garbage can. Comput. Math. Organ. Theory (2023). https://doi.org/10.1007/s10588-023-09378-3
12. Carley, K.: On the evolution of social and organizational networks. In: Andrews, S., Knoke, D., Bacharach, S. (eds.) Research in the Sociology of Organizations: Networks in and Around Organizations. Emerald Publishing Ltd., Somerville (1999)
13. Hollingshead, A., Fulk, J., Monge, P.: Fostering intranet knowledge sharing: an integration of transactive memory and public goods approaches. In: Hinds, P., Kiesler, S. (eds.) Distributed Work, pp. 335–355. MIT Press, Cambridge (2002)
14. Mortimore, D., Aten, K., Buettner Jr., R.: A new technology rises: non-human knowledge workers and decision-making in a system of complex systems. In: Proceedings of the 18th Annual System of Systems Engineering Conference, Villeneuve d'Ascq, France (2023). https://doi.org/10.1109/SoSE59841.2023.10178624
15. Perrow, C.: A framework for the comparative analysis of organizations. Am. Sociol. Rev. **32**(2), 194–208 (1967). https://doi.org/10.2307/2091811
16. Ren, Y., Argote, L.: Transactive memory systems 1985–2010: an integrative framework of key dimensions, antecedents, and consequences. Acad. Manage. Ann. **5**(1), 189–229 (2011). https://doi.org/10.1080/19416520.2011.590300
17. Thompson, J.: Organizations in Action. Routledge, New York (2003)
18. Levitt, R.: Organizational Design as "Virtual Adaption": Designing Project Organizations Based on Micro-Contingency Analysis (2005)
19. Jin, Y., Levitt, R.: The virtual design team: a computational model of project organizations. Comput. Math. Organ. Theory **2**(3), 171–196 (1996)
20. Levitt, R.: Computational modeling of organizations comes of age. Comput. Math. Organ. Theory **10**(2), 127–145 (2004)
21. *POW-ER*. (Version 3.4a): Collaboratory for Research on Global Projects
22. Collaboratory for Research on Global Projects: *POW-ER* Documentation for *POW-ER* 2.0. Stanford University, Stanford (2006)
23. Carley, K., Behrens, D.: Organizational and individual decision making. In: Sage, A., Rouse, W. (eds.) Handbook of Systems Engineering and Management, 1st edn. John Wiley and Sons, Inc., New York (1999)

Modeling Human Actions in the Cart-Pole Game Using Cognitive and Deep Reinforcement Learning Approach

Aadhar Gupta$^{(\boxtimes)}$, Mahavir Dabas$^{(\boxtimes)}$, Shashank Uttrani$^{(\boxtimes)}$,
Sakshi Sharma$^{(\boxtimes)}$, and Varun Dutt$^{(\boxtimes)}$

Applied Cognitive Science Lab, Indian Institute of Technology Mandi, Kamand
175005, Himachal Pradesh, India
{aadhar.innovate,mahavirdabas18,shashankuttrani,sakshi28720}@gmail.com,
varun@iitmandi.ac.in

Abstract. Designing optimal controllers still poses a challenge for modern Artificial Intelligence systems. Prior research has explored reinforcement learning (RL) algorithms for benchmarking the cart-pole control problem. However, there is still a lack of investigation of cognitive decision-making models and their ensemble with the RL techniques in the context of such dynamical control tasks. The primary objective of this paper is to implement a Deep Q-Network (DQN), Instance-based Learning (IBL), and an ensemble model of DQN and IBL for the cart-pole environment and compare these models' ability to match human choices. Forty-two human participants were recruited to play the cart-pole game for ten training trials followed by a test trial, and the human experience information containing the situations, decisions taken, and the corresponding reward earned was recorded. The human experiences collected from the game-play were used to initialize the memory (buffer) for both the algorithms, DQN and IBL, rather than following the approach of learning from scratch through environmental interaction. The results indicated that the IBL algorithm initialized with human experience could be proposed as an alternative to the Q-learning initialized with human experience. It was also observed that the ensemble model could account for the human choices more accurately compared to the Q-learning and IBL models.

Keywords: Instance-Based Learning · Cognitive Modeling · Reinforcement Learning · Q-Learning · DQN · cart-pole · Ensemble

1 Introduction

Reinforcement Learning (RL) is a paradigm of machine learning where the agent learns by indirect supervision signal in the form of rewards [17]. Contrary to supervised learning, RL is used when the target outputs are unknown, so the agent needs to interact with the environment to gather information [17]. The agent heads towards optimal behavior by exploring the rewards associated with various actions under various situations and exploiting the hence-gained knowledge of the goodness of actions to maximize the cumulative reward for an entire sequence of actions [17]. With the advent of Deep RL (DRL) [10], it has become possible to apply RL to complex problems, earlier considered to be intractable [2]. The recent success of RL in tasks like playing Atari games at a superhuman level [11] has demonstrated the capability and robustness of RL algorithms.

DRL suffers from certain shortcomings, such as reward shaping, sample inefficiency, and local optima [5]. Learning from human behavior offers an alternative to achieving intelligent behavior. Imitation Learning (IL) [14] is a branch of AI where the agent tries to mimic human behavior. Similarly, Cognitive Science is another branch of Artificial Intelligence (AI) that uses human behavior and aims at creating techniques as robust, insightful, and adaptive as human intelligence [7]. Prior research has contributed to more than a hundred cognitive architectures, including production rule-based, psychology-based, and a combination of neural networks with cognitive psychology, to mention a few [7]. Adaptive Control of Thought-Rational (ACT-R) [1] is a psychologically motivated cognitive model that combines AI, cognitive psychology, and some components of neurobiology. Many researchers have extended upon the principles of ACT-R yielding architectures avoiding the high complexity yet retaining the efficiency, such as Instance-Based Learning (IBL) [6].

The cart-pole problem [3,9] provides a simple and cost-effective platform to test AI algorithms for control. It consists of a pole attached to a cart like an inverted pendulum, and the player needs to balance the pole by moving the cart. Prior research has investigated a wide range of techniques for the cart-pole problem [4,8,12,13,15,16,18,19], with a major focus on RL and DRL [10, 11]. However, little is known about the capability of RL techniques to account for human choices in these games. The learning in RL techniques examined in the literature so far, with regard to control problems like cart-pole, is purely mathematical. It doesn't incorporate human intuition. To address this literature gap, we have made a two-fold attempt to give a human touch to RL: by building it over human behavior data and by developing a cognitive model to work in an ensemble with RL.

The upcoming sections include the background on the cart-pole problem, followed by the detailed methodology of this study. Next, the results are presented, followed by the conclusion of our findings with a brief analysis.

2 Background

The cart-pole problem seems to be introduced in [9] and popularized by [3]. Since then, literature has witnessed a plethora of experimentation on this problem, mostly focused on the RL techniques [4,8,12,13,15,16,18,19]. The algorithms of Q-Learning [20] and deep Q-learning [10] have been thoroughly investigated, along with a few others. [12] examined Q-Learning and SARSA for playing cart-pole and found that both performed quite well. [13] examined a variety of algorithms, including Policy Gradient (PG), Temporal difference (TD), and DQN, and found TD to perform the best while PG displayed better stabilization and faster convergence than Q-Learning. While [18] examined Deep Q-Learning over cart-pole, [16] examined the Baseline PG and the Reinforce PG and found Reinforce PG to outperform in cumulative reward while Baseline PG outperformed in episode speed. [19] proposed novel variants of DQN and other advanced algorithms and found the rewards to increase with a reduced need for training. [4,8] examined various advanced algorithms, including DQN, and found PER with DQN to perform remarkably well. [15] investigated the difference in the performance of Q-learning and DQN over cart-pole but didn't observe any significant difference. However, the Q-learning algorithm was found to train the agent significantly faster than DQN. However, the aforementioned techniques investigated on cart-pole don't take into consideration the human aspect of decision making. Moreover, prior research has also lacked an investigation of ensemble techniques that combine the RL and cognitive paradigms.

In this study, we began by developing a virtual cart-pole game, followed by the implementation of a DRL and a cognitive algorithm for the agent to balance the pole. Among the DRL techniques, we developed a deep Q-learning network (DQN), and the cognitive model was based on IBL. The goal of the agent trained on DQN, IBL, and the ensemble of these two algorithms was to keep balancing the pole by moving the cart left or right. The study began with collecting the game-play data of human participants, with multiple trials in the training phase and a single trial in the testing phase. Next, the DQN and IBL were applied to the agent to play the cart-pole game in the same way human players did. Furthermore, an ensemble model was developed to combine the IBL cognitive architecture and DQN. Finally, the results for IBL, DQN, and their ensemble were observed and compared.

3 Methodology

3.1 Game Design

A cart-pole game was developed. The task was to balance the pole on the cart for as long as possible. The cart-pole system dynamics were completely governed by pre-defined equations [3,9] for the horizontal motion of the cart and the angular displacement of the pole. The cart and the pole were assigned a virtual weight of 1 kg each, and a left or right action exerted a force of 10N on the cart. Hence a keypress in either direction caused the cart to accelerate, either increasing the speed in that direction or reducing the speed if the cart was moving in the reverse direction. On initialization of the game, the cart appeared vertically above the horizontal center of the platform. The initial angle of the pole with the vertical was obtained randomly between 0.05 rad (approximately 2.86°) to the left and 0.05 rad to the right side of the vertical. There were two terminating conditions for the game: the angle of the pole with the vertical axis exceeding a threshold of 30° and the cart falling off the platform. A reward of 0.1 and -5 was given for non-terminating and terminating actions, respectively. The game-play was divided into two phases: the training phase, with ten trials per player, and the testing phase, with a single trial per player. The situation of the cart-pole was defined by four values: cart position, cart velocity, pole angle to the vertical axis, and pole angular velocity. There were two possible actions for a participant to be taken in the game: move left, and move right.

3.2   Participants

42 participants were enlisted from the Indian Institute of Technology, Mandi, to collect human data after approval from the ethics committee. There were 76.18% males and 23.82% females(mean $= 25$, sd $= 3$). 93% of the participants belonged to STEM, and the rest belonged to the humanities.

3.3   Procedure

The experiment began with instructing the participants on the game's rules, along with a collection of the demographic details. No time limitation was set for either of the phases. The actions taken and the corresponding situation vector were recorded. The recorded data was fed into the DQN, IBL, and ensemble model of DQN and IBL (more details ahead), which were then made to act in the environment, and the observations were collected.

IBL Model

Conceptual Details. IBL [9] works similarly to how humans make judgments by gathering and refining memory experiences. The past experiences are stored as situation-decision-utility(SDU) tuples called instances. Given a situation, the most similar situations are retrieved and are used to compute the goodness score for each decision, called blended value (BV). The decision with maximum BV is executed. IBL uses the formulations of the Activation, Probability of retrieval(PR), and BV, given as:

$$A_{i,t} = \sigma \ln(\frac{\gamma_{i,t}}{1 - \gamma_{i,t}}) + \ln(\sum_{t_p=1}^{t-1} (t - t_p)^{-d}) + \mu(S) \tag{1}$$

where d, σ and γ represent the parameter for memory decay, cognitive noise, and a random draw from a uniform probability distribution, respectively. t_p and S for instance i, represent the timestamp and the similarity measure with the current test situation, respectively, while μ is the scaling factor.

$$P_{i,t} = \frac{e^{A_{i,t}/\tau}}{\sum_j e^{A_{j,t}/\tau}}) \tag{2}$$

where τ represents the random noise and $A_{i,t}$ represents the activation of the instance i.

$$V_j = \sum_{i=1}^{n} p_i x_i \qquad (3)$$

where j is the concerned action, while x_i and p_i are the utility and PR of the instance i.

Implementation. The IBL agent's memory was initialized with the SDU instances of a human participant's ten training trials of game-play. All the instances were timestamped with 0; hence, base activation was not used. The IBL agent was evaluated on situations from the participant's test session data. The memory instances with cosine similarity greater than 0.85 for the current situation were shortlisted and used to compute the activation and PR, and BV. The model's performance was measured by comparing predicted decisions with the human participants. Considering the class imbalance, the F1 score was opted to evaluate the model's ability to mimic human decision making. Each of the 42 participant's data was used to initialize an IBL model. Hence there were 42 distinct instances of the IBL model. The F1 score over all the model instances was averaged to give a generalized metric for IBL's human behavior-mimicking ability.

Hyper-parameters. In the IBL model, the hyper-parameters used were the cognitive noise and the similarity threshold, as mentioned in Table 1a. Cognitive noise (CN) is added to capture the variability in decisions from one agent to another, while the similarity threshold controls the memory instances that are allowed to contribute to the decision-making.

DQN Model

Conceptual Details. Q-learning is a model-free RL algorithm [20] that uses a trial-and-error approach to learn via environmental interaction [20]. The algorithm aims to determine the State-Action values (Q-Value) and store it in a table called the Q-table [20]. The Q-values are updated using the Bellman equation, given as:

$$Q(s_t, a_t) = (1 - \alpha)Q(s_t, a_t) + \alpha(r_t + \gamma \max_a [Q(s_{t+1}, a)]) \qquad (4)$$

where $Q(s_t, a_t)$ represents the Q-Value for state s_t and action a_t. s_t and s_{t+1} stand for the current and the next state, respectively, and r_t represents the reward on the transition from the state s_t to the state s_{t+1} on taking action a_t, α represents the learning rate that controls the amount of updation in the Q-values and γ represents the discounting factor.

Implementation. The experience replay buffer [20] of DQN was initialized with the quadruplets of 'State, Decision, Feedback, and Next state' to enable it to learn from human behavior rather than environmental interactions. The model was trained for a maximum of 10 epochs with early stopping. The model predicted Q values corresponding to the actions left and right. The cosine distance between the predicted Q-values and the target Q-values, along with the validation loss, was computed. A distinct model instance was trained on each participant's data. The overall performance of the DQN model was computed by averaging the F1 score between predicted and actual human decisions for all the model instances.

Network Architecture. The neural network architecture (16-32-2) comprised two fully connected hidden layers with 16 and 32 units and the output layer with two units corresponding to the two actions. Rectified linear activation followed by Dropout with a rate of 0.1 was used for both hidden layers.

Hyper-parameters. Hyper-parameters can play a significant role in the learning of a neural network. The hyper-parameters used for DQN in this study are presented in Table 1b.

Ensemble Model

Conceptual Details. The Ensemble model was obtained by performing weighted addition of the cognitive model's BVs and the DQN model's State-Actions Values (SAV) for the corresponding decisions. The BV of the IBL model represents the experienced utility (experienced reward) for the current action, and the State-Action value predicted by the DQN approximates the cumulative future rewards. With the aim of attaining more informed decision making, these two values were brought together. The values were normalized to bring the BV and the SAV to the same scale. A weight variable was used to determine the contribution of each approach in the ensemble value corresponding to each decision alternative. The decision against the higher ensemble value was chosen. For the weight value, 'x' multiplied by the IBL BV V_j, a weight of '1-x' was multiplied with the DQN SAV $Q(s_t, j)$, before adding, given as:

$$EnsembleValue = x * V_j + (1 - x) * Q(s_t, j) \qquad (5)$$

Hyper-parameter. The weight combinations for IBL and DQN varied from 0.1 to 0.9 in steps of 0.1. In this case, as well, a distinct ensemble model was created, corresponding to each participant, with the IBL and DQN model initialized with that participant's data. For each weight combination in the ensemble model, the generalized performance was computed by averaging over the F1 score for the ensemble model for each participant's data.

4 Results

Figure 1 shows the total number of human choices for the left and the right action in the cart-pole game for the ten training trials and the single test trial for the 42 human participants. Figure 2a and b shows the confusion matrix for the IBL and the DQN model, respectively, taking into consideration the actions predicted by all the model instances (each uniquely trained on the data of one participant, where the number of model instances equaled the number of human participants).

As shown in Table 2, the average F1 score for the DQN model and the IBL model was observed to be 0.829 and 0.809, respectively. Table 3 shows the F1 score of the Ensemble model for each weight combination of the IBL and DQN model. The highest F1 score was achieved for the weight of 0.2 and 0.8 for IBL and DQN, respectively. Notably, the F1 scores of all the weights combinations for the Ensemble exceeded the F1 scores of the individual DQN and IBL models. However, as shown in Table 3, the F1 score is found to decline with the increase in weight of IBL beyond the value of 0.2, indicating the inclination of optimal decision-making toward the long-sighted RL approach.

graph total actions 42 all train session test session nowhere avg.png

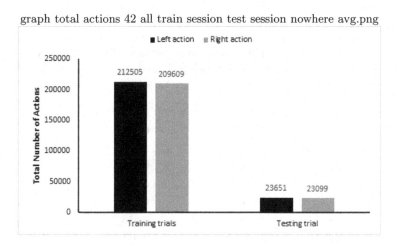

Fig. 1. The total number of human choices for the left and the right action by the 42 players in the cart-pole game, corresponding to all 10 training trials and the single testing trial.

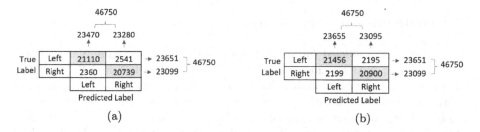

Fig. 2. The confusion matrix for the total actions taken by all the 42 model instances (each uniquely trained on the data of one participant) for a) the IBL model and, b) the DQN model.

Table 1. Hyper-parameters used for the IBL and DQN model.

Name	Value
Cognitive Noise	0.25
Similarity Threshold	0.85

(a) IBL

Name	Value
Weight initialization	Xavier Uniform
Batch size	32
Optimizer	Adam
Learning rate	0.001
Discounting factor	0.95
Dropout rate	0.1

(b) DQN

Table 2. Average F1 score of the DQN and IBL model.

Model	Average F1 score
DQN	0.8228
IBL	0.8091

Table 3. F1 score of the Ensemble model, averaged over the model instances corresponding to all human participants, for each weighted combination of the IBL and DQN

Weight IBL (w)	Weight DQN (w)	Mean F1 score
0.1	0.9	0.859
0.2	0.8	0.860
0.3	0.7	0.859
0.4	0.6	0.858
0.5	0.5	0.859
0.6	0.4	0.858
0.7	0.3	0.855
0.8	0.2	0.848
0.9	0.1	0.836

5   Discussion and Conclusion

In this study, we modeled human behavior in the cart-pole game via an IBL cognitive model, a DQN model, and their ensemble. The IBL and DQN models were initialized with human behavior data via memory pre-population and experience replay initialization, respectively. The IBL model performed moderately well in matching human choices with an F1 score of 80%. A possible explanation of why the model fell short of a 100% F1 score might be that the model could recognize the frequent states but not the rarely occurring states. The DQN model outperformed the IBL model in matching human choices, the likely reason being that the DQN approach of maximizing the cumulative reward fits human decision-making better than the IBL approach of maximizing the current reward. However, the F1 scores of both the models lay in the interval of 80-82%, indicating their inability to model a significant portion of the human decisions.

The ensemble of the IBL and the DQN models was developed to bring the principles of the cognitive and RL approaches under one roof. Results revealed that the ensemble models could predict human choices with greater accuracy than the standalone cognitive and RL models. The likely reason could be that the human decision-making process is based on a trade-off between immediate short-term and long-term goals, more accurately modeled by the ensemble. The optimal weights of IBL and DQN were found to be 0.2 and 0.8, respectively. This points to the trade-off being inclined towards the far-sighted approach.

The limitation of this research is that in the process of data extraction from the recorded human game-play, only two actions, left and right, were considered, but for a human, another outcome of 'no action' occurred for some situations while switching between the left and right action key. Dropping the 'no action' action might prevent capturing actual human behavior. Additionally, a very simple architecture was used for DQN, and there may be a possibility to push the DQN results a little further through more complex networks.

There is a broad scope of future work based on this study. Various modifications could be done to the IBL models by importing concepts from other cognitive mechanisms [7], which could increase the match with human behavior. More complex network architectures could be examined for improving DQN performance. It would be interesting to observe the models' performance if different rewards are associated with a win or loss in the episode, and each action's reward is obtained via discounting, unlike predefined rewards for each step, as in this study. Apart from DQN, other more advanced state-of-the-art algorithms could also be investigated. Moreover, the approaches possible to achieve an ensemble of these two techniques from such different paradigms are limited only by one's imagination. Instead of addition, multiplication of the corresponding action scores to give a final measure of an action's goodness, weighted multiplication, and a hybrid mechanism to merge the working principle of the two algorithms, to mention a few.

References

1. Anderson, J., Bothell, D., Byrne, M., Douglass, S., Lebiere, C., Qin, Y.: An integrated theory of mind. Psychol. Rev. **111**, 1036–1060 (2004)
2. Arulkumaran, K., Deisenroth, M.P., Brundage, M., Bharath, A.A.: A brief survey of deep reinforcement learning. arXiv preprint arXiv:1708.05866 (2017)
3. Barto, A.G., Sutton, R.S., Anderson, C.W.: Neuronlike adaptive elements that can solve difficult learning control problems. IEEE Trans. Syst. Man Cybern. SMC-**13**(5), 834–846 (1983). https://doi.org/10.1109/TSMC.1983.6313077

4. Duarte, F.F., Lau, N., Pereira, A., Reis, L.P.: Benchmarking deep and non-deep reinforcement learning algorithms for discrete environments. In: Silva, M.F., Luís Lima, J., Reis, L.P., Sanfeliu, A., Tardioli, D. (eds.) ROBOT 2019. AISC, vol. 1093, pp. 263–275. Springer, Cham (2020). https://doi.org/10.1007/978-3-030-36150-1_22

5. Fgadaleta: top 4 reasons why reinforcement learning sucks (ep. 83) (2019). https://datascienceathome.com/what-is-wrong-with-reinforcement-learning/

6. Gonzalez, C., Dutt, V.: Instance-based learning models of training. In: Proceedings of the human factors and ergonomics society annual meeting, vol. 54, pp. 2319–2323. SAGE Publications Sage CA: Los Angeles, CA (2010)

7. Kotseruba, I., Tsotsos, J.K.: A review of 40 years of cognitive architecture research: core cognitive abilities and practical applications. arXiv preprint arXiv:1610.08602 (2016)

8. Kumar, S.: Balancing a cartpole system with reinforcement learning-a tutorial. arXiv preprint arXiv:2006.04938 (2020)

9. Meltzer, B., Michie, D.: Machine intelligence 4 (1970)

10. Mnih, V., et al.: Playing Atari with deep reinforcement learning. arXiv preprint arXiv:1312.5602 (2013)

11. Mnih, V., et al.: Human-level control through deep reinforcement learning. Nature **518**(7540), 529–533 (2015)

12. Mothanna, Y., Hewahi, N.: Review on reinforcement learning in cartpole game. In: 2022 International Conference on Innovation and Intelligence for Informatics, Computing, and Technologies (3ICT), pp. 344–349. IEEE (2022)

13. Nagendra, S., Podila, N., Ugarakhod, R., George, K.: Comparison of reinforcement learning algorithms applied to the Cart-Pole problem. In: 2017 International Conference on Advances in Computing, Communications and Informatics (ICACCI), pp. 26–32. IEEE (2017)

14. Schaal, S.: Is imitation learning the route to humanoid robots? Trends Cogn. Sci. **3**(6), 233–242 (1999)

15. Sunden, P.: Q-learning and deep Q-learning in OpenAI gym cartpole classic control environment (2022)

16. Surriani, A., Wahyunggoro, O., Cahyadi, A.I.: Reinforcement learning for cart pole inverted pendulum system. In: 2021 IEEE Industrial Electronics and Applications Conference (IEACon), pp. 297–301. IEEE (2021)

17. Sutton, R.S., Barto, A.G.: Reinforcement learning: an introduction MIT press. Cambridge, MA 22447 (1998)

18. Tan, Z., Karakose, M.: Optimized deep reinforcement learning approach for dynamic system. In: 2020 IEEE International Symposium on Systems Engineering (ISSE), pp. 1–4. IEEE (2020)

19. Wang, X., Gu, Y., Cheng, Y., Liu, A., Chen, C.P.: Approximate policy-based accelerated deep reinforcement learning. IEEE Trans. Neural Netw. Learn. Syst. **31**(6), 1820–1830 (2019)

20. Watkins, C.J.C.H., Dayan, P.: Q-learning. Mach. Learn. **8**, 279–292 (1992). https://doi.org/10.1007/BF00992698

CCTFv1: Computational Modeling of Cyber Team Formation Strategies

Tristan J. Calay[1], Basheer Qolomany[2], Aos Mulahuwaish[1(✉)],
Liaquat Hossain[2], and Jacques Bou Abdo[3]

[1] Saginaw Valley State University, University Center, Michigan, USA
`amulahuw@svsu.edu`
[2] University of Nebraska at Kearney, Kearney, USA
[3] University of Cincinnati, Cincinnati, USA

Abstract. Rooted in collaborative efforts, cybersecurity spans the scope of cyber competitions and warfare. Despite extensive research into team strategy in sports and project management, empirical study in cybersecurity is minimal. This gap motivates this paper, which presents the Collaborative Cyber Team Formation (CCTF) Simulation Framework. Using Agent-Based Modeling, we delve into the dynamics of team creation and output. We focus on exposing the impact of structural dynamics on performance while controlling other variables carefully. Our findings highlight the importance of strategic team formations, an aspect often overlooked in corporate cybersecurity and cyber competition teams.

Keywords: Cybersecurity · Cyber Competition · Collaborative Teams

1 Introduction

Teams, more than just collectives engaged in mutual tasks, present complex dimensions studied across various fields, including organizational sociology, anthropology, organizational behavior, industrial psychology, sociometry, and social network analysis [8,17,31]. Echoing Aristotle's philosophy that the whole is greater than the sum of its parts, comprehending teams is key to understanding the evolution and future trajectory of industrial revolutions [29]. Their potential manifests in enhanced performance, improved quality through error identification and recovery [28], increased productivity via labor distribution, and fostering innovation [25]. As such, optimizing team performance becomes a significant factor for the success of organizations, enterprises, large-scale endeavors, and competitive teams [9,11].

While teamwork can bolster performance, it may also introduce errors and dysfunction [28]. Factors such as power dynamics, leadership, culture, organizational processes, communication, cohesion, and team composition influence team performance [18]. This was demonstrated in FIFA's 2002 World Cup when the 1998 winning French team, despite retaining over 80% of its members [2,3], and led by the same captain, Zinedine Zidane, failed to win any match. This downfall is attributed to a change in team formation from a defensive 4-3-3 in 1998 to a

R. Thomson et al. (Eds.): SBP-BRiMS 2023, LNCS 14161, pp. 199–208, 2023.
https://doi.org/10.1007/978-3-031-43129-6_20

more balanced 4-2-3-1 in 2002, catching opponents off guard [1,4]. Such strategies of role configuration are ubiquitous in professional team formation and have been extensively studied in competitive sports and project management [10,21]. However, such exploration is scant in the cybersecurity domain.

This study aims to fill this gap and hypothesizes that team formation strategies significantly affect performance and victory probability. Utilizing computational modeling, we simulated a Blue team/Red team cyber competition, where teams were given two unique roles. The performance was assessed based on role configuration. To eliminate skill biases, we developed roles using Agent-Based Modeling, focusing on the impact of team formation strategies and network properties on performance.

This work serves as an initial stride in comprehending the performance of cyber teams, laying the groundwork for the development of the Collaborative Cyber Team Formation (CCTF). CCTF is a three-pillared framework that we are building to dissect the performance of cyber teams. The first pillar of CCTF is rooted in theoretical research that explores collaborative systems, particularly through the lenses of game theory, social learning, and complex systems. The second pillar hinges on the empirical analysis of cyber competitions coupled with subject matter expertise. The third pillar, to which this work contributes, draws on computational modeling.

2   Implementation Methodology

This work aims to assess the influence of team formation strategies on performance in cyber competitions, considering several influencing factors:

- **Team Size:** Larger teams tend to yield greater collective performance [23], indicating a direct proportionality between team size and performance.
- **Skills:** Teams with higher skills, encompassing talent [19] and training [27], typically outperform those with lesser skills, suggesting a direct correlation between members' skills and team performance.
- **Collaboration:** More collaborative teams generally perform better [20], implying a direct relationship between collaboration and performance. However, talent and collaboration may sometimes be inversely proportional [30].
- **Leadership:** Teams exhibiting superior leadership and trust usually perform better than those facing leadership crises [13].
- **Social Influence:** Performance can be swayed by social pressures within the team, like interpersonal dynamics [22], or external pressures, such as crowd influence [16]. This encompasses factors like motivation, stress, and discrimination.

To ascertain the impact of team formation distinct from these factors, the study design neutralizes or, when impractical, fixes these influences as detailed in Sect. 2.1. Moreover, performance is contingent upon specific goal attainment, thus, it is evaluated based on key predefined objectives, as elaborated in Sect. 2.1.

2.1 Study Design

In our study, a cyber competition is modeled as a defender team safeguarding a network against an attacker team. Following the models established in [14,15], the network is constituted of interconnected nodes, with only the peripheral nodes (internet-facing) being accessible to attackers initially. On successfully overtaking a peripheral node, attackers can target linked nodes. Figure 1(a) depicts such a network, where peripheral nodes in the yellow region are exposed to attack attempts (black arrows). Non-peripheral nodes in the green region are unreachable directly but become accessible (blue arrows) once a peripheral node is compromised (red arrow).

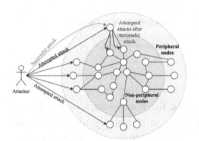

Attacker Roles	Hacking Steps as defined by EC-Council	Incident Handling Steps as defined by EC-Council	Defender Roles
	Step 5: Clearing Track		
Role 2: Exploiter	Step 4: Maintaining Access	Step 4: Act	Role 2: Interceptor
	Step 3: Gaining Access	Step 3: Decide	
Role 1: Scout	Step 2: Scanning	Step 2: Orient	Role 1: Detector
	Step 1: Reconnaissance	Step 1: Observe	

(a) Modeling the network and attackers

(b) Attacker and defender roles matched to EC-Council's hacking and incident handling steps

Fig. 1. Modeling study agents (Color figure online)

In our model, each attacker assumes one of two distinct roles: a "scout" or an "exploiter". The scout scans accessible nodes to detect vulnerabilities, with the probability of successful detection-given a node is vulnerable-set by the user and defined as:

$$P(detect \mid node\ is\ vulnerable) = P_{scout}$$

Once a scout identifies a vulnerability, exploiters are notified to attempt to exploit this vulnerability and control the compromised node. The probability of successful exploitation—given a node is vulnerable—is set by the user as:

$$P(exploit \mid node\ is\ vulnerable) = P_{exploiter}$$

Systematically, the scout performs the initial two hacking steps as defined by EC-Council [5], as illustrated in Fig. 1(b). In parallel, the exploiter is systematically characterized as the role executing steps 3 through 5. The attacker team consists of s scouts and $N-s$ exploiters, where N denotes the total team size. To mitigate the influence of "team size," as outlined in Sect. 2, this study maintains constant and equivalent team sizes for both attacker and defender teams.

Agent-Based Modeling (ABM) is employed to curtail the impact of "skills", "collaboration", and "social influence" (detailed in Sect. 2.2). Identical agents mitigate "skills", while broadcasting communication and anonymous interactions limit "collaboration" and "social influence", respectively. An ad-hoc team arrangement, with a user-determined size (N), helps manage the "leadership" factor. Success in a cyber-attack hinges on its objectives; this study outlines three unique objectives and corresponding metrics (Table 1).

Table 1. Measured Objectives

Index	Objective	Metric
1	Overtake as many nodes as possible	Portion of the network overtaken by the attackers. This metric is a decimal number ranging between 0 and 1.
2	Overtake the whole network	Whether the whole network was overtaken by the attackers. This metric is a Boolean.
3	Overtake the central nodes representing core services such as the database	Whether the central nodes were overtaken by the attackers. This metric is a Boolean

In our model, each defender can assume one of two distinct roles: a "detector" or an "interceptor". The detector scans the network for compromised (exploited or overtaken) or vulnerable nodes. The probability of successfully identifying an infected node—given that it is indeed infected—is set by the user and defined as:

$$P(detect \mid node\ is\ infected) = P_{detector-exploited}$$

while the probability of successfully identifying a vulnerable node-given that it is indeed vulnerable-is set by the user and defined as:

$$P(detect \mid node\ is\ infected) = P_{detector-vulnerable}$$

Upon detection of an infected node, interceptors are alerted to defend the network. The interceptor can either flag the infected node as untrusted—thus isolating it [12, 24, 26]—or recover the node based on user-defined action. The interceptor requires time $\Delta_{interceptor}$, set by the user, to carry out this action. The detector is systematically aligned with the first three steps of incident handling, as defined by EC-Council [6], as illustrated in Fig. 1(b). Similarly, the exploiter is systematically defined as the role that performs step 4. The team of defenders is composed of d detectors and $N - d$ interceptors. The same measures implemented to limit the impact of the factors defined in Sect. 2 on attacker team's performance are implemented to limit the impact on defender team's performance.

Hypothesis. As noted in Sect. 2.1, team performance, gauged by the three metrics, is influenced by the team formation strategies of both attackers and defenders. In this section, we posit that team formation strategies notably impact performance and the likelihood of victory. Given that team formation invariably precedes performance and potential confounding factors have been controlled, correlation analysis remains necessary for deciding whether to accept or reject the proposed hypothesis.

2.2 Study Parameters

Our methodology utilizes an Agent-Based Modeling approach featuring five agent types: Router, Scout, Exploiter, Detector, and Interceptor. These are illustrated in our NetLogo simulation, as shown in Fig. 3. While the network size and structure can be user-specified, for the rest of this paper, we are considering a scale-free network with 30 routers. Each input parameter can accommodate a broad spectrum of values; however, for demonstration, we have used incremental values iteratively as enumerated in Table 2.

It is notable that every tick triggers packet generation and routing table broadcasts, with the simulation offering real-time status updates to emulate a realistic network model. The workflow, outlined in Fig. 2, involves offline routers becoming active again once the simulation surpasses their shutdown delay. Mean-

Table 2. Input Parameters

Input Parameter	Description	Simulation Values
N	Total number of team members (N) for each side	10
S	Number of attacker scouts (S). $1 \leq S < N$. Exploiters: $N - S$	1–9
d	Number of defender detectors (d). $1 \leq d < N$. Interceptors: $N - d$	1–9
Vul_rate	Chance of a router becoming vulnerable when vulnerabilities are generated	2%
P_{scout}	Probability of attacker scout discovering a vulnerable router	100%
$P_{exploiter}$	Probability of exploiter to exploit an exploitable router	2%
$P_{detector-vulnerable}$	Probability of defender detector discovering a vulnerable router	25%, 50%, 75%, 100%
$P_{detector-exploited}$	Probability of defender detector discovering an exploited router	25%, 50%, 75%, 100%
$\Delta_{interceptor}$	Time needed to restore a router	10 ticks

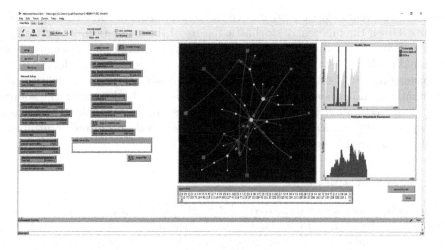

Fig. 2. NetLogo simulation snapshot

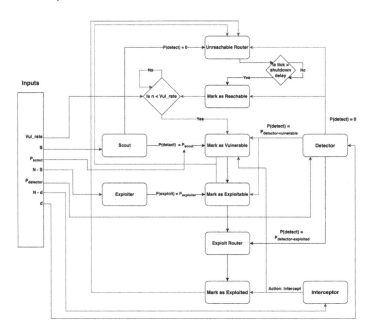

Fig. 3. NetLogo simulator workflow diagram

while, scouts and detectors can identify vulnerable routers, exploiters have the capacity to compromise routers, and interceptors can both shut down and rehabilitate routers.

3 Results and Analysis

Our study accommodates 1296 unique input combinations and corresponding outputs. Given the inherently stochastic nature of the simulation, we performed five trials for each setup to capture possible output variations. Each result is evaluated against the three objectives, as described in Sect. 2.1, employing three established metrics. As a result, we derived a dataset comprising 6480 distinct combinations and corresponding outcomes. This dataset, along with the source code [7] for reproducibility and validation, is publicly accessible. The results are represented graphically in Sect. 3.1. We subsequently posit and scrutinize a hypothesis using this assembled dataset in Sect. 3.2.

3.1 Sample Results

In this section, we showcase the derived dataset via graphical representation. All diagrams, as depicted in Figs. 4 and 5, employ the number of Exploiters $(N - s)$ as the y-axis and the number of Interceptors $(N - d)$ as the x-axis. The z-axis corresponds to the utilized metric for all configurations and tests.

A router is deemed compromised from the instant it is exploited until recovery initiation by an interceptor. Similarly, a router is classified as offline from

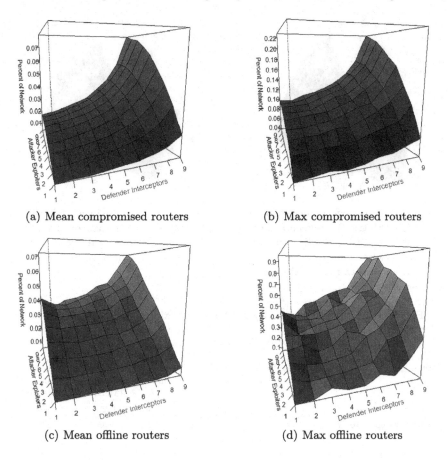

(a) Mean compromised routers   (b) Max compromised routers

(c) Mean offline routers   (d) Max offline routers

Fig. 4. Team performance measured following "Metric 1"

the commencement of its recovery (or that of its root router) until successful recuperation. Tracking both compromised and offline routers is paramount, as the former represents the infiltrated network and the success of the attackers, while the latter signifies the network's containment and the defenders' success in thwarting the attack. However, both are critical for "Metric 1" (defined in Sect. 2.1). Figure 4(a) illustrates the mean percentage of the network compromised throughout the test, while Fig. 4(b) displays the maximum percentage. Likewise, Figs. 4(c) and 4(d) represent the mean and maximum percentages of the offline network over the test duration, respectively. These four figures all utilize the same axes: number of interceptors $(N-d)$, number of exploiters $(N-s)$, and network percentage. These visuals collectively indicate that, per metric 1, the defenders' optimum performance is achieved with a formation of 8 detectors and 2 interceptors. However, this team structure doesn't consistently yield superior performance in other configurations.

In accordance with metrics 2 and 3, depicted in Figs. 5(a) and 5(b), respectively, the defenders' formation of 8 detectors and 2 interceptors does not consis-

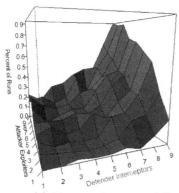
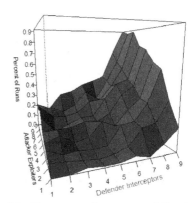

(a) Team performance following "Metric 2", percentage of the runs that resulted in the attackers compromising more than two-thirds of the network

(b) Team performance following "Metric 3", percentage of the runs that resulted in the attackers compromising the central router

Fig. 5. Team performance measured following "Metric 2" and "Metric 3"

Table 3. Performance Metrics

R Values	Mean Compromised	Max Compromised	Mean Offline	Max Offline	2/3 Network (Metric 2)	Center Wins (Metric 3)
Attacker Strategy	−0.47	−0.54	−0.74	−0.43	−0.28	−0.28
Defender Strategy	−0.67	−0.54	−0.27	−0.39	−0.36	−0.36

tently exhibit top performance. The ideal formation for defenders is contingent on the attackers' formation, an outcome anticipated from a game theory standpoint. It's also important to acknowledge that these results are based on the sample configuration and are furnished for illustrative purposes. A more comprehensive analysis will follow.

3.2 Hypothesis Testing

In Sect. 2.1, we hypothesized that team formation strategies impact performance and the likelihood of victory. To test this hypothesis, we compute the correlation coefficient (R values) to gauge the strength of the relationship between two variables, thereby allowing us to understand and evaluate their statistical association independent of any extraneous variables. Table 3 presents the correlation between the performance metrics (defined in Sect. 2.1, extracted from the generated dataset, and depicted in Figs. 4 and 5) and the Attacker/Defender strategies. In this study, our hypothesis is validated, as there exists a significant negative correlation between the strategies of the attackers/defenders and the

designated metrics. As the quantity of Attacker exploiters or Defender interceptors escalates, we observe a corresponding decline in metrics, and vice versa.

4 Conclusion

In this paper, we proposed that the formation strategy plays a crucial role in
the performance of cyber teams. To investigate this hypothesis, we adopted a
computational approach, and the results yielded significant findings that support
the acceptance of the hypothesis. Moving forward, our next endeavor involves
validating and potentially calibrating this computational model using empirical
data collected from a competition. To ensure the integrity of the study, the
competition will be organized as a controlled experiment, effectively minimizing
the influence of other factors that could introduce biases. This work serves as a
foundational component in the development of the Collaborative Cyber Team
Formation (CCTF) framework, which aims to integrate theoretical, empirical,
and computational approaches to comprehensively understand the performance
dynamics of cyber teams.

References

1. France 1998–2006. https://footballsgreatest.weebly.com/france-1998-2006.html
2. French national team 1998. https://www.fifa.com/tournaments/mens/worldcup/1998france/teams/43946
3. French national team 2002. https://www.fifa.com/tournaments/mens/worldcup/2002korea-japan/teams/43946
4. How France really won the world cup. https://www.aspeninstitute.org/blog-posts/how-france-really-won-the-world-cup/
5. What is ethical hacking?. https://www.eccouncil.org/cybersecurity/what-is-ethical-hacking/
6. What is incident response? https://www.eccouncil.org/cybersecurity/what-is-incident-response/
7. Collaborative Cyber Team Formation (CCTF) Simulation Framework. github repository (2023). https://github.com/Starwhip/CCTF-Framework
8. Bär, M., Kempf, A., Ruenzi, S.: Is a team different from the sum of its parts? Evidence from mutual fund managers. Rev. Finan. **15**(2), 359–396 (2011)
9. Bavelas, A.: A mathematical model for group structures. Hum. Organ. **7**(3), 16–30 (1948)
10. Budak, G., Kara, İ, İc, Y.T., Kasımbeyli, R.: New mathematical models for team formation of sports clubs before the match. Central Eur. J. Oper. Res. **27**, 93–109 (2019)
11. Cadima, R., Ojeda Rodríguez, J., Monguet Fierro, J.M.: Social networks and performance in distributed learning communities. Educ. Technol. Soc. **15**(4), 296–304 (2012)
12. Callaway, D.S., Newman, M.E., Strogatz, S.H., Watts, D.J.: Network robustness and fragility: Percolation on random graphs. Phys. Rev. Lett. **85**(25), 5468 (2000)
13. Dirks, K.T.: Trust in leadership and team performance: evidence from NCAA basketball. J. Appl. Psychol. **85**(6), 1004 (2000)

14. Dobson, G.B., Carley, K.M.: Cyber-FIT: an agent-based modelling approach to simulating cyber warfare. In: Lee, D., Lin, Y.-R., Osgood, N., Thomson, R. (eds.) SBP-BRiMS 2017. LNCS, vol. 10354, pp. 139–148. Springer, Cham (2017). https://doi.org/10.1007/978-3-319-60240-0_18

15. Dobson, G.B., Carley, K.M.: Cyber-fit agent-based simulation framework version 4. In: Center for the Computational Analysis of Social and Organizational Systems (2021)

16. Friesen, A.P., Wolf, S.A., van Kleef, G.A.: The social influence of emotions within sports teams. In: Feelings in sport, pp. 49–57 (2020)

17. Greco, L.M., Porck, J.P., Walter, S.L., Scrimpshire, A.J., Zabinski, A.M.: A meta-analytic review of identification at work: relative contribution of team, organizational, and professional identification. J. Appl. Psychol. **107**(5), 795 (2022)

18. Grove, J.R., Fish, M., Eklund, R.C.: Changes in athletic identity following team selection: self-protection versus self-enhancement. J. Appl. Sport Psychol. **16**(1), 75–81 (2004)

19. Gula, B., Vaci, N., Alexandrowicz, R.W., Bilalić, M.: Never too much-the benefit of talent to team performance in the national basketball association: comment on swaab, schaerer, anicich, ronay, and galinsky (2014). Psychol. Sci. **32**(2), 301–304 (2021)

20. Harris, C.M., McMahan, G.C., Wright, P.M.: Talent and time together: the impact of human capital and overlapping tenure on unit performance. Pers. Rev. **41**(4), 408–427 (2012)

21. Jahanbakhsh, F., Fu, W.T., Karahalios, K., Marinov, D., Bailey, B.: You want me to work with who? Stakeholder perceptions of automated team formation in project-based courses. In: Proceedings of the 2017 CHI Conference on Human Factors in Computing Systems, pp. 3201–3212 (2017)

22. Johnson, A.R., Van de Schoot, R., Delmar, F., Crano, W.D.: Social influence interpretation of interpersonal processes and team performance over time using Bayesian model selection. J. Manag. **41**(2), 574–606 (2015)

23. Mao, A., Mason, W., Suri, S., Watts, D.J.: An experimental study of team size and performance on a complex task. PLoS ONE **11**(4), e0153048 (2016)

24. Moore, C., Newman, M.E.: Epidemics and percolation in small-world networks. Phys. Rev. E **61**(5), 5678 (2000)

25. Osburg, T., Schmidpeter, R.: Social innovation. In: Solutions for a Sustainable Future, p. 18 (2013)

26. Sahimi, M.: Applications of Percolation Theory, vol. 213. Springer, Cham (2023). https://doi.org/10.1007/978-3-031-20386-2

27. Salas, E., et al.: Does team training improve team performance? A meta-analysis. Hum. Factors **50**(6), 903–933 (2008)

28. Sasou, K., Reason, J.: Team errors: definition and taxonomy. Reliabil. Eng. Syst. Saf. **65**(1), 1–9 (1999)

29. Senge, P.M., Carstedt, G., Porter, P.L.: Next industrial revolution. MIT Sloan Manag. Rev. **42**(2), 24–38 (2001)

30. Swaab, R.I., Schaerer, M., Anicich, E.M., Ronay, R., Galinsky, A.D.: The too-much-talent effect: team interdependence determines when more talent is too much or not enough. Psychol. Sci. **25**(8), 1581–1591 (2014)

31. Xu, H., Bu, Y., Liu, M., Zhang, C., Sun, M., Zhang, Y., Meyer, E., Salas, E., Ding, Y.: Team power dynamics and team impact: new perspectives on scientific collaboration using career age as a proxy for team power. J. Am. Soc. Inf. Sci. **73**(10), 1489–1505 (2022)

Simulating Transport Mode Choices
in Developing Countries

Kathleen Salazar-Serna[1,2]([⊠]) [iD], Lorena Cadavid[2] [iD], Carlos J. Franco[2] [iD],
and Kathleen M. Carley[3] [iD]

[1] Pontificia Universidad Javeriana, Calle 18 No. 118-250, Cali, Colombia
kathleen.salazar@javerianacali.edu.co
[2] Universidad Nacional de Colombia, Av. 80 No. 65-223, Medellin, Colombia
{kgsalaza,dlcadavi,cjfranco}@unal.edu.co
[3] Carnegie Mellon University, 4665 Forbes Ave, Pittsburgh 15213, USA
kathleen.carley@cs.cmu.edu

Abstract. Agent-based simulations have been used in modeling transportation systems to gain deeper understanding of travel behavior and transport mode choices. This study focuses on analyzing the factors that influence transportation mode decisions specifically in developing countries. As motorcycles are the preferred mode of transport in these economies, we have developed an agent-based model that includes two-wheeler vehicles as a transport alternative for commuters. Our model represents individuals who must make decisions regarding their choice of transport mode for their daily commute to work or school. These decisions are influenced by a combination of factors, including personal satisfaction, uncertainty about trying a new transport solution, and social comparisons of these two aspects within their social network. The model was ran using data from a Colombian city. The results show that our model represents the behavior of the system, which means in the absence of any policy intervention, the number of motorcycles and private cars will continue to increase in the coming years. This growth exacerbates negative impacts such as traffic congestion and road accidents, presenting significant challenges to the transportation system. Several key factors emerge as influential in the decision-making process. The time of travel and personal security considerations play a significant role, leading individuals to favor private transport alternatives over public transit. These findings underscore the need for targeted interventions that address these factors and promote sustainable and efficient modes of transportation.

Keywords: Agent-based simulation · Motorcycles · Social influence

1 Introduction

Agent-based simulation is a powerful modeling technique that takes a bottom-up approach, representing individual agents as interdependent decision-makers. These agents autonomously interact with each other and their environment,

R. Thomson et al. (Eds.): SBP-BRiMS 2023, LNCS 14161, pp. 209–218, 2023.
https://doi.org/10.1007/978-3-031-43129-6_21

adapting their behaviors and decisions accordingly. The collective actions and interactions of these individual agents within an agent-based system give rise to crowd behavior [1]. This modeling approach has gained widespread adoption globally, as it proves effective in capturing the complexities of systems such as transportation and studying individual travel behavior [2]. While a significant amount of research has been conducted in developed economies, it is important to note that motorcycles are often overlooked in favor of cars, bicycles, and public transportation [3,4]. However, in developing countries, motorcycles serve as the preferred mode of transportation for middle- and lower-class individuals due to their affordability, autonomy, and speed [5,6]. The increasing number of motorcycles on the roads in these countries contributes to issues such as traffic congestion, accidents, and pollution, necessitating the implementation of effective public policies [7]. It is crucial to note that policies derived solely from studies conducted in developed territories may not effectively address the dynamics of the system (the environment), since socioeconomic and cultural conditions significantly influences decision-making patterns. Therefore, further research is needed to understand how individuals make decisions regarding their choice of transportation mode and the social factors that impact those decisions. This paper presents results obtained with an agent-based simulation developed by Salazar et al. [7]. Although the model is not presented in this paper in detail, it represents the interactions of urban travelers as they make decisions regarding their primary mode of transport. The model includes motorcycles and incorporates parameters that capture realistic human behavior patterns within the specific socioeconomic context. A Colombian city is selected as a case study to analyze commuter behavior in developing countries. The objective of this research is to provide a testbed to identify the factors that influence transport mode choice, considering motorcycles as a full-fledged mode of transport. Ultimately, the study aims to inform the development of policies that can mitigate the negative impacts of the increasing motorcycle circulation on mobility.

2   Related Works

Agent-based simulations have emerged as a valuable tool for investigating a wide range of transportation issues. These applications consider driver-vehicle elements to simulate and assess decision-making behaviors under various traffic conditions, such as congestion, lane changes, and traffic light coordination [1]. In the realm of transport mode choice research, there are a few studies applying agent-based modeling; for instance, Faboya [9], introduces the Modal Shift (MOSH) framework, which analyzes the adaptive travel behavior of individuals commuting to and from a university. This framework considers multiple user groups, including public transit users, car drivers, bicyclists, and pedestrians. The findings highlight the significant impact of comfort on transport mode selection and emphasize the role of social interactions in influencing mode adoption. Another study by Kangur et al. [10] presents an agent-based model that explores consumer behavior and the large-scale interactions between consumers and the system in the context of electric vehicle adoption. The analysis encompasses gasoline, hybrid, and electric cars. Importantly, both of these studies incorporate the CONSUMAT approach [12], a consumer behavior model that considers agents' social and personal needs, as well as their levels of satisfaction and tolerance for uncertainty. By assessing satisfaction and uncertainty levels, individuals in these studies can employ various decision strategies, such as repetition, imitation, inquiry, and deliberation. This approach acknowledges the rational and calculative nature of human decision-making and emphasizes the engagement of individuals with similar travelers to gather information and reduce decision uncertainties. Nevertheless, the approach does not propose a specific way to calculate the satisfaction or the uncertainty and it is a generic consumer model that needs to be adapted to the transportation context.

Our model similarly to the MOSH approach, integrates the CONSUMAT to capture the interactions of interdependent agents who employ diverse decision-making strategies to select their transportation mode. In addition, we are considering motorcycles as an alternative to commute in a whole city and including additional influencing factors i the mode shift. In order to incorporate social influence into the decision-making process, we establish a social network that connects individuals with similar attributes, enabling them to gather information and make decisions in uncertain situations. As Carley [11] suggests, the cultural context plays a significant role in shaping the social structure and the likelihood of interactions among individuals. Our study extends previous works by proposing an approach that combines personal and collective experiences, weighted by the level of importance individuals assign to social engagement. This level of importance is parameterized based on the cultural dimensions studied by Hofstede et al. [14], specifically individualism and uncertainty avoidance. Additionally, our research includes motorcycles as a transportation alternative and examins travel behavior within the context of a developing country.

3   Simulation Model

In this section, we provide a high-level view of our agent-based model. The model simulate the transportation choices made by commuters during the peak hour as they travel from their homes to work. The available transport modes include motorcycles, cars, and public transit. Agents have attributes such as gender, age, and socioeconomic status. Individuals are distributed in the environment (the city) and located in one neighborhood that align with their socioeconomic condition and need to travel to the assigned workplace. The model is implemented with NetLogo 6.3.0 [15], in which time steps are counted in ticks. One tick corresponds to one minute of a peak hour in the real world.

3.1   Decision-Making Module

The conceptual model that guides agents in their mode choice decisions is explained in Fig. 1. Transport users select a mode of transportation for their journey from home to work. Subsequently, they assess their satisfaction with the journey based on a list of influencing factors. Simultaneously, agents check the uncertainty about the satisfaction that will obtain with their current transport mode. This uncertainty is a combination of their personal user experience and the experiences of their contacts within the social network, reflecting how individuals can be influenced by the experiences of their immediate peers. As agents accumulate more experience, their level of uncertainty decreases. Agents utilize both satisfaction and uncertainty to evaluate a mental model, which determines the strategy for the decision-making process and, consequently, the mode of transportation for the subsequent period. To implement the mental model, we employ the CONSUMAT approach [12] and the MOSH framework [9].

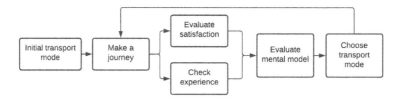

Fig. 1. Conceptual model of transport mode choice.

Similar to previous studies [9,13], our agents take into account various factors or needs to calculate their satisfaction. These factors encompass the cost of acquisition, operating cost, road safety, personal security during travel, travel comfort, commuting time, and pollution generated. Each of these factors holds a varying degree of importance for individuals. Consequently, the overall satisfaction is determined by a weighted sum of satisfaction with each specific need. The values associated with these needs are influenced by the state of the system, which is a result of individual and aggregate decisions made by the agents at each tick. Weights are established according to the socioeconomic and cultural context. For this purpose, we applied a survey in our case study city (See Supplementary Materials section).

3.2 Social Influence

The decision-making process of individuals is often influenced by their social connections or social network ties [9]. When faced with uncertain situations, people often seek information from others within their social network who have similar experiences. Previous research suggests that large social networks exhibit a power-law distribution, and this characteristic can be extended to the dynamics of transportation [10]. In our model, we establish connections between individuals who share similar conditions, forming a scale-free network within socioeconomic groups. Additionally, across these groups, some individuals are connected using the concept of a small world network [17]. Figure 2 illustrates the social network structure, using parameters specific to Cali, a selected Colombian city serving as a case study.

3.3   Validation

Data from Cali city was used to parameterized the model. Cali is located in the southern part of Colombia and has a population of 1.8 million people. The population of Cali is concentrated mainly in the middle class (41%) and low-income class (35%). Due to the impact of the pandemic on public transit usage, historical data of transport mode users prior to 2019 were utilized. The validation of the model was conducted using the "validation in parts" technique, as suggested by Carley [18]. This approach involves separately validating inputs, processes, and outputs. Inputs were validated to ensure they followed the distributional properties observed in the real system. Census data and official records were employed to parameterize the initial distribution of accident rates, fleet distribution, and agent properties such as age, gender, and socioeconomic status. The internal processes within the model were designed to resemble real-world processes, with qualitative alignment between model elements and their real-world counterparts. A conceptual model was initially defined based on the literature on transport modeling and human behavior theories. The code procedures were then incrementally validated in each module, including setup and go procedures. Validation techniques such as dimensional consistency, extreme value analysis, and structure validation, were employed. Furthermore, experts in agent-based simulation and transportation studies were consulted to validate the model's assumptions conceptually. For outputs, the model's mean predictions were compared to the general patterns of behavior observed in the real system or historical examples. Pattern modeling validation, was performed by configuring a tailored scenario that represented the conditions of the system in 2017. The proportion of people using private vehicles and public transportation was analyzed. As demonstrated in Fig. 3, the simulation results from 100 runs indicated that the emerging macro-level outputs were comparable to real-world patterns observed in subsequent years.

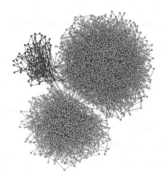

Fig. 2. Social network connecting agents.

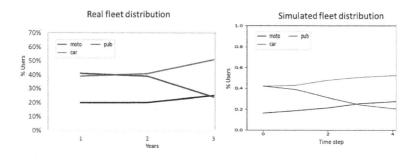

Fig. 3. Social network connecting agents.

4   Virtual Experiment

We conducted a virtual experiment using data from Cali city to analyze the shift in transport modes over time and the factors influencing this change. The model employs a scale of 1:1000 and simulates a synthetic population of 1,121 individuals, each characterized by their socio-economic attributes and an initial transport mode. These attributes follow distributions derived from our case study. Each agent is assigned thresholds for satisfaction and uncertainty, which are normally distributed, and they are connected to other individuals within the virtual society. The independent variables introduced into the model are listed in Table 1. The transport modes are parameterized based on technical specifications such as average efficiency, average CO2 emissions, average possible speed, and average costs. Historical data is utilized to calculate accident rates. The analysis period spans 10 years, equivalent to 600 ticks in NetLogo, where each tick represents one typical peak hour. Agents evaluate information annually to make decisions regarding mode changes. The simulation is run 100 times, and the average results are calculated.

Table 1. Model parameters.

Parameters
Socio-demographics
Initial fleet distribution
Satisfaction and Uncertainty thresholds
Mode costs
Accident and incident rates
Mode speeds
Mode emissions

Figure 4 illustrates the proportion of users by mode over the 10 analyzed periods. In the absence of any changes or policies introduced in the system, an increase in private vehicle usage and a continued decrease in public transit are expected. Two-wheeler registrations are projected to increase by 95%, resulting in an additional 215,000 motorcycles in the city. Car registrations would increase by 50%, adding approximately 228,000 more cars. As a consequence, the share of public transit would decline from 41% to 14%. This validates that the model represents the expected behavior from the real system and that later implementation of policies in the agent-based simulation can provide reliable results.

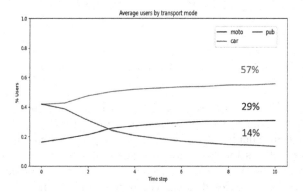

Fig. 4. Average users by transport mode.

5   Discussion and Future Work

As a result of the evaluation of the mental model, individuals employ different strategies to make decisions. In our model, agents are categorized into four groups based on the CONSUMAT approach. When agents' uncertainty fall below the threshold, the decision-making process is conducted individually, and depending on whether the satisfaction level is high or low, they may repeat the selection from the previous period. High levels of uncertainty prompt agents to engage with individuals in their social network. For agents with high satisfaction, the social comparison results in an imitation process where they adopt the most popular mode among their contacts. High uncertainty and low satisfaction values mean agents inquire in their network and compare their own satisfaction to the expected satisfaction if using the others' transport modes. If any of those expected satisfactions surpasses their current satisfaction level, the agent will opt for that alternative. Figure 5 demonstrates how individuals are classified into groups of decision-makers based on the strategies they employ to select a transport mode over time. As people begin transitioning from public transit to private alternatives, uncertainty increases (as shown in Fig. 6b), leading to a greater number of people seeking information from their peers. In contrast,

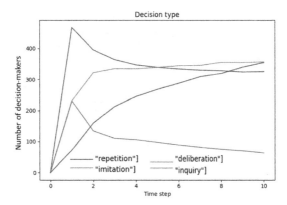

Fig. 5. Average satisfaction by transport mode.

motorcyclists tend to be more satisfied than users of other modes (see Fig. 6a), resulting in their tendency to repeat their mode selection.

The satisfaction calculated in every decision period depends on the values that each agent obtains for the transport mode attributes; some of them have a greater influence, for instance, the travel time and the personal security. As seen in Fig. 7, those variables disfavor the public transit. In contrast, accidents has a big impact on motorcyclist safety, but this factor is less important for individuals.

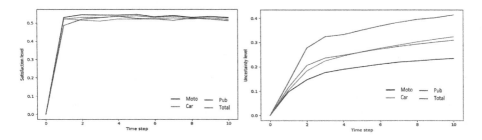

Fig. 6. a) Average satisfaction by transport. b) Average uncertainty by transport mode.

For future work, we intend to analyze the dynamics of mode shift within demographic groups. This facilitates the identification of factors for intervention through segmented policies. Based on this analysis, we will select some policies to implement in the model and study their impact on the system. Currently, the model does not include the road capacity, this limitation will be addressed in later work. This model can be parameterized with information for different cities, taking into account their economic, social, and cultural conditions. We hope this research can be a reference point for policymakers and contribute to the design of better transport systems in developing countries.

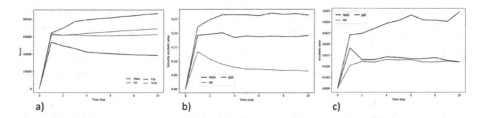

Fig. 7. a) Average travel time. b) Average insecurity rate. c) Average accident rate.

6   Conclusions

In this paper, we present results from an agent-based model that simulates the decision-making process of urban commuters and analyze the factors that influence their choice of transportation mode. Our study examines how commuters assess their satisfaction and uncertainty by comparing themselves to their social network, ultimately leading them to select one of three alternatives: cars, motorcycles, or public transit. The findings from our case study reveal that motorcycle users report higher levels of satisfaction and their numbers are projected to continue increasing at the highest rate among the alternatives examined. The primary reason behind this preference is the swift commuting experience offered by motorcycles, which is highly valued by users. Furthermore, the migration of public transit users to private transportation is driven by two key factors: the prolonged travel time and the heightened perception of personal insecurity. Despite the significant impact of road accidents, this factor is generally undervalued, particularly among motorcyclists, who exhibit the highest accident rate. By shedding light on these dynamics, our study serves as a starting point for the development of targeted public policies aimed at addressing mobility issues. Policymakers can leverage these insights to formulate strategies that cater to specific commuter groups based on their individual behaviors and preferences.

Acknowledgement. We thank CASOS at Carnegie Mellon University and the CBIE at Arizona State University for providing support during Kathleen Salazar's visit as a PhD student at Universidad Nacional de Colombia. Alike to the Fulbright Scholarship Program for making her visit possible.

   Supplementary materials Additional information can be found at:
   https://github.com/Kathleenss/SBP-Brims23-SupplementaryMaterials

References

1. Chen, B., Cheng, H.H.: A review of the applications of agent technology in traffic and transportation systems. IEEE Trans. Intell. Transp. Syst. **11**(2), 485–497 (2010)
2. Kagho, G.O., Balac, M., Axhausen, K.W.: Agent-based models in transport planning: current state, issues, and expectations. In: The 9th International Workshop on Agent-based Mobility, Traffic and Transportation Models, Methodologies and Applications, Proceedings, pp. 726–722 (2020)

3. Bakker, S.: Electric two-wheelers, sustainable mobility and the city. Sustainable Cities - Authenticity, Ambition and Dream (2019). https://doi.org/10.5772/intechopen.81460

4. Cadavid, L., Salazar-Serna, K.: Mapping the research landscape for the motorcycle market policies: sustainability as a trend-a systematic literature review. Sustainability **13**(19), 10813 (2021). https://doi.org/10.3390/su131910813

5. Eccarius, T., Lu, C.-C.: Adoption intentions for micro-mobility - insights from electric scooter sharing in Taiwan. Transp. Res. Part D: Transp. Environ. **84**, 102327 (2020). https://doi.org/10.1016/j.trd.2020.102327

6. Tanabe, R., Asahi, Y.: Analysis of trends of purchasers of motorcycles in Latin America. In: The 20th International Conference on Human Interface and the Management of Information, Las Vegas, Nevada, pp. 136–144 (2018)

7. Suatmadi, A.Y., Creutzig, F., Otto, I.M.: On-demand motorcycle taxis improve mobility, not sustainability. Transp. Res. Part D: Transp. Environ. **7**(2), 218–229 (2019). https://doi.org/10.1016/j.cstp.2019.04.005

8. Salazar, K., Cadavid, D., Franco, C.: Mode choice agent-based model. Universidad Nacional de Colombia. https://github.com/Kathleenss/SBP-Brims23-SupplementaryMaterials/blob/main/MotosV31-Apr23.nlogo. Accessed 4 Jul 2023

9. Ng, L.H.X., Carley, K.M.: Is my stance the same as your stance? A cross validation study of stance detection datasets. Inf. Process. Manage. **59**(6), 103070 (2022)

10. Faboya, O.T., Ryan, B., Figueredo, G.P., Siebers, P.-O.: Using agent-based modelling for investigating modal shift: The case of university travel. Comput. Ind. Eng. **139** (2020). https://doi.org/10.1016/j.cie.2019.106077

11. Kangur, A., Jager, W., Verbrugge, R., Bockarjova, M.: An agent-based model for diffusion of electric vehicles. J. Environ. Psychol. **166**(182) (2020). https://doi.org/10.1016/j.jenvp.2017.01.002

12. Jager, W., and Janssen, M.: An updated conceptual framework for integrated modeling of human decision making: The Consumat II. In: Workshop Complexity in the Real World @ECCS, pp. 1–18, Brussels, (2012). https://doi.org/10.1016/j.cie.2019.106077

13. Carley, K.: A theory of group stability. Am. Sociol. Rev. **56**(3), 331–354 (1991). https://doi.org/10.2307/2096108

14. Hofstede, G., Hofstede, G.J., Minkov, M.: Cultures and Organizations: Software of the Mind, vol. 2. Mcgraw-hill, New York (2005)

15. Wilensky.NetLogo, Center for Connected Learning and Computer-Based Modeling, Northwestern University. http://ccl.northwestern.edu/netlogo/. Accessed 4 Mar 2023

16. Wangsness, P., Proostb, S., Løvold, K.: Vehicle choices and urban transport externalities. Transp. Res. Part D. **86**, 102384 (2020). https://doi.org/10.1016/j.trd.2020.102384

17. Doerr, B., Fouz, M., Friedrich, T.: Why rumors spread so quickly in social networks. Commun. ACM **55**(6), 70–75 (2012)

18. Carley, K.: Validating Computational Models. CASOS technical report (2017). http://www.casos.cs.cmu.edu/publications/papers/CMU-ISR-17-105.pdf

User Identity Modeling to Characterize Communication Patterns of Domestic Extremists Behavior on Social Media

Falah Amro[(✉)] and Hemant Purohit

George Mason University, Fairfax, VA 22030, USA
famro@gmu.edu

Abstract. Years of online manifestos and mass killings by Domestic Violent Extremists (DVEs) in the U.S. and other nations indicate a grave global crisis. DVEs employ online propaganda to maximize their influence during societal crises and become the conduits to amplifying hate, organizing, and mobilizing to possibly influence non-extremists. User behavior modeling provides a promising approach to characterize behavioral patterns of DVEs and model the evolving identities of users. However, empirical research falls short to create explanatory user behavior modeling techniques for learning the classes of DVE user identity. Specifically, existing techniques lack measures to make sense of DVEs' online behavior relating to the well-known factors such as social relationships and beliefs that contribute to shaping one's identity. Guided by social science theories, primarily Social Identity Theory (SIT), this study proposes an explanatory and scalable user modeling approach that leverages unsupervised machine learning to analyze and model the identity classes of users. We model the identity classes of DVEs through a set of derived behavioral attributes–sociability, reach, and subjectivity from user profile metadata available in streaming social media data that helps capture the factors of social relations and attitudes partially but rapidly at scale and in near real-time. Using the data of Twitter conversations leading up to the 2021 January 6th U.S. Capitol attack, this study presents novel insights on how DVEs identities can be discovered and present better distinct analyses to understand their differences.

1 Introduction

The U.S. Office of the Director of National Intelligence (ODNI) released an unclassified intelligence report on Domestic Violent Extremists (DVEs) assessing that they pose a heightened threat in 2021 and the years to follow [1]. DVEs are not only motivated by ideologies, social relationships, and opinions, but they're galvanized by socio-political developments such as the narrative of fraud in the 2020 general election in the U.S., or the breach of the U.S. Capitol on January 6, 2021 [1]. The ODNI report shows that they have proliferated on social media and fomented socially divisive rhetoric. Furthermore, a recent report by the U.S. Congress [30] reveals that the actions of individuals who rise to violence are "the product of years of incitement, spread with stunning speed, scope, and impact on social media" (p.23). These warning signs from government agencies suggest that social media such as Twitter could provide a fertile environment for DVEs

© The Author(s), under exclusive license to Springer Nature Switzerland AG 2023
R. Thomson et al. (Eds.): SBP-BRiMS 2023, LNCS 14161, pp. 219–230, 2023.
https://doi.org/10.1007/978-3-031-43129-6_22

to gradually spread their toxic narrative, mobilize, and plan for extremist events, often unnoticed. According to a recent ADL survey [8], social media platforms such as Twitter enabled radicalized individuals to disseminate and consume radicalized content and amplify hate. In addition to amplifying hate, social media like Twitter has also given a platform for amplifying the voice of extremists to communicate, organize, and mobilize where they leverage the social platform to recruit non-extremists [3, 5].

Although past research on detecting extremist content and predicting DVEs behavior on social media is plentiful [7, 9], little empirical efforts are made to holistically model the identity of DVEs using explanatory modeling techniques. Furthermore, prior research not only lacks the psychological measures to make sense of DVEs' online behavior but also failed to investigate the potential association of their online activities to offline actions- links to real life and offline social crises (e.g., U.S capitol riot, January 6th). As DVEs threats are increasing, amongst other agencies, the U.S. Department of Homeland Security warned of repercussions of inadequate resources, including early warning systems supported by reliable and explanatory algorithms to distinguish DVE behaviors for intelligence analysts, to minimize the threat landscape from evolving on social media, and to potentially mobilize DVEs offline [1, 23]. To that end, modeling user identity using user behavior modeling techniques can be an important tool to rapidly distinguish behavioral patterns of DVEs based on the measures of social relationships and opinions, relying upon users' network structure, interaction behavior, and conversational behavior on social media.

Social Identity Theory (SIT), first developed by Henri Tajfel and John Turner [10] to examine intergroup relations and conflicts can be used to illuminate user behavior and communication patterns on social media. In describing how DVEs identity is shaped in the context of this study, one can argue that social categorization, the extent to which an individual self-identification with social groups determines the degree to which his behavior aligns with those within that social group [10]. Similarly, those who express positive emotions towards the group are more likely to express hostility towards outgroup members. Out-group hostility towards other groups usually results from intergroup extreme identification with a group [10]. Drawing from the SIT (Sect. 3.2), we hypothesize that the user network can be explained by the extent to which an individual network structure overlaps with social movement groups in which his or her behavior aligns with those within the group. SIT can also help operationalize interaction behaviors of DVEs where users who seek out violent behavior are more likely to act on content from their group by sharing it with others. Finally, SIT can also explain the conversational content of DVEs which can be measured by positive or negative emotions toward groups. If an individual shares positive emotions towards his group, he or she is likely to be subjective towards outgroup members. Our research questions are:

1. How can we model user identity based on interaction behavior in social media?
   a. What characterizes the patterns of different users' identity?

2 Literature Review

2.1 Unsupervised Learning for Identifying Extremists Users

Few studies used clustering algorithms to identify individual extremist behavior on social media. For example, in mapping ISIS ideology, Benigni [5] employs a graph-based vertex clustering algorithm to detect extremist individuals from Twitter data collected on ISIS supporters. A dataset of 22,000 Twitter accounts whose online behavior shows support for ISIS propaganda was used in the study. The study reveals a diverse range of user types including *fighters, propagandists, recruiters, religious scholars, and unaffiliated sympathizers*. The authors also classified user accounts into two categories; *member* and *non-member, and suspended*. Similarly, to detect extremist content and identify extremist users, Kursuncu et al. [37] also used clustering approaches such as hierarchical clustering over 538 users in the extremist dataset. The authors reveal two main clusters, one of which is where the majority of the users fall-they call this class of users *likely extremist users*. The other small cluster is called *likely outliers*. Finally, in an exhaustive analysis of extremist behavior study, Gaikwad et al. [39] applied clustering in part of their methodology to categorize topics of extremist content and then predicted the type of content. While the three studies attempt to identify extremist behavior, none relied on theory-driven approaches in identifying key behavioral dimensions nor were they contextualized in the Western social media context.

2.2 Supervised Learning for Detecting Hateful Rhetoric

Other studies used supervised machine learning to detect hateful rhetoric on social media like Twitter. For example, Agarwal et al. [12] attempted to classify content into hateful or not-hateful content from a labeled corpus of 10,486 tweets. The authors used hashtags such as #extremists and #Islam, and #islamophobia to collect tweets of extremist hate speech. The authors classify content into hate-supporting only if the data fits the hate class, otherwise, it is labeled unknown. Another study by Araque et al. [13] proposed a sentiment-based approach to classify text as *radical, non-radical*, or *neutral* to characterize ISIS extremist content. Finally, the authors use Linear SVM to classify the text into positive and negative. Ferrara et al. [14] applied supervised machine learning models, Logistic Regression, and Random Forest, on millions of tweets to characterize interactions of extremist users and identify extremist content. A set of 20,000 tweets of extremist users' accounts that were later suspended by Twitter were used in the experiment. The model achieved 93% accuracy in detecting extremist users, identifying users with extremist content, and finally predicting whether a user will interact with extremist content. Within this landscape of analysis, other studies focused on assessing the extremist ideology of Twitter users through employing features sharing, interaction, and influence. For example, Rowe et al. [15] used expert annotators to label tweets into *pro-ISIS* or *anti-ISIS*. In addition to content classification, the authors revealed further insights on the temporal radicalization behavior by measuring the time at which a user exhibited a radicalized behavior through the identification of pathways to radicalization, the radicalization peak, and post-radicalization stage. Although the study attempted to capture the temporal aspects of extremist identities, it does not associate online behavior with offline activities. These findings beg the need for further research to determine whether online social dynamics contribute to offline activities during societal crises.

3 Methodology

This section outlines the methodological approaches used in this study to address our research objective: analyze and model the identity classes of DVEs relying upon the user's network structure, interaction behavior, and conversational behavior on social media. Figure 1 below illustrates our proposed methodology for the study.

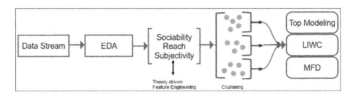

Fig. 1. A schematic representation of the user identity modeling approach.

3.1 Data Collection and Processing

We first leveraged the GDELT project [16] to identify active social movements in North America in the years 2020 and 2021. Supported by Google Jigsaw [17], the GDELT project monitors the world's broadcast, print, and web news in over 100 languages. We use the CAMEO codes (1451: Engage in violent protest for leadership change). From the preliminary analysis of the events, we identify the January 6, 202 riots on the U.S. Capitol as our context. Subsequently, we extracted Twitter data using the Twitter Academic API (Twitter API for Academic Research Product Track, 2021). We used the keyword-based crawling approach with the hashtag #StoptheSteal as our input for the data query. The data collection was constrained between 2020-12-28 and 2021-01-18 and resulted in 57,056 tweets. #StoptheSteal was the top used hashtag that covered the entire timeframe.

3.2 Theory-Driven Attributes of User Behavior

By building on the literature review and integrating insights from the SIT, we formulate the following attributes representing each user:

3.2.1 Sociability

The ratio of the number of following to followers of a user, to inform the user's social network structure, and is computed by:

$$\log\left(1 + \frac{1 + Following_{count}}{1 + Follower_{count}}\right) \tag{1}$$

The desire for sociability motivates users to interrelate with others on social media platforms. This desire explains why users seek to belong to other groups [21]. In fact, fulfilling the socializing desire is the primary reason why people use social media [22]. Additionally, users' sociability preserves and expands their social networks, where users can feel a sense of belonging [24]. On Twitter, a user can follow any other user of interest. The number of followers reflects the sociability and likeability of a user and is another manifestation of his or her status in the real world [25].

3.2.2 Reach

The reputation and influence of a user based on interactions from the reply and post counts, and is computed by:

$$\log\left(1 + \frac{(1 + a.RT_{count} + b.REPLY_{count})}{(1 + Post_{count})}\right) \quad (2)$$

where a and b are weights where we considered 1 for experiments. Influential users desire to be correct, which emboldens others and strengthens their feelings and morals [24]. According to Baek et al. [28, 29], contrary to the common belief that users share or retweet because they are stimulated by the content presented in front of them, users often react to the reputation of the source user. On Twitter, users can interact with each other in many ways such as commenting on others' tweets, directly messaging others, or disseminating others' tweets. Given that retweeting is not only disseminating information but also presenting one's self-beliefs to others, we argue that retweeting a friend is a strong indicator of their relationship, implying trust, endorsement, and support of claims made by that friend.

3.2.3 Subjectivity

Subjectivity analysis of the text is a part of sentiment analysis, where using Natural Language Processing (NLP), text was classified as opinionated or not opinionated. The subjectivity of every tweet is calculated using the sentiment analyzer tool in nltk (Natural Language Toolkit).

$$\log\left(1 + \frac{(1 + POS_{count} - NEG_{count})}{(1 + TOTAL_{LEN})}\right) \quad (3)$$

For users with multiple tweets, we log-transformed the sentiment score; then used aggregation to give an average subjectivity score for each user. Subjectivity analysis helps us determine whether users were tweeting emotional subjective opinions. Consistent with prior work [25, 27, 28], unless users are motivated by another user's influence, one way to amplify their messages is through new content. VaderSentiment provides an outcome that represents both polarity and subjectivity [29]. In examining how message content affects users' interactions online, the characteristics of the content, such as the amount of sentiment, argument quality, and information quality, have significant impacts on retweeted times of a message or individual retweeting decisions [25, 28, 30].

We experimented with several other combinations of additional theory-based attributes to identify the best input combination that yields the highest Silhouette score.

Namely, *Influence*: the number of retweeted times to represent other people's opinions about the tweet. [31], *Favorability*: ratio of the number of favorites to the total number of tweets; it informs higher engagement in contrast to just posting tweets [31–35], *Survivability*: the potential active existence on the platform over time, and it is measured as the difference between current timestamp and the timestamp at which a tweet is created [31], *Information Richness*: number of URLs and Hashtags in the tweet [31, 35, 36], and *Source Attractiveness*: number of followers the tweet's author has [31]. Subsequently, we determine that the best attributes in terms of producing the most efficient and rapid clustering outcome are *sociability, reach, and subjectivity*. These attributes allow us to effectively capture the similarities and differences between DVE users and cluster them based on their network structure, interaction behavior, and conversational behavior.

3.3 Identifying Behavior Classes via Unsupervised Learning

We experimented with different types of clustering algorithms; namely, DBSCAN [21] and K-means [11] where the optimal number of clusters is determined by the Elbow method [11]. Based on Calinski-Harabasz, we determine KMeans as the most suitable technique. We feed the KMeans algorithm with our theory-based behavioral attributes relevant to interactions of DVEs (sociability, reach, and subjectivity). KMeans clustering can group users based on these attributes and identify whether they have similar identities.

4   Results and Discussion

4.1   Cluster Performance Evaluation

We evaluated the performance of KMeans clustering using three metrics: Silhouette coefficient (0.36), Calinski-Harabaz index (1791.40), and Davies-Bouldin index (0.61).

4.2   Visualization of User Behavior Categories

We use our pre-identified key behavior dimensions to visualize the user behavior space as shown in Fig. 2, representing how the user features can vary by cluster.

Fig. 2. Visualization of user behavior classes (Cl1 n = 10274, Cl2 n = 4726, Cl3 n = 7446).

4.3 Cluster Topical Analysis

We first apply topic modeling based on Latent Dirichlet Allocation (LDA) to the entire dataset to better understand the semantic meaning of the obtained clusters. Based on the perplexity metric (0.39), the number of suitable topics for our analysis was 3. We also evaluated results from 3 to 9 latent topics and found that topics become similar and redundant after 3. First, we discuss the dominant topics for the entire dataset as presented in Table 1 below.

Table 1. Dominant topics

Topic	Top Keywords
Topic 1	support, america, like, know, let, going, need, please, want, back
Topic 2	trump, realdonaldtrump, election, president, fraud, maga, fightfortrump, january, votes, get
Topic 3	people, biden, fight, one, time, joebiden, would, patriots, see, world

In topic 1, we observe discourse representative of users' patriotism. In topic 2, we observe a heavy focus on Trump, realdonlad-trump, and the president. Contextual analysis of the terms election, fraud, and votes reveals that users started casting doubt on the election results after President Trump lost the election. Similarly, the rhetoric framing of the rallies also appears in terms fightfortrump, january, get, maga. In topic 3, discourse appears to be directed at Biden and JoeBiden. These discussions reveal users' attempts to disrupt the voting approval of Biden, and prevent him from being sworn as the next president.

Next, we focus our attention on examining how users in each cluster discuss the aforementioned topics and determine whether divergence in terms used existed in the discourse across clusters. Table 2 depicts the comparative analysis of the topical terms, based on term frequency and distribution, for the entire dataset separated by clusters. We observe different patterns in the terms used by users within each cluster to discuss the overarching topics. Users in Cl1 predominantly use terms related to the election corruption-oriented rhetoric, most represented by right-wing news channels. In Cl2 we observe users focus on voting fraud-oriented rhetoric, discussing alleged election fraud and corruption, expressing belief in supporting evidence, and emphasizing the need to fight for Donald Trump. Finally, in Cl3, users adopt a rallies-oriented rhetoric, calling for rallies against perceived rigged elections.

4.4 Moral Values Analysis

We leverage the Moral Foundations Dictionary (MFD) [35] to determine if motivational differences exist between our clusters. We evaluate clusters based on the dimension of *authority*. That is, how do users in each cluster prioritize virtues related to social order and the obligations of hierarchical leadership, including deference to legitimate authority and respect for traditions [35]? We conducted a t-test to compare the mean

Table 2. Topical cluster of terms extracted from messages across clusters in discussing dominant topics. Cl1 uses terms related to election corruption, Cl2 uses terms related to voting fraud, and Cl3 uses terms related to framing rallies.

	Cluster 1 (Cl1)	Cluster 2 (Cl2)	Cluster 3 (Cl3)
Topic 1	corrupt, corruption, daily, entire, newsmax	anything, communist, democrat, donald, elected, elections, evidence, family, fighting, god, going, great, hawleymo, hear, knows, lost, making, marshablackburn, money, proof, ros, wants	america, anyone, bill, bless, care, dead, democrats, gatewaypundit, gopleader, government, lead, lying, nation, officials, old, party, patriot, patriots, power, thanks, veto, voter, work
Topic 2	across, scotus, trying, world	actually, agree, allow, americafirst, around, biden, cheaters, covid, dems, joe, justice, look, made, numbers, office, point, put, real, rigged, scott, Santacruz, sure, tedcruz, trumps, win	best, better, calls, campaign, done, enough, erictrump, even, fightback, free, friends, latest, political, protest, rigge delection, run, senategop, trumpwon, united, usa, use
Topic 3	amacforamerica	ballots, country, election, electors, fightfortrump, first, fraudulent, history, hope, join, landslide, law, lindseygrahamsc, man, martial, must, needs, never, people, please, president, rally, realdonaldtrump, right, see, senator, special, stand, states, take, tell, think, time, true, vote, voting, watch, wethepeople	big, christmas, congress, fair, fight for america, fight like aflynn, fraud, media, retweet, senators, thank

differences in authority between the three clusters. Overall, as shown in Fig. 3, the three clusters exhibit statistically different activation levels of authority ($p < .000$). The results revealed significant differences between Cl1 and Cl2 ($p < .000$), Cl1 and Cl3 ($p < .000$), and moderate differences between Cl2 and Cl3 ($p = 0.051$).

4.5 Psychometric Analysis

We investigate if different psychological patterns exist in the identified behavioral classes. We apply the Linguistic Inquiry Word Count (LIWC) [36] to collect the LIWC category scores for each cluster's tweets on the category of affiliation and drives. LIWC counts words in psychologically meaningful categories for a given text. Figure 4 below,

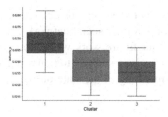

Fig. 3. Average authority foundation per cluster.

users in Cl1 tend to use messages with lesser drives and affiliation-related terms in general. Second, unlike Cl1, users in Cl2 tend to write with greater expressions of drives and affiliation terms in their messages. According to Fig. 4 below, messages of users in Cl2 tend to have greater expressions of anger than those in Cl1 but lower than those in Cl3. Finally, expressions of affiliation and drives are strongly manifested in Cl3 users.

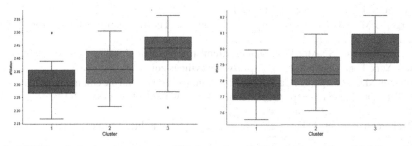

Fig. 4. LIWC analysis showing average affiliation values (left) and drives values (right) per cluster.

Using a t-test, we observe a statistically significant difference between clusters in both categories ($p < .000$) (Fig. 4). The t-test results using cluster pairwise comparison indicate the differences in the "drives" values between different clusters. Specifically, the p-values for the t-tests between Cl1 and Cl2, Cl1 and Cl3, and Cl2 and Cl3 are 0.024, 0.000, and 0.000, respectively. Similarly, the t-test results using cluster pairwise comparison for the "affiliation" values also show statistically significant differences between different clusters. The p-values for the t-tests between Cl1 and Cl2, Cl1 and Cl3, Cl2 and Cl3 are 0.026, 0.000, and 0.000, respectively.

5   Conclusion

We presented a novel explanatory approach to model the identity classes of DVEs through a set of derived behavioral attributes - sociability, reach, and subjectivity. Using KMeans clustering, we demonstrated that DVEs behavior can be explained by three distinctive clusters, namely, election corruption rhetoric, voting fraud-oriented rhetoric, and rallies-oriented rhetoric. Results for the DVEs identity classes indicate that a user's sociability contributed most to the prediction of its DVE identity class, followed by reach. The application of the Moral Foundations and LIWC determined motivational differences and psychological patterns exist between our clusters, suggesting DVEs identities can be discovered efficiently relying upon their social network structure, interaction behavior, and conversational behavior. Our future work will incorporate large language models and additional behavioral dimensions such as toxicity and semantic network analysis to provide a holistic identity mapping of the state of extremism and the mind frame of extremists today.

Acknowledgment. This work was supported by the Office of the Director of National Intelligence (ODNI) under Award No. ICPD-2021-01. Opinions, interpretations, conclusions, and recommendations are those of the authors.

References

1. U.S. Department of Homeland Security, 18 October 2020. https://www.dhs.gov
2. Gupta, A., Joshi, A., Kumaraguru, P.: Identifying and characterizing user communities on Twitter during crisis events. In: DUBMMSM (2012)
3. Berger, J.M.: Extremist Construction of Identitys. The International Center for Counter-Terrorism – The Hague 8, no. 7 (2017)
4. Cai, W., Landon, S.: Attacks by white extremists are growing. So are their connections. The New York Times, 3 April 2019. https://www.nytimes.com
5. Benigni, M.C., Joseph, K., Carley, K.M.: 'Online extremism and the communities that sustain it: Detecting the ISIS supporting community on Twitter.' PLoS ONE **12**(12), 1–23 (2017). https://doi.org/10.1371/jour-Nal.pone.0181405
6. MacFarquhar, N., et al.: Far-Right Group That Trades in Political Violence Gets a Boost. New York Times, 30 September 2020
7. Fernandez, M., Asif, M., Alani, H.: Understanding the roots of radicalization Twitter. In: Proceedings of the 10th ACM Conference on Web Science, pp. 1–10, May 2018
8. ADL Survey: On-line Hate. https://www.adl.org/online-hate-2020
9. Mashechkin, I.V., Petrovskiy, M.I., Tsarev, D.V., Chikunov, M.N.: 'Machine learning methods for detecting and monitoring extremist information on the Internet. Program. Comput. Softw. **45**(3), 99–115 (2019). https://doi.org/10.1134/S0361768819030058
10. Tajfel, H., Turner, J.C.: The social identity theory of intergroup behavior (1986). https://student.cc.uoc.gr/uploadFil_SIT_xs.pdf. Accessed 10 Jul 2022
11. Likas, A., Vlassis, N., Verbeek, J.J.: The global k-means clustering algorithm. Pattern Recogn. **36**(2), 451–461 (2003)
12. Kursuncu, U., Purohit, H., Agarwal, N., Sheth, A.: When the bad is good and the good is bad: understanding cyber social health through online behavioral change. IEEE Internet Comput. **25**(1), 6–11 (2021). https://doi.org/10.1109/MIC.2020.3045232
13. Araque, O., Iglesias, C.A.: An approach for radicalization detection based on emotion signals and semantic similarity. IEEE Access **8**, 17877–17891 (2020). https://doi.org/10.1109/ACCESS.2020.2967219
14. Ferrara, E., Wang, W.-Q., Varol, O., Flammini, A., Galstyan, A.: Predicting online extremism, content adopters, and interaction reciprocity. In: International Conference on Social Informatics (2016)
15. Rowe, M., Saif, H.:. Mining pro-ISIS radicalisation signals from social media users (2016). https://www.aaai
16. Amazon Web Services AWS (n.d.): Global Database of Events, Language and Tone (GDELT). https://registry.opendata.aws/gdelt

17. Google (n.d.). Jigsaw. https://jigsaw.google.com/
18. Purohit, H., Pandey, R.: Intent mining for the good, bad, and ugly use of social web: concepts, methods, and challenges. In: Agarwal, N., Dokoohaki, N., Tokdemir, S. (eds.) Emerging Research Challenges and Opportunities in Computational Social Network Analysis and Mining. LNSN, pp. 3–18. Springer, Cham (2019). https://doi.org/10.1007/978-3-319-941 05-9_1
19. Mirrlees, T.: The alt-right's discourse on "cultural marxism": a political instrument of intersectional hate. Journals.msvu.ca (2018). https://journals.msvu.ca/index.php/atlantis/article/view/5403. Accessed 18 Jul 2021
20. López-Sánchez, D., Corchado, J.M., González Arrieta, A.: Dynamic detection of radical profiles in social networks using image feature descriptors and a case-based reasoning methodology. In: Cox, M.T., Funk, P., Begum, S. (eds.) ICCBR 2018. LNCS (LNAI), vol. 11156, pp. 219–232. Springer, Cham (2018). https://doi.org/10.1007/978-3-030-01081-2_15
21. Marutho, D., Handaka, S.H., Wijaya, E.: The determination of cluster number at k-mean using elbow method and purity evaluation on headline news. In: Proceedings of the 2018 International Seminar on Application for Technology of Information and Communication, pp. 533–538 (2018). https://doi.org/10.1109/ISEMANTIC.2018.8549751
22. Liu, Z., Liu, L., Li, H.: Determinants of information retweeting in microblogging. Internet Res. **22**(4), 443–466 (2012)
23. Ravndal, J.A., Bjorgo, T.: Investigating Terrorism from the Extreme Right (2018). https://www.universiteitleiden.nl
24. Suh, B., Hong, L., Pirolli, P., Chi, E.H.:. Want to be retweeted? Large scale analytics on factors impacting retweet in Twitter network. In: IEEE Second International Conference on Social Computing (Socialcom), pp. 177–184 (2010). https://doi.org/10.1109/SocialCom.201 0.33
25. Lazarsfeld, P.F., Merton, R.K.: Friendship as a social process: a substantive and methodological analysis. Freedom Control Mod. Soc. **18**(1), 18–66 (1954)
26. Hutto, C.J., Gilbert, E.E.: VADER: a parsimonious rule-based model for sentiment analysis of social media text. In: Proceedings of the Eighth International Conference on Weblogs and Social Media (ICWSM 2014), vol. 8(1), pp. 216–225 (2014). https://doi.org/10.1609/icwsm.v8i1.14550
27. Stieglitz, S., Dang-Xuan, L.: Emotions and information diffusion in social media sentiment of microblogs and sharing behavior. J. Manag. Inf. Syst. **29**(4), 217–248 (2013)
28. Baek, K., Holton, A., Harp, D., Yaschur, C.: The links that bind: uncovering novel motivations for linking on Facebook. Comput Hum Behav. **27**(6), 2243–2248 (2011)
29. Shi, J., Hu, P., Lai, K.K., Chen, G.: Determinants of users' information dissemination behavior on social networking sites: an elaboration likelihood model perspective. Internet Res. **28**(2), 393–418 (2018). https://doi.org/10.1108/IntR-01-2017-0038
30. Gaikwad, M., Ahirrao, S., Phansalkar, S., Kotecha, K.: Online extremism detection: a systematic literature review with emphasis on datasets, classification techniques, validation methods, and tools. IEEE Access **9**, 48364–48404, (2021). Congress.gov. https://www.congress.gov/event/117th-congress/house-event/LC65965/text?
31. Ma, L., Sian Lee, C., Hoe-Lian Goh, D.: Understanding news sharing in social media. Online Inf. Rev. **38**(5), 598–615 (2014)
32. Chu, S.-C., Kim, Y.: Determinants of consumer engagement in electronic word-of-mouth. Int. J. Advert. **30**(1), 47–75 (2011)
33. Nagarajan, M., Purohit, H., Sheth, A.P.: A qualitative examination of topical tweet and retweet practices. In: Proceedings of the Fourth International AAAI Conference on Weblogs and Social Media, vol. 2(10), pp. 295–298 (2010). https://doi.org/10.1609/icwsm.v4i1.14051

34. An, D., Zheng, X., Rong, C., Kechadi, T., Chen, C.: Gaussian mixture model based interest prediction in social networks. In: Proceedings of the 2015 IEEE 7th International Conference on Cloud Computing Technology and Science, pp. 196–201 (2015)

35. Graham, J., Haidt, J., Nosek, B.A.: Liberals and conservatives rely on different sets of moral foundations. J. Pers. Soc. Psychol. **96**(5), 1029–1046 (2009). https://doi.org/10.1037/a00 15141

36. Francis, M.E., Booth, R.J.: LIWC: Linguistic Inquiry and Word *Count*. Southern Methodist University (1993)

37. Kursuncu, U., et al.: Modeling islamist extremist communications on social media using contextual dimensions: religion, ideology, and hate. Proc. ACM Hum.-Comput. Interact. CSCW **3**, Article 151 (2019). 22 p. https://doi.org/10.1145/3359253

Understanding Clique Formation in Social Networks - An Agent-Based Model of Social Preferences in Fixed and Dynamic Networks

Pratyush Arya$^{(\boxtimes)}$ and Nisheeth Srivastava

Indian Institute of Technology Kanpur, Kanpur 208016, India
`parya22@iitk.ac.in`

Abstract. This paper presents results from *in silico* experiments trying to uncover the mechanisms by which people both succeed and fail to reach consensus in networked games. We find that the primary cause for failure in such games is preferential selection of information sources. Agents forced to sample information from randomly selected fixed neighborhoods eventually converge to a consensus, while agents free to form their own neighborhoods and forming them on the basis of homophily frequently end up creating balkanized cliques. We also find that small-world structure mitigates the drive towards consensus in fixed networks, but not for self-selecting networks. Preferentially attached networks appear to show the highest convergence to one color, thereby showing resilience to balkanization of opinion in self-selecting networks. We conclude with a brief discussion of the implications of our findings for the representation of behavior in socio-cultural modeling.

Keywords: Social preference · preference learning · agent-based modeling · clique formation · balkanization · filter bubbles · polarization

1 Introduction

In the ever-evolving digital landscape of the 21st century, we confront socio-cognitive divides shaped by polarization, filter bubbles, and clique formation. Polarization reflects the increasing divergence of societal and political viewpoints, fragmenting ideological landscapes into opposing extremes [1]. In the realm of social media, this phenomenon is amplified by algorithmic personalization, transforming it into a potent force that drives societies towards divisiveness [2]. In the same vein, filter bubbles, a term birthed by Eli Pariser [3], encapsulate the unsettling reality of intellectual isolation, which now pervades the World Wide Web. Algorithmically generated digital echo chambers present users with content that aligns with their preexisting preferences, reinforcing self-confirming information loops. Clique formation materializes as individuals, sharing common attributes or beliefs, clustering together in cyberspace, resulting in islands of homogeneity [4]. Homophily, an age-old sociological phenomenon, has experienced an exponential surge due to the reduction of friction in communication in cyberspace, exerting profound influences on society, politics, and cognitive processes [5].

Previous efforts to comprehend and address these phenomena have predominantly adopted social and cultural perspectives, examining how societal structures, media environments, and cultural contexts shape their manifestations [1,4]. However, there has been a noticeable lack of focus on how these phenomena impact and are influenced by individual cognitive and information processing mechanisms. This gap in the literature signals an uncharted frontier in our understanding of polarization, filter bubbles, and clique formation. The intricate interaction between external stimuli and internal cognitive processes is at the heart of how individuals navigate their social and informational environments. As such, understanding these phenomena from an information processing standpoint is crucial to understand why and when polarization is likely to result in networks of individuals.

In the context of this paper, we operationalize networks of individuals as graphs produced by three different mechanisms, two of which make sociological assumptions: Erdos-Renyi (ER), Barabasi-Albert (BA), and Watts-Strogatz (WS). ER graphs, characterized by random connections between nodes, offer a baseline mathematical graph model, with no socio-cultural appurtenances, for studying network dynamics. On the other hand, BA graphs, generated via a preferential attachment mechanism, exhibit power-law degree distributions, meaning the probability of encountering highly connected nodes is relatively higher. This aligns with the structure of social media and other digital networks, where influential individuals gather larger followings. BA graphs thus have a clear sociological connotation and provide insights into the dynamics of online communities wherein low social friction easily permits large inequalities in the degree distribution of connectivity between individuals. WS graphs, with their small-world architecture, strike a balance between local clustering and global connectivity, reflecting networks in the real-world wherein higher social friction reduces the range of degree distributions accessible to individuals. The small-world property further reflects the interconnected nature of real-world social networks, wherein individuals can establish connections with others through short paths, akin to the "six degrees of separation" concept, without requiring to make a large number of direct connections personally. We examine how social preferences vary over the course of networked consensus games in all three categories of graphs in this paper.

2   Empirical Background for this Work

The motivation, and the empirical background, for this work comes primarily from Michael Kearns' paper, "Behavioral Experiments on a Network Formation

Game" [6]. The paper talks about a series of behavioral experiments where 36 human participants had to solve a competitive coordination task (of biased voting) for monetary compensation. Communication, in these games, happens only via the game GUI, and only with individuals in one's assigned social neighborhood. It has been found that in such cases, where the social neighborhoods are explicitly fixed, and participants are then asked to achieve a collective goal, human participants tend to perform well - subjects are able to extract almost 90% of the value that is available to them in principle. This has led researchers to conclude that humans are quite good at solving a variety of challenging tasks from only local interactions in an underlying network [7].

However, when Kearns made a slight change to the game, human performance deteriorated. The slight change entailed participants having to build the network during the experiment, via individual players purchasing links whose cost is subtracted from their eventual task payoff. A striking finding is that the players performed very poorly compared to behavioral experiments in which network structures were imposed exogenously. Despite clearly understanding the biased voting task, and being permitted to collectively build a network structure facilitating its solution, participants instead built very difficult networks for the task. This finding is in contrast to intuition, case studies and theories suggesting that humans will often organically build communication networks optimized for the tasks they are charged with, even if it means overriding more hierarchical and institutional structures [8,9].

These results suggest that humans are able to achieve a collective goal if a network structure is imposed on them, and they are restricted to communicating within the fixed neighborhood itself; however, when they are free to choose people to communicate with, instead of selecting people that will maximize the chances of global coordination, human participants end up building sub-optimal networks and fail to coordinate effectively.

3 Social Preference Formation

Central to our model is the assumption that the inference of social preferences occurs through the same information processing mechanisms as the inference of individual preferences. Building upon this assumption, our account relies on two specific information-processing assumptions.

Firstly, we embrace the principle of inductive inference, which posits that individuals make decisions by inferring what to do based on their past choices involving similar options. In our model, agents exhibit this inductive reasoning by updating their color preferences based on previous interactions and outcomes, thereby gradually adjust their preferences over time, resulting in the emergence of distinct color clusters.

Secondly, our model incorporates the concepts of memory growth and memory decay. Inspired by the workings of human memory, we assume that agents' memories of past interactions can both strengthen and fade. Memory growth reflects the reinforcement of memory traces associated with interactions that led to similar color preferences, promoting the formation of social ties with like-minded individuals. On the other hand, memory decay represents the natural process of forgetting, allowing agents to adapt and respond to changing social dynamics. These memory dynamics contribute to the evolution of the network structure and the emergence of distinct color clusters in the dynamic network case.

By integrating inductive inference, memory growth, and memory decay into our model, we aim to provide a more comprehensive understanding of how cognitive processes shape social behavior. While our model is a simplified representation of complex human decision-making, it offers insights into the mechanisms underlying social preferences and network dynamics.

3.1   Preference Inference per Iteration

There is now substantial evidence to believe that inductive inference underpins the construction of several (if not all) mental attributes [10]. This Bayesian approach to cognition was recently applied to the problem of preference learning [11]. Following their notation, an agent's preference for an option is identical to the probability that it is desirable, $p(r|x)$, and can be calculated by summing out across evidence of desirability observed in multiple contexts,

$$p(r|x) = \frac{\sum_{c \in C} p(r|x, c)p(x|c)p(c)}{\sum_{c \in C} p(x|c)p(c)} \tag{1}$$

Here C is the set of all contexts offering x as a possible choice. The desirability probability $p(r|x, c)$ simply considers the frequency with which the agent had previously preferred option x in context c, the option probability $p(x|c)$ expresses the frequency with which the option x is observed in context c, and the context probability $p(c)$ expresses the base rate of context c in the agent's environment.

3.2   Memory Growth and Memory Decay Through Iterations

In the context of the model, memory decay and memory growth are parameters that control how the memory matrix evolves over time. The role of these parameters comes in particularly in the case of dynamic network.

Memory decay signifies the gradual decrease in the strength of an agent's memory of past interactions. It models the natural forgetting process in human memory. A higher memory decay rate means that memories of past interactions fade more quickly, while a lower decay rate means that memories persist for a longer time.

$$new\_memory = memory \times (1 - memory\_decay) \tag{2}$$

Memory growth, on the other hand represents the strengthening of an agent's memory of past interactions that have led to similar color preferences. It captures the idea that repeated experiences of similarity reinforce memory traces. A higher memory growth rate means that agents are more likely to remember and interact with agents who have similar color preferences, while a lower growth rate means that memory is less influenced by past interactions.

$$new\_memory[i,j] = memory[i,j] +$$
$$(similar\_preferences[i,j] \times memory\_growth) \quad (3)$$

where:

- new_memory[i, j] is the updated memory value for agent i's memory of agent j,
- memory[i, j] is the previous memory value for agent i's memory of agent j,
- similar_preferences[i, j] is a measure of the similarity between agent i's and agent j's color preferences,
- memory_growth is a parameter controlling the rate at which memory is reinforced.

We introduce an exponential decay factor to the memory distances, which represents the influence of memory decay. The memory weights are then calculated as the product of the exponential decay factor and the corresponding memory values between agents. This way, we emphasize stronger memories while accounting for the decay process.

The use of the exponential decay factor ensures that closer memory distances and stronger memory values lead to higher memory weights, indicating a higher probability of selecting an agent as a neighbor. The normalization step ensures that the memory weights sum up to 1, providing a valid probability distribution for neighbor selection. In doing so, the neighborhood selection process takes into account both memory growth and memory decay, resulting in the formation of connections based on the strength and recency of agents' memories.

4   Demonstrations and Results

In a typical consensus game, members of a group are permitted to preferentially assign themselves one of a small set of colors, but the entire group is rewarded if it eventually converges to one color. Kearns [6] finds that people are very good at maximizing the group's welfare across a variety of network structures and incentives, so long as the set of their neighbors is held constant: human subjects achieved approximately 90% of the theoretically maximum payout attainable by a perfectly coordinated group.

To assess the behavior of our social preference learning agents, we simulated an environment containing 36 agents, each randomly endowed with one of four color preferences. In other words, for a given agent i, the initial $p_i(r|x) = 1$ for one x, and $= 0$ for the three other xs (colors). The agents could interact with any of the other agents in a sequence. The possible agents with which the initiator i interact with are, from his perspective, the context; thus, interaction partners (responders) are considered c and the interaction is selected by sampling the available neighbors. For simulations using fixed networks, each agent's

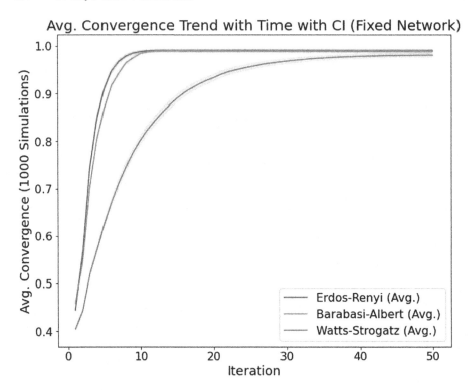

Fig. 1. The plot shows convergence over time for all the fixed network simulations for all three types of graphs. Shaded area represents 95% CI after 1000 simulations. We see convergence for all three graph types.

neighborhood was specified and it could not be changed during the course of the iterations. During an interaction, the responder indicates to the initiator his preferred color ($\arg\max[p(r|x,c)]$), and the responder received no information. At each time step, the initiator updates their own color preferences by marginalizing across the preferences expressed by their neighbors using the preference inference computation mentioned earlier.

We simulate neighborhoods randomly using all three types of graphs - ER, BA, and WS - 1000 times, and report results using the average convergence (the greatest number of nodes converging to a particular color divided by the total number of nodes in the graph at any point in time) obtained for 50 iterations of the consensus game played on each graph for all three categories of graphs in Fig. 1. Even in the absence of an explicitly specified reward for group consensus, our simulation results show that individual agents use the preferences of their neighbors to change their personal preferences, until consensus is reached.

Consistent with the existing literature [12], we find that the color with the greatest representation in the initial condition of each graph wins most frequently (this result simply verifies that under a fixed network structure, our model appropriately propagates beliefs). We also find that the rate of convergence to consensus is directly proportional to the degree of nodes on average in all three types of graphs.

Erdos-Renyi (Fixed) - Initial State   Erdos-Renyi (Fixed) - Final State

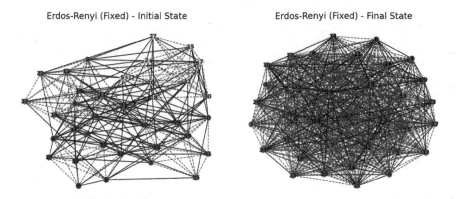

Fig. 2. On the left is the initial network for the dynamic network condition in one of the trials, using the ER graph. On the right is the final network after 50 iterations. We see balkanization and cliques formation based on color preference similarities.

But what happens when agents are free to choose their neighbors? When Kearns [7] relaxed the fixed network structure, such that subjects could select which of their neighbors they wished to receive information about, they found that coordination suffered massively, with efficiency dropping to about 40%. It turns out that while humans are extremely good at adapting their preferences to existing network structures, something about the process of social link formation causes this facility of coordination to break down.

We find similar results from our simulation experiment across a broad range of parameter values for memory growth and memory decay. Since network connections were now permitted to be dynamic, agents updated their neighborhoods using encounter information throughout the simulation. At each model iteration, the propensity for interacting with other agents changed, and so did their current preference, using the computation for $p(r|x)$ as above. See Fig. 2 - agents start out with a fixed network, and are then allowed to sample from other agents to update their neighborhood and connections. As a result of this, the final network state (on the right of Fig. 2) turns out to be balkanized.

When updating preferences in fixed network conditions, agents performed the computation as suggested by Eq. 1, and that was enough to get them to global convergence - even in absence of any specified rewards. However, in the case of dynamic networks, when agents were free to choose their neighbors in every iteration, agents retain memories of past interactions, enabling them to recall and potentially favor agents with whom they have had shared color preferences in the past. This memory retention allows for the persistence of social ties and the potential formation of clusters based on shared preferences. This contributes

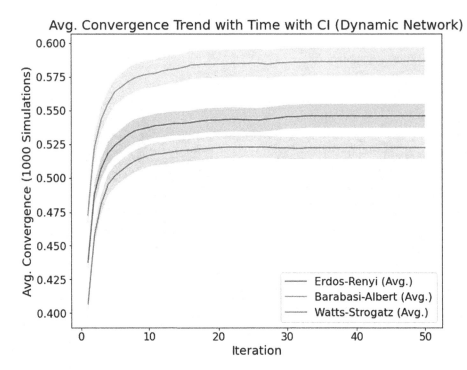

Fig. 3. The plot shows convergence over time for all the fixed network simulations for all three types of graphs. Shaded area represents 95% CI after 1000 simulations.

to the reinforcement of existing social ties, potentially leading to the emergence of cohesive clusters of agents with similar color preferences. This is the case for ER, BA, as well as WS graphs. However, there is a curious differentiation that can be observed when we look at the convergence asymptote value for the three types of graphs across all simulations and all iterations - see Fig. 3 above.

We see that Barabasi-Albert networks show convergence to a higher asymptotic value compared to Watts-Strogatz as well as Erdos-Renyi networks. Considering the structural differences in how the three graphs are generated, we find an interesting explanation for this difference. What makes the BA graph different from the other two is its degree distribution, which follows a power law - thereby increasing the probability of finding nodes that are thickly connected with many neighbors, compared to ER graphs, where the degree distribution is binomially (approximately normally) distributed. Likewise, with WS, we have a small world structure, yielding a close to uniform degree distribution.

For the consensus game, all that matters is the local neighborhood - so, if a node is thickly connected, there is a high chance that it is connected to nodes that have varying colors. If such a node switches over, it's going to have a lot of impact on the rest of the graph. Since we are more likely to see this sort of highly trusted or highly influential node in a BA network than in ER or WS graphs, we see a higher convergence asymptote for BA than ER or WS graphs.

Thus, we find that the same algorithm, when allowed to work with a fixed network structure, performs information coordination efficiently, whereas when allowed freedom to preferentially create local network neighborhoods, agents behave in locally optimal ways that reduce global coordination. We believe these findings explain to a considerable extent the mysterious gap in coordination performance in Kearns' networked game experiments: Agents, and likely humans, assure themselves that they have equilibrated to the consensus preference through sampling the preference of their neighbors. When forced to consider all neighbors, they must necessarily engage with all the information present in their neighborhood; when free to choose, they end up restricting communication with neighbors who share their preference.

5   Conclusion

In this paper, we used a memory-based model of social preference learning to reproduce both the success and failure of agents to attain consensus in a networked game, based on whether agents were permitted to select their social neighborhood. We showed that networks of agents forced to play with neighborhoods assigned to them nearly always converged to a consensus color in the game, although this process was slower for Watts-Strogatz small-world neighborhoods. We also showed that networks of agents permitted to create their own neighborhoods failed to converge to a consensus, with Barabasi-Albert style preferentially attached networks reaching more majority consensus than alternative types.

One alternative to the memory model we used is instance-based learning (IBL). IBL assumes that decision making is based on remembering past experiences and generalizing from these to new situations [13]. The Adaptive Control of Thought - Rational (ACT-R) model is another alternative. ACT-R posits that cognition is composed of a set of basic modules (e.g., visual and auditory), a single production system that coordinates interactions among the modules, and a single declarative memory system that stores factual knowledge [14]. However, while each of these models focuses on different aspects of cognition, they are ultimately just vehicles for the assumptions - and it is these assumptions that determine how accurately the model can predict phenomena in the real world. The choice of model does not fundamentally change our conclusions, so long as the assumptions that guide our model are valid and are themselves representative of the phenomena we seek to understand.

Our findings have theoretical as well as practical implications for enhancing group efficiency and cohesion, particularly in addressing the challenges posed by clique formation and balkanization. By understanding the mechanisms underlying network dynamics and their impact on group behavior, we can also design social media platforms and online communities that foster a less balkanized environment. In particular, our results show that it is not necessary to impose fixed networked structure to prevent balkanization. The presence of highly connected nodes in networks also protects communities from failures in consensus, so long as these nodes are open to changing their colors based on observing their local neighborhood's majority view. Interestingly, these results are consistent with recent empirical work showing that the effect of filter bubbles in large-scale social media may be overstated [15].

Naturally, our current model is highly simplified, and ignores the possibility of alternative reward structures influencing the opinions of individual nodes in the graph. Exploring these possibilities constitutes a clear direction for future work in this project.

References

1. DiMaggio, P., Evans, J., Bryson, B.: Have American's social attitudes become more polarized? Am. J. Sociol. **102**(3), 690–755 (1996)
2. Borgesius, F.J.Z., et al.: Should we worry about filter bubbles? Internet Policy Rev. **5**, 1–16 (2016)
3. Pariser, E.: The Filter Bubble: What the Internet Is Hiding from You. Penguin Press, New York (2011)
4. McPherson, M., Smith-Lovin, L., Cook, J.M.: Birds of a feather: homophily in social networks. Ann. Rev. Sociol. **27**, 415–444 (2001)
5. Sunstein, C.R.: #Republic: Divided Democracy in the Age of Social Media. Princeton University Press, Princeton (2017)
6. Kearns, M., Judd, S., Vorobeychik, Y.: Behavioral experiments on a network formation game. In: Proceedings of the ACM Conference on Electronic Commerce (2012)
7. Kearns, M., Judd, S., Tan, J., Wortman, J.: Behavioral experiments on biased voting in networks. Proc. Natl. Acad. Sci. U.S.A. **106**, 1347–1352 (2009)
8. Burns, T., Stalker, G.M.: The Management of Innovation. Oxford University Press, Oxford (1994)
9. Nonaka, I., Nishiguchi, T.: Fractal design: self-organizing links in supply chain management. In: Knowledge Creation: A Source of Value, pp. 199–230. Ed. St. Martin's Press (2009)
10. Tenenbaum, J.B., Kemp, C., Griffiths, T.L., Goodman, N.D.: How to grow a mind: Statistics, structure, and abstraction. Science **331**, 1279–1285 (2011)
11. Srivastava, N., Schrater, P.: Rational inference of relative preferences, Proceedings of Advances in Neural Information Processing Systems 25, vol. 26 (2012)
12. Tang, J., Wu, S., Sun, J.: Confluence: conformity influence in large social networks. In: Proceedings of the 19th ACM SIGKDD International Conference on Knowledge Discovery and Data Mining, pp. 347–355 (2013)
13. Gonzalez, C., Lerch, F.J., Lebiere, C.: Instance-based learning in dynamic decision making. Cogn. Sci. **27**, 591–635 (2005)
14. Anderson, J.R., Bothell, D., Byrne, M.D., Douglass, S., Lebiere, C., Qin, Y.: An integrated theory of the mind. Psychol. Rev. **111**, 1036 (2004)
15. Dahlgren, P.M.: A critical review of filter bubbles and a comparison with selective exposure. Nordicom Rev. **42**(1), 15–33 (2021)

A Bayesian Approach of Predicting the Movement of Internally Displaced Persons

Obed Domson[1], Jose J. Padilla[2(✉)], Guohui Song[1(✉)],
and Erika Frydenlund[2(✉)]

[1] Department of Mathematics and Statistics, Old Dominion University,
Norfolk, VA 23529, USA
[2] Virginia Modeling, Analysis and Simulation Center, Old Dominion University,
Suffolk, VA 23435, USA
{odoms001,jpadilla,gsong,efrydenl}@odu.edu

Abstract. This paper proposes an approximate Bayesian model to predict the number of internally displaced people arriving to a location. Locations are characterized by their elevation, distance from point of departure, and land cover. The model is applied to the population and terrain data of the North Kivu province in the Democratic Republic of Congo (DRC). Results suggest that distance captures about 67% of the influence on the choice of destination; elevation captures 9%, and land cover 24%.

Keywords: human migration · Internally Displaced Persons (IDPs) · approximate Bayesian computation (ABC)

1 Introduction

Global human migration is on the rise, both voluntary migration like tourism and business as well as forced migration where people are displaced from their homes by conflict, environmental disasters, and political and economic turmoil [7]. The United Nations High Commissioner for Refugees (UNHCR) estimates that there were 89.3 million forcibly displaced persons worldwide as of the end of 2021, of which 53.2 million were internally displaced persons (IDPs) [2]. Internal displacement refers to those who are forced to leave their homes but remain inside their country of origin, in contrast to refugees and asylum seekers who have crossed international boundaries. Although they remain in their home countries, IDPs often experience social, political, and economic exclusion but may be less nationally and internationally visible than refugees. This can lead to difficulties for humanitarian aid agencies to identify IDPs who require assistance and to anticipate where IDPs may end up in order to set up the appropriate humanitarian response.

R. Thomson et al. (Eds.): SBP-BRiMS 2023, LNCS 14161, pp. 241–250, 2023.
https://doi.org/10.1007/978-3-031-43129-6_24

In this study, we use Approximate Bayesian Computation (ABC) to predict the movement of IDPs given factors such as location of inhospitable terrain or locations of ongoing conflict. There are usually many factors migrants consider when choosing a potential movement place such as the distance, the socio-political context of the destination, and their individual and group health and mobility levels.

Other modeling approaches have been used to study similar phenomena, attempting to predict the movement or final destinations of forced migrants. Agent-based models (ABMs), for instance, have been widely used to study the distribution of migrants across terrains and country borders [8,9,12,14]. ABMs have been used because of their decentralized approach, allowing a heterogeneous mix of agents to act and interact autonomously to make decisions about where to go in the environment [8,16]. Depending on the decision rules of agents and ways in which they interact with each other and the modeled environment, computational costs for ABMs can rise exponentially as a function of the number of factors/parameters included. This can make it impractical for modeling a large number of factors and/or a large number of agents directly. It would be more efficient if we could combine some factors into a single quantity to measure the "attractiveness" of potential movement places and then use this quantity rather than model all factors individually. This is where mathematical modeling approaches can benefit human migration models.

Markov Chains and Bayesian techniques have also been used to model refugee migration. For instance, Huang and Unwind [11] used a Markov process to predict the local movement of refugees in near real time. Also, Singh et al. [15] applied a hierarchical Bayesian approach to estimate the number of families moving from one location to another. More recently, data availability and computational tools have aided in constructing machine learning models for predicting displacement [10]. In the refugee migration models, it can be challenging to specify an analytical likelihood function that accurately represents the underlying data-generating process. Approximate Bayesian Computation (ABC) allows for the relaxation of this requirement by approximating the likelihood indirectly through simulations. This makes ABC more flexible and applicable to a wider range of problems, including cases with complex or intractable likelihoods.

In this paper, we focus on predicting the movement of internally displaced persons (IDPs) in the Democratic Republic of Congo (DRC) in response to conflict. The DRC has one of the highest numbers of IDPs globally, with an estimated four million internally displaced from conflict and environmental disaster in 2022 [3]. The large number of IDPs; high number of ongoing civil conflicts; dense, mountainous, and often inhospitable terrain; and access to internal migration data over time makes Eastern DRC an interesting test case for developing an Approximate Bayesian Computation model of human movement.

Specifically, we will investigate the combination of three factors that influence the movement of IDPs: the distance to the destination; the highest elevation along the path; and the forest cover along the path. We propose a model to quantify location attractiveness as a weighted combination of these factors. The parameters/weights in such a weighted combination could be viewed as the (relative) importance of these factors. We employ a Bayesian approach to estimate such parameters from historical data. We then use the estimated parameters to predict the movement of IDPs. We will also use the estimated parameters to identify the potential settlement locations of IDPs within Eastern DRC and predict the number of individuals likely to end up at such places.

This paper is organized as follows. In Sect. 2, we describe the problem situation in more detail. In Sect. 3, we describe the model and the Bayesian approach to estimate the parameters. In Sect. 4, we present the results. In Sect. 5, we discuss the results and conclude the paper.

2 Problem Description

2.1 Background

The Democratic Republic of Congo experiences adversities related to IDPs and refugees. Chaos in the nation dates to the colonial era and poor leadership. The colonizers extracted a vast majority of wealth from the nation's inhabitants. While many sought freedom, the leaders that took over were driven by their selfish interests hindering development. By 1993, income levels had shrunk to 35 percent of their pre-independence level; inflation reached 23,000 percent in 1995; foreign debt ballooned to fourteen billion; and only 15 percent of the roads inherited in 1960 remained passable [13].

Those who reside on the eastern side of Congo are mostly affected due to inadequate resources - some of which are legacies of Colonialism - as well as the chaos of large numbers of armed groups that fill leadership voids left by a weak and physically distant central government [17]. Spillover conflict from the Rwandan genocide and ongoing struggles to access eastern DRC's vast natural resources has left eastern provinces struggling with civil conflict, sparse health access, and human rights violations. While this has driven millions - half a million in 2022 alone - into neighboring countries, over 5 million are internally displaced [1]. Regular reports from eastern provinces of the DRC continue to illustrate the instabilities that regularly and repeatedly uproot people across the region [4].

The movement of IDPs is a complex process that is influenced by various factors including geographic features of the destination, the living environment, and the health conditions and opportunities elsewhere. Distance to the destination has been considered in the literature to model refugee movement [11,16]. The North Kivu province, eastern DRC is a mountainous area with uneven terrain and is densely forested. As the elevation range of the region spans from 568 to 4671 m and movement of IDPs is mostly on land, elevated areas along potential escape routes appear to pose a challenge to movement [6]. Forests within the region under study are reported hideouts for armed groups [5], and the presence of these hideouts is likely to influence the movements of the IDPs as well. With such geographical complexity, we examine the influence of the following three factors on the destinations of IDPs in that region: distance, elevation, and land cover. We use these three terrain factors combined with some partial data to predict the movement of IDPs through the implementation of a Bayesian method described below.

2.2   Data

We use OCHA (United Nations Office for the Coordination of Humanitarian Affairs): *Sites de displacement au 25 avril 2017* data on displacement in North Kivu province in eastern DRC updated as of April 25, 2017. It includes the number of internally displaced persons (IDPs) at each settlement, the type of site, location, and other relevant information.

Figure (1) shows a map of GIS data of the IDP settlements and Armed groups' areas of activity in the region.

- ▲ : Deserted sites - refer to places which were formerly camps but had people moving away during a conflict [2].
- ●, ◐ : Displacement sites such as IDP camps.
- Pink regions represent the areas controlled by armed groups.

Fig. 1. Displacement Sites and Armed groups

2.3   Problem Formulation

This dataset only provides the number of internally displaced persons (IDPs) at specific displacement sites or settlements. It does not include information about the number of IDPs settled in other potential locations. Based on conversations with experts in refugee and IDP management roles within humanitarian response organizations, it is possible to predict the locations where displaced persons might settle using subject matter expertise. However, there are exceptions to these predictions, particularly in eastern DRC. The region's mountainous terrain, dense forests, complex ethnic affiliations, and areas controlled by armed groups influence the direction people flee and ultimately settle. In such cases, self-settlements can occur spontaneously and remain undetected by humanitarian organizations for extended periods. Therefore, there is a need for models that can predict the locations of self-settled populations, enabling quicker identification and effective delivery of humanitarian aid to them.

We focus on the prediction of the movement of IDPs to other potential places outside the official IDP camps run by humanitarian agencies like OCHA. Specifically, we utilize the data at existing IDP camps to gain insight into how geographic factors impact the settlement patterns of IDPs and then use such understanding to estimate the number of IDPs settled at other potential locations.

3 Model and Bayesian Technique

3.1 Model

Consider people moving from the red places that are M deserted areas in Fig. (1) to C campsites and P other potential destinations. We will model the movement of IDPs from each starting place to every ending place. Suppose we have M starting places (the deserted areas/villages) and $N = C + P$ ending places (the camps and other potential settled places). We use $\theta_{m,n}$ to denote the chance of the people at the m-th starting place moving to the n-th ending place. We need to model this probability $\theta_{m,n}$ for all $m = 1, \ldots, M$ and $n = 1, \ldots, N$.

As discussed in the previous section, we will consider three factors that influence the movement of IDPs: $D_{m,n}$, $E_{m,n}$ denote the distance and the highest elevation from village m to a settlement site or potential destination n, and χ_n denotes whether the place n where people might end up is in the forest or not. That is

$$\chi_n = \begin{cases} 0, & \text{if the place } n \text{ is in forest;} \\ 1, & \text{otherwise.} \end{cases}$$

We will use the following model of $\theta_{m,n}$:

$$\theta_{m,n}(\boldsymbol{w}) = \frac{w_1 \frac{1}{D_{m,n}} + w_2 \frac{1}{E_{m,n}} + w_3 \chi_n}{\sum\limits_{n=1}^{N} \left(w_1 \frac{1}{D_{m,n}} + w_2 \frac{1}{E_{m,n}} + w_3 \chi_n \right)}, \qquad 1 \le m \le M; 1 \le n \le N,$$

$$(1)$$

where $\boldsymbol{w} = (w_1, w_2, w_3)$ denote the weights of the factors satisfying $w_1 + w_2 + w_3 = 1$. From equation (1), the probability is set to be inversely proportional to the distance between their initial and final destinations since people fleeing conflicts are known to prefer moving shorter distances (Ravenstein's Laws of Migration). Similarly, we set the probability to be inversely proportional to the elevation since the higher the elevation, the less the likelihood of people traversing it.

We emphasize that the probability $\theta_{m,n}$ could also be viewed as the "attractiveness" of the n-th destination for the people at m-th town. We will need to identify such weights to obtain the "attractiveness". In particular, we will use the Approximate Bayesian Computation (ABC) described below to obtain a posterior distribution of the weights.

3.2   Approximate Bayesian Computation (ABC)

Suppose $L(w; y) = p(y|w)$ is the likelihood of the observed data given the parameters w and $p(w)$ is the prior distribution of the parameter w. We will use the Bayesian technique to estimate the posterior distribution $p(w|y)$:

$$p(w|y) \propto L(w; y) p(w).$$

The prior distribution typically comes from other studies or subjective evaluation. To obtain the posterior distribution, we either have to get the likelihood or estimate the posterior directly.

The human movement process is a complex process and its likelihood function is usually impossible to obtain. In this study, we estimate the posterior directly using the Approximate Bayesian Computation (ABC). The idea of ABC is to replace the evaluation of the likelihood with a 0–1 indicator, describing whether the model outcome is close enough to the observed data. It works as follows

– A candidate vector w^c is drawn from a prior distribution.
– A simulation is run with parameter vector w^c to obtain simulated data y^c .
– The candidate vector is either retained or dismissed depending on whether the simulated data y^c is close enough to the observed data y .

Figure (2) below illustrates the proposed model.

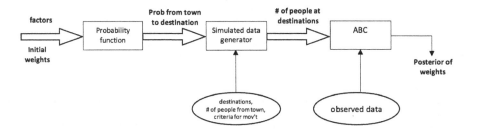

Fig. 2. Schematic diagram of the model

4   Results

4.1   Setup

To evaluate the model, potential destinations are identified with the following criteria:

– The destination should not be at a conflict zone
– The destination comes before any settlement site with the assumption that the IDPs will not move from camps once they get there.

Figure (3) shows the factors - elevation and land cover - as well as the selected potential destinations. There are 25 camps represented by blue and green circles. Additionally, we selected 5 potential destinations depicted by yellow ovals. Furthermore, there are 6 villages denoted by red triangles, namely Bibwe, Bweru, Kitso, Kivuye, Nyange, and Mpati. The population estimates for individuals originating from these villages were taken to be

$$[35\,K, 29\,K, 40\,K, 37\,K, 40\,K, 23\,K]$$

respectively. Distances and highest elevations between departing villages and destinations were obtained using GIS. While it is important to acknowledge that actual ground distances and elevations may vary, this approach provides a representation that is relatively more realistic on a comparative scale.

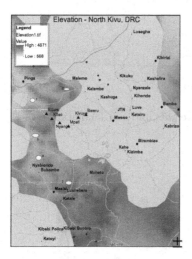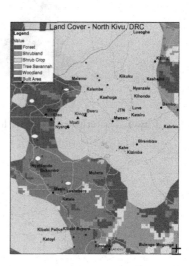

Fig. 3. Elevation and Land Cover of North Kivu, DRC (Color figure online)

4.2   Weights Estimation

The prior distribution of the weights w is set to be uniform on the simplex. For each candidate vector w^c, we compute the probability of people migrating from a specific village to a particular destination in (1). These probabilities are then used to simulate the number of people that end up at each destination. We then compare the number of IDPs at each settlement with the observed data. If the difference is less than a threshold, the candidate vector is retained. Otherwise, it is dismissed. The posterior distribution is then estimated using the retained candidate vectors.

With 150 weights randomly generated from a uniform prior distribution, the ABC technique is used to obtain the posterior distribution of the weights as demonstrated in Fig. (4). That is, with all weights initially having equal chance, the posterior illustrates the updated chance of each weight given the sample data. Figure (5) below shows the weights whose counts are more than 30 and points out the weight having the most count. The three sets of weights with the highest chances are:

$$[0.6742, 0.0869, 0.2389], \quad [0.6403, 0.0475, 0.3123], \quad [0.6247, 0.0299, 0.3455].$$

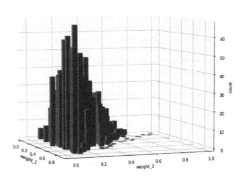

Fig. 4. Posterior of weights, for 150 weights and 100 simulations

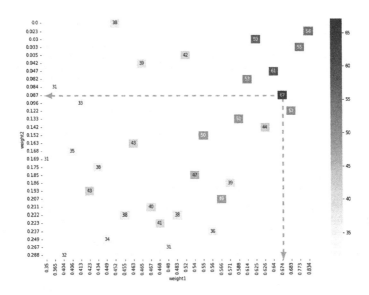

Fig. 5. Heat map of weights having counts greater than 30

4.3   Prediction of IDPs at Potential Destinations Other than Camps

For the weights with the largest probability, $w = (0.6742, 0.0869, 0.2389)$, we estimate the number of persons at these five potential destinations (yellow ovals in Fig. 3) outside the camps: $2400, 7400, 7100, 3300, 1800$.

5   Discussion

We have proposed a Bayesian model to predict the number of IDPs at potential destinations outside the camps. The proposed model is based on the "attractiveness" of the destination determined by the distance, elevation and land cover of the destination. The weights of these factors are estimated using the ABC technique. The proposed model is applied to the population and terrain data of North Kivu province in the DRC. The results indicate that the distance has about a 67% influence on the choice of where people end up or settle. Elevation, on the other hand, had a 9% influence, with land cover having an influence of 24%. It is useful to note that this is particular to this migration context only. Different factors such as different geographical locations and dispersion from more villages than those used in this study may yield different results.

In conclusion, information from this study can help determine which locations may attract the greatest number of persons given geographic driving factors. Additionally, the computational models can help determine or predict the most influential elements when considering a location preference for people.

We point out that this proposed model demonstrates promising results on modeling the human migration. However, there are still some limitations. First, we only consider three factors for deciding the movement direction. More factors could be easily included in the proposed model. This computational cost of the proposed model is only linear on the number of factors, while regular ABM has a geometric computational cost on the number of factors. Second, we assume the relative "attractiveness" (1) to be a (weighted) linear combination of these factors. More general models could also be considered, such as a generalized linear combination. Third, we only consider the IDP migration in a static setting. In reality, the IDP migration is a dynamic process. The proposed model could be extended to a dynamic setting to incorporate the changes over time, including the daily travel distance, new conflicts, and moving direction changes. With the consideration of time, we would model the dynamic movement of IDPs between their origins and potential destinations. Furthermore, the proposed method could also be extended to study other human movement contexts, such as predicting the travel directions of people in different cities/countries in response to environmental change or other large-scale movements.

Acknowledgements. This research was supported by the National Science Foundation DMS-1939203. Additionally, this research was supported by a grant from the Office of Naval Research (N000141912624) through the Minerva Research Initiative. None of the views reported in the study are those of the funding organizations.

References

1. The DRC: Regional refugee response plan - 2023. https://data.unhcr.org/en/documents/details/98918. Accessed 17 Jul 2023
2. Fresh fighting drives displacement in eastern DR congo. https://www.unhcr.org/en-us/news/latest/2016/4/570dfb126/fresh-fighting-drives-displacement-eastern-dr-congo.html
3. IDMC DRC Country Profile. https://www.internal-displacement.org/countries/democratic-republic-of-the-congo. Accessed 17 Jul 2023
4. No escape for civilians trapped in eastern DRC. https://bit.ly/3pHglzt. Accessed 17 Jul 2023
5. UN. Sustainable Forest Management for Peace Building. https://www.un.org/esa/forests/wp-content/uploads/2015/06/SFM-for-PeaceBuilding.pdf. Accessed 17 Jul 2023
6. UNHCR. Stories. Families fleeing DRC violence. https://bit.ly/3pTxK8i. Accessed 17 Jul 2023
7. World migration report 2022. https://worldmigrationreport.iom.int/wmr-2022-interactive/. Accessed 17 Jul 2023
8. Frydenlund, E., Foytik, P., Padilla, J.J., Ouattara, A.: Where are they headed next? Modeling emergent displaced camps in the DRC using agent-based models. In: 2018 Winter Simulation Conference (WSC), pp. 22–32. IEEE (2018)
9. Groen, D.: Simulating refugee movements: where would you go? Procedia Comput. Sci. **80**, 2251–2255 (2016)
10. Hoffmann Pham, K., Luengo-Oroz, M.: Predictive modeling of movements of refugees and internally displaced people: Towards a computational framework. arXiv preprint arXiv:2201.08006 (2022)
11. Huang, V., Unwin, J.: Markov chain models of refugee migration data. IMA J. Appl. Math. **85**(6), 892–912 (2020)
12. Johnson, R.T., Lampe, T.A., Seichter, S.: Calibration of an agent-based simulation model depicting a refugee camp scenario. In: Proceedings of the 2009 Winter Simulation Conference (WSC), pp. 1778–1786. IEEE (2009)
13. Kaplan, S.: The wrong prescription for the Congo. Orbis **51**(2), 299–311 (2007)
14. Kniveton, D., Smith, C., Wood, S.: Agent-based model simulations of future changes in migration flows for Burkina Faso. Glob. Environ. Change **21**, S34–S40 (2011)
15. Singh, L., et al.: Blending noisy social media signals with traditional movement variables to predict forced migration. In: Proceedings of the 25th ACM SIGKDD International Conference on Knowledge Discovery & Data Mining, pp. 1975–1983 (2019)
16. Suleimenova, D., Bell, D., Groen, D.: A generalized simulation development approach for predicting refugee destinations. Sci. Rep. **7**(1), 1–13 (2017)
17. Verweijen, J.: From autochthony to violence? Discursive and coercive social practices of the Mai-Mai in Fizi, eastern DR Congo. Afr. Stud. Rev. **58**(2), 157–180 (2015)

Social-Cyber Behavior Modeling

Few-Shot Information Operation Detection Using Active Learning Approach

Meysam Alizadeh[1]([✉]) and Jacob N. Shapiro[2]

[1] Department of Political Science, University of Zurich, 8050 Zurich, Switzerland
alizadeh@ipz.uzh.ch
[2] School of Public and International Affairs,
Princeton University, Princeton, NJ 08544, USA
jns@princeton.edu

Abstract. Previous research suggested that supervised machine learning can be utilized to detect information operations (IO) on social media. Most of the related research assumes that the new data will always be available in the exact timing that models set to be updated. In practice, however, the detection and attribution of IO accounts is time-consuming. There is thus a mismatch between the performance assessment procedures in existing work and the real-world problem they seek to solve. We bridge this gap by demonstrating how active learning approaches can extend the application of classifiers by reducing their dependence on new data. We evaluate the performance of an existing classifier when it gets updated according to five active learning strategies. Using state-sponsored information operation Twitter data, the results show that if querying from Twitter is possible, the best active learning strategy requires 5–10 times less tweets than the original model while only showing 1–3% reduction in the average monthly F1 scores across countries and prediction tasks. If querying from Twitter is not possible, the corresponding active learning strategy requires 5–10 times less tweets while showing 1–9% reduction in the average monthly F1 scores. Depending on the country, a hand-full to few hundred new ground-truth examples would suffice to achieve a reasonable performance.

Keywords: Information operation · Active learning · Text classification

1 Introduction

Detecting inauthentic behaviors on social media has always been a challenge for platforms. The most recent case of such problems is that of online *information operations* (a.k.a. political astroturfing, influence operation, coordinated networks, or inauthentic coordinated behavior) [1,8,10]. Information operation (IO) refers to coordinated campaigns by one organization, party, or state to

© The Author(s), under exclusive license to Springer Nature Switzerland AG 2023
R. Thomson et al. (Eds.): SBP-BRiMS 2023, LNCS 14161, pp. 253–262, 2023.
https://doi.org/10.1007/978-3-031-43129-6_25

impact one or more specific aspects of politics in domestic or another state through social media, by producing content designed to appear indigenous to the target audience or state [7]. This literature is different from bot detection. Recent reports suggest that in most cases the accounts used in such campaigns are either human-operated or hybrid human-automated accounts [4].

Previous research on detecting IO proposed machine learning frameworks that assume new ground-truth data are available on a regular weekly or monthly basis and they can fine-tune their model with the new data released by social media platforms (e.g. [1]). However, this does not match the reality where discovery and attribution of information operations are complicated and time-consuming efforts. This raises questions of what is the best model updating strategy when obtaining new ground-truth data is limited or even not possible, and how much performance we loose by implementing that strategy? To answer these practical questions, we pick an existing detection model by Alizadeh et al. [1] and implement five 'active learning' sampling strategies on it (including a baseline updating scheme) and compare the performance of the corresponding classifiers with that of having all new ground-truth data for updating the model overtime, which were reported in [1].

We use the exact data, features, and model settings as used in Alizadeh et al. [1] and follow their approach in testing the performance of classifiers on a monthly basis in which the classifier gets updated with ground-truth data from previous month to detect troll activity in the current month. However, instead of assuming the availability of all ground-truth data at the beginning of each month, we implement four different active learning and a simple baseline strategy to reduce the dependence of classifiers on new ground-truth data.

2   Related Work

2.1   Characterizing Information Operations

Early research on IO characteristics used natural language processing and machine learning to assess whether Russian Internet Research Agency (IRA) trolls were creating unique content [2]. Comparing the syntactic patterns of IRA trolls' tweets to those written by a sample of random English-speaking Twitter users revealed clear differences. They interpret this as an indicator that there was no serious attempt to obfuscate the non-nativity of IRA content. However, the authors did not test the predictive value of that content distinction or how it will change over time.

A related effort used publicly available Facebook, Reddit, and Twitter data of IRA trolls and a conflict-oriented events data to test hypotheses on temporal dynamics of IRA activities [6]. They found that "IRA Reddit activity granger caused IRA Twitter activity within a one-week lag", but that there were no similar relationships with Facebook activity [6]. In addition, the author found no similar Granger relationship between IRA activity and (1) Russian threat or military posturing toward U.S., (2) U.S. threat or military posturing toward Russia, and (3) approval of Donald Trump.

2.2   Troll Detection

Previous research developed a machine learning classifier to predict whether a Twitter account was one of the IRA accounts and then use it to identify current active IRA accounts on Twitter [5]. The model is trained on data which combine the most-recent 200 posts by the first set of IRA accounts released by Twitter on October 2018, with the most-recent 200 posts of 171,291 random American Twitter users. While similar in spirit to our analysis, this approach relies on historical behavioral data of users over a long time-period, and does not show variation in performance of the approach over time. In addition, the study uses a random sample of American users as control data, which is potentially an easier classification task compared to distinguishing coordinated IO campaign content from that of politically-engaged users.

Early related work used textual features representing thematic information such as emotions, morality, and profile traits to train classifiers for detecting trolls based on their tweets content [3]. The positive cases are IRA Twitter accounts, and the negative cases are a random sample of American users who posted at least 5 election-related tweets. The final data is highly imbalanced with IRA users being only 2% of users. More recently, researchers showed that classifiers trained on content-based features can perform well on detecting out-of-sample post-URL pairs that are part of already known IOs [1].

3   Data and Methods

In this paper, we use the exact same data, features, model settings, and two prediction tasks as in Alizadeh et al. [1] and compare the performance of classifiers trained on early ground-truth data and getting updated with new data according to four different active learning sampling strategies and one simple baseline sampling method.

3.1   Troll and Control Data

Since 2018, Twitter has released reports and data on IOs conducted by various countries. Following Alizadeh et al. [1] we use English-language tweets from China, Russia, and Venezuela campaigns. These include tweets from 2,660 Chinese, 3,722 Russian, and 594 Venezuelan accounts (Table 1). For replication purposes we focus on the same time period as in [1] which is from January 1, 2015 to February 29, 2019.

We use the same control data that Alizadeh et al. [1] used which include 5,000 random and 5,000 politically-engaged US Twitter accounts. We made this decision so that our classification results are comparable to those of them. The random US accounts were sampled by generating random numeric user IDs and checking the location to be in the United States. The politically-engaged Twitter users were defined as those who follow at least 5 American politicians. Table 1 summarizes the number of accounts and tweets in each dataset.

3.2   Active Learning Strategies

Active learning is a subfield of machine learning in which an algorithm inter-
actively query an oracle to label new data points. The key idea behind active
learning is that in situation where data labelling is expensive, we can obtain
better results if we allow the learning algorithm to choose what new data-points
should be labeled [9]. We consider five active learning scenarios to simulate various real-world situations in which one received data about an IO and is then
able to manually investigate a subset of post-URL pairs on a monthly basis:

Table 1. Summary of Twitter Trolls and Control Data

Type	Dataset	# Users	# Tweets
Troll	China	2,660	1,940,180
Troll	Russia	3,722	3,738,750
Troll	Venezuela	594	1,488,142
Control	US Political	5,000	22,977,929
Control	US Random	5,000	20,935,038

1. *Least Confident Sampling*: Query for those data points for which the classifier
   is the least confident. For each month t, train a classifier on all previous
   months (i.e. $t - 1$, $t - 2$, ...) and test on current month t. Then, using the
   result of the classifier on month t, query the highly uncertain tweets (p(troll)
   $\in [.3, .7]$) and add them to the pool of labeled data for predicting $t + 1$ with
   their true labels.
2. *Entropy Sampling*: The most common uncertainty sampling strategy uses
   Entropy as the uncertainty measure. For each month t, we train a classifier
   on all previous months' labeled data ($t - 1$, $t - 2$, ...) and test on month t.
   Then we measure the entropy of class probabilities for all data-points, query
   the top decile value data, and add them to the pool of labeled data
3. *Certainty Sampling*: For each month t, train a classifier on all previous months
   labeled data ($t - 1$, $t - 2$, ...) and test on current month t. Then select highly
   certain tweet-URL pairs (p(troll) $> .8$) and add them to the pool of labeled
   data
4. *Hybrid Sampling*: For each month t, train a classifier on all previous months
   labeled data (i.e. $t - 1$, $t - 2$, ...) and test on current month t. Query half of
   the least confident tweet-URL pairs and add their true labels to the training
   data for predicting $t + 1$. Also select half the highly certain tweet-URL pairs
   and add their model-assigned labels to the training data.
5. *Baseline*: Starting at month 4, we train a classifier on labeled data from all
   previous months (i.e. $t - 1$, $t - 2$, and $t - 3$) and test it on the current month
   t. Then, for each of the next months, we randomly select 2.5% of the output
   of the classifier in the previous month and add it to the pool of train data for
   the next month (2.5% is chosen to make the query size comparable to other
   sampling strategies).

3.3 Prediction Tasks

Out of the four prediction tasks reported in [1], we focus on two of them that are relevant to our purposes: (1) find content in month t from users who were not active in the training period, which simulates situation in which the already identified accounts were shut down by platforms and have not been replaced by new troll accounts; (2) considering all post-urls of month t but randomly shuffling the accounts creation dates for users who were active in the training period. Here, the assumption is that previously identified accounts were shut down by platforms, however, they have been replaced by a pool of new unused troll accounts who were created on the same dates as identified accounts.

3.4 Modeling and Evaluation

Following Alizadeh et al. [1], we train *Random Forests* classifiers using the *scikit-learn* library for Python in each month using the exact same features and settings as in [1] for all of our hyperparameters (i.e. *scikit-learn*'s default setting except for *n_estimators = 1000*). We report macro-weighted precision, recall, and F1 scores using the default classification threshold of 0.5. We use three months of 'ground-truth' data and start the simulation at 2015/4 for Russian activity, 2016/04 for Chinese activity, and 2016/12 for Venezuelan activity.

We avoid any model comparison or grid/random search on hyperparameters. Since we test the performance overtime, this decision ensures we have the same parameters for all classifiers. It also makes our results comparable to those of Alizadeh et al. [1] as they have not performed any hyperparameter tuning as well. The results, therefore, represent a lower-bound on the performance of active learning strategies with content-based classifiers.

4 Results

4.1 Classification Performance

Table 2 and Table 3 report mean and standard deviation of F1 scores for monthly classifiers for detection of tweets posted by new (those who were not active on previous months) and all (those who were active on previous months but their account creation times have been randomly replaced) trolls respectively. We also report the corresponding results from Alizadeh et al. [1] in the last row of each table for comparison purposes. Six points stand out: (1) when querying from Twitter is possible (i.e. least confident and entropy methods), we only see 1–3% reduction in average monthly F1 scores across different countries and prediction tasks, (2) when querying from Twitter is not possible, we see 1–9% reduction in average monthly F1 scores across different countries and prediction tasks, (3) Uncertainty sampling (i.e. least confident and entropy) is always the best strategy; (4) hybrid sampling always decreases performance compared to least confident strategy; (5) certainty sampling performs better than baseline when

Table 2. Monthly Macro-Averaged F1-Scores For Detecting New Users' Content

	China	Russia	Venezuela
Duration (# Months)	33	42	14
Sampling Strategy			
Baseline	0.57	0.70	0.90
	(0.01)	(0.02)	(0.000)
Least Confident	0.86	**0.79**	**0.92**
	(0.02)	**(0.02)**	**(0.000)**
Entropy	**0.86**	0.78	**0.92**
	(0.01)	(0.02)	**(0.000)**
Certainty	0.80	0.76	0.91
	(0.02)	(0.03)	(0.000)
Hybrid	0.81	0.76	0.91
	(0.02)	(0.03)	(0.000)
Alizadeh et al. [1]	0.89	0.81	0.92
	(0.12)	(0.13)	(0.15)

detecting tweets written by new trolls. However, baseline model slightly outperforms certainty sampling when detecting all users' content; and (6) certainty sampling, in which we totally rely on model output and never query from platforms, usually perform well enough. We discuss each in details below.

If it is possible to query highly uncertain data points from platforms, uncertainty sampling is always the best strategy for both new and all users experiments. For detecting new users in a given test month, we obtain fairly stable prediction performance across countries (Table 2), with minimum average monthly F1 score of 0.79 for Russian operation and maximum of 0.99 for Venezuelan operation. For the prediction task of detecting all users, we achieve stable and usually better performance compared to detecting only new users' tweets across countries (Table 3).

When querying from platforms is accessible, pooling platform-labeled tweets with model-labeled tweets (i.e. hybrid strategy) decreases prediction performance for Chinese and Russian IOs (Table 2). In fact, for the prediction task of detecting all users content, we see 8–18% reduction in average monthly F1 scores when pooling platform-labeled tweets with model-labeled tweets (Table 3).

When querying new observations from platforms is not possible, our classifiers still show reasonable prediction performance, specially for Chinese and Venezuelan operations. For the prediction task of detecting new users content, we obtained average monthly F1 scores of 0.80, 0.76, and 0.99 for Chinese, Russian, and Venezuelan trolls respectively (Table 2). For he prediction task of detecting all users content, we achieved average monthly F1 scores of 0.83, 0.70, and 0.97 for Chinese, Russian, and Venezuelan campaigns respectively (Table 3). This is very promising as it shows one can obtain fairly stable prediction performance over time (i.e. 2–3 years) just by having an initial set of platform-labeled data on nation-state IOs.

Table 3. Monthly Macro-Averaged F1-Scores For Detecting All Users' Content

	China	Russia	Venezuela
Duration (# Months)	33	42	27
Sampling Strategy			
Baseline	0.85	0.73	0.98
	(0.008)	(0.01)	(0.000)
Least Confident	**0.92**	**0.79**	**0.99**
	(0.002)	**(0.002)**	**(0.000)**
Entropy	**0.92**	**0.79**	**0.99**
	(0.002)	**(0.002)**	**(0.000)**
Certainty	0.83	0.67	0.97
	(0.01)	(0.02)	(0.003)
Hybrid	0.84	0.70	0.97
	(0.01)	(0.02)	(0.003)
Alizadeh et al. [1]	0.93	0.81	0.99
	(0.04)	(0.07)	(0.002)

Indeed, a machine learning approach could have tracked Chinese and Venezuelan campaigns through time without any human intervention using simple baseline or certainty sampling strategies. Only the Russian information operation from 2015–18 was so dynamic that it was hard to follow through time using a content-based machine learning approach. But even there our results suggest that limited additional investigations of highly-uncertain posts can yield strong performance over a long timeframe.

4.2 Query Size

As mentioned in the introduction, our main goal is to test whether content-based machine learning approaches with active learning can help identifying current and future information operations with minimum dependence on ongoing investigations. Since the results show the superiority of uncertainty sampling strategies, one might wonder what are the actual sizes of their queries.

Table 4 summarizes the average number of queried tweets and their equivalent number of users for our prediction task of new users. Three observations stand out: (i) querying only a handful of users each month is enough to track Venezuelan activity over time; (ii) comparing the average query sizes of the least confident and entropy sampling methods with the actual number of ground-truth data used by Alizadeh et al. [1] in their simulations overtime shows that these two active learning sampling methods require almost 5–10 times less tweets than their classifiers; and (iii) while having similar classification performances, the entropy sampling requires fewer queries than the least certain sampling.

Table 4. Average Number of Monthly Queried Tweets and Users by Active Learning Strategies For Detecting New Users' Content

		China	Russia	Venezuela
Duration (# Months)		33	42	14
Sampling Strategy	Query			
Baseline	Tweets	0	0	0
	Users	0	0	0
Least Confident	Tweets	495	1,267	3
	Users	190	328	2
Entropy	Tweets	425	1,028	7
	Users	172	311	4
Certainty	Tweets	0	0	0
	Users	0	0	0
Hybrid	Tweets	495	1,267	3
	Users	190	328	2

4.3   Important Features

To formally examine the relative importance of various types of features, we categorize our features into 4 groups: content, meta-content, content timing, and account timing. We consider the model trained on content features alone as a baseline and compare prediction performances by adding each group of features across two tests and four strategies. Due to space limit, we only demonstrate the results for the Russian Twitter campaign (Table 5). Compared to baseline, adding meta-content features on average increases the F1 score by 7.5% point across our two tests and four active learning strategies. Content timing features are not effective and add little to the performance. Account timing features, however, increases the F1 score by 3.3% point on average across various tests and strategies. Finally, including network features (e.g. various attributes of the co-shared and co-occurring hashtags network) has mixed effects on the prediction performance.

5   Discussion

Previous research used supervised learning approaches to classify malicious coordinated activity and normal behavior based only on content, though assuming the availability of entire data at once or over time. In this paper, we relaxed this assumption and studied the performance of active learning strategies to asses how well a classifier can perform in detecting information operations content over time without depending on new investigations. We evaluate this approach on a monthly basis across three different IOs on Twitter, two distinct tests, and four active learning strategies. Overall, simulation efforts to track three major

Table 5. Mean of Monthly F1-Scores with Varying Predictor Sets for Detection of Tweets Written by New Russian Trolls in Test Month t

	Content	(1) + Meta-Content	(2) + Content Timing	(3) + Account Timing	(4) + Network
Model Number	(1)	(2)	(3)	(4)	(5)
Sampling Strategy					
Baseline	0.59	0.65	0.65	0.70	0.71
Least Confident	0.68	0.76	0.76	0.79	0.78
Entropy	0.68	0.74	0.75	0.78	0.76
Certainty	0.66	0.73	0.73	0.76	0.74
Hybrid	0.68	0.72	0.73	0.76	0.76

IO campaigns over time shows that active learning strategies can help content-based classifiers to follow coordinated information operations on social media over long periods of time.

More particularly, three results stand out. First, being able to query a small portion of hard-to-classify new data on a monthly basis always leads to higher prediction performance compare to other active learning strategies. Second, using high-confidence output of the model as input for the next month alone does a pretty good job of following IO content for dozens of months across all three campaigns, and performs exceptionally well against Venezuelan IO. Third, classifiers almost always exhibit higher precision than recall, with near perfect precision scores in most cases.

The fact that our simple baseline works well across multiple tasks and campaigns over such a long time-period is strong evidence that content-based approaches could: (1) support public-facing dashboards alerting polities to the extent of foreign disinformation campaigns; (2) drive recommender systems to alert users when they are seeing promoted content or inadvertently spreading it themselves; and (3) cue investigations on platforms where user data must be hidden for privacy reasons. Future work should focus on improving prediction performance by utilizing features extracted from images and videos shared by trolls, develop richer features sets, and implement classification approaches that leverage longer histories.

References

1. Alizadeh, M., Shapiro, J.N., Buntain, C., Tucker, J.A.: Content-based features predict social media influence operations. Sci. Adv. **6**(30), eabb5824 (2020)
2. Boyd, R.L., et al.: Characterizing the internet research agency's social media operations during the 2016 us presidential election using linguistic analyses (2018)
3. Ghanem, B., Buscaldi, D., Rosso, P.: TexTrolls: Identifying Russian trolls on twitter from a textual perspective. arXiv preprint arXiv:1910.01340 (2019)

4. Grimme, C., Assenmacher, D., Adam, L.: Changing perspectives: is it sufficient to detect social bots? In: Meiselwitz, G. (ed.) SCSM 2018. LNCS, vol. 10913, pp. 445–461. Springer, Cham (2018). https://doi.org/10.1007/978-3-319-91521-0_32

5. Im, J., et al.: Still out there: Modeling and identifying Russian troll accounts on twitter. arXiv preprint arXiv:1901.11162 (2019)

6. Lukito, J.: Coordinating a multi-platform disinformation campaign: Internet Research Agency activity on three US social media platforms, 2015 to 2017. Polit. Commun. 37(2), 238–255 (2020)

7. Martin, D.A., Shapiro, J.N.: Trends in Online Foreign Influence Efforts. ESOC Report. Empirical Studies of Conflict Project, Princeton University, Princeton (2019)

8. Schoch, D., Keller, F.B., Stier, S., Yang, J.: Coordination patterns reveal online political astroturfing across the world. Sci. Rep. 12(1), 1–10 (2022)

9. Settles, B.: Active Learning Literature Survey. University of Wisconsin-Madison Department of Computer Sciences, Tech. rep. (2009)

10. Uyheng, J., Cruickshank, I.J., Carley, K.M.: Mapping state-sponsored information operations with multi-view modularity clustering. EPJ Data Sci. 11(1), 25 (2022)

Dynamic Modeling and Forecasting of Epidemics Incorporating Age and Vaccination Status

Nitin Kulkarni[✉], Chunming Qiao, and Alina Vereshchaka

Department of Computer Science and Engineering, State University of New York
at Buffalo, Buffalo, USA
{nitinvis,qiao,avereshc}@buffalo.edu

Abstract. Controlling the spread of infectious diseases poses a significant challenge, necessitating a thorough understanding of the interplay between human behavior and disease transmission. This research introduces an age and vaccination stratified epidemiological model called SIHRD^{a,v}, which considers the differential impact of diseases on various age and vaccination groups. The model's parameters are estimated using real-world data. We demonstrate that our model is well suited for forecasting the spread of epidemics such as COVID-19. Through the evaluation of different scenarios, we project the impact on infection rates, hospitalizations, and fatalities. These analyses and simulations provide valuable insights for effectively managing and preventing disease outbreaks.

Keywords: Dynamic modeling · Curve fitting · SIHRD model · Pandemic · COVID-19 · Epidemiological model · Mitigation regulations

1 Introduction

The outbreaks of diseases such as H3N2, Ebola, and COVID-19 in recent times have sparked significant interest among researchers in the field of epidemiology. Consequently, the development of epidemiological models is of great importance, as these models play a crucial role in assisting the prevention and control of diseases. Many diseases have varying degrees of impact based on an individual's age and vaccination status. Therefore, it is essential to develop an epidemiological model stratified by age and vaccination groups to accurately predict such differential effects. Such a model can assist policymakers in identifying more targeted measures to protect the most vulnerable population groups.

In this paper, we introduce a novel time-varying compartmental epidemiological model [1], called SIHRD^{a,v} which is stratified by age and vaccination groups (Sect. 3.1). This model is an extension of the widely used SIRD model [2] (Sect. 2.1). Compartmental epidemiological models have proven effective in studying and simulating the transmission dynamics of various diseases [3–5]. By incorporating age and vaccination groups into our model, we aim to enhance its accuracy and provide a more comprehensive understanding of disease spread.

In addition, we describe our methodology for estimating model parameters by leveraging real-world data using the Levenberg-Marquardt algorithm [6] (Sect. 3.2). This algorithm, renowned for its effectiveness in curve fitting [7], has been successfully applied in various domains, including nuclear physics [8], spectroscopy [9], and electron optics [10]. Additionally, we have demonstrated the efficacy of our approach by successfully applying it to COVID-19 data from the United States and obtaining accurate fits to the observed trends. (Section 4.2). This validation demonstrates the robustness and applicability of our method in accurately capturing the dynamics of the epidemic.

The ultimate objective of this work is to provide accurate epidemic forecasts that can assist policymakers in devising effective mitigation strategies. We perform a comparative analysis between our model's predictions for the COVID-19 pandemic and the data collected from the United States (Sect. 4.2). Furthermore, we evaluate a range of scenarios encompassing common government interventions (Sect. 4.3). By employing our approach, policymakers can gain valuable insights into the progression of the epidemic, enabling them to make informed decisions regarding the implementation of appropriate measures.

2 Background

2.1 SIRD Epidemiological Model

Compartmental epidemiological models divide the population under study into distinct groups (compartments) with underlying dynamics describing the transition of individuals between these compartments [1,11].

The SIRD epidemiological model divides the population under study (N) into four compartments: susceptible (S): individuals susceptible to infection; infected (I): currently infected individuals; recovered (R): individuals who have recovered from the infection; deceased (D): individuals who have died as a result of infection [2,12]. An individual can only occupy a single compartment at any given time.

The dynamics of the infection are mathematically represented by the following set of differential equations [13]:

$$\frac{dS}{dt} = -\beta\frac{S}{N}I, \quad \frac{dI}{dt} = \beta\frac{S}{N}I - \gamma I - \mu I \quad \frac{dR}{dt} = \gamma I, \quad \frac{dD}{dt} = \mu I \tag{1}$$

Where, β represents the infection rate, γ corresponds to the recovery rate, and μ signifies the mortality rate.

In Sect. 2.2, we provide an overview of relevant prior research, while in Sect. 3.1, we present our extended epidemiological model, SIHRD^{a,v}.

2.2 Related Works

Previous studies have extensively utilized the SIRD epidemiological model to model various diseases, including COVID-19 and Ebola [2,12,14]. However, these studies were conducted either prior to the development of a vaccine or did not consider the impact of vaccination. Subsequent research efforts focused on incorporating the impact of vaccination into epidemic models by introducing an additional compartment, denoted as 'Vaccinated', to account for individuals who had received the vaccine. Usherwood [15] and Nastasi [16] made notable contributions by developing the SIRDV model, which incorporated the concept of vaccine efficacy to account for the possibility of vaccinated individuals getting infected.

Furthermore, prior research has investigated the effects of various interventions on curbing the spread of the epidemic. Giordano [17] and Reiner [18] conducted exemplary studies evaluating the effectiveness of interventions such as lockdowns, mask mandates, population-wide testing, and contact tracing. Their work provided policymakers with a valuable tool to assess the impact of different mitigation strategies.

Thus, previous works primarily focused on the efficacy of vaccines, but did not take into consideration the potential influence of vaccination on the severity of symptoms, resulting in varying mortality and hospitalization rates [19,20]. Additionally, these studies did not consider the differential impact of the disease across different age groups.

To address these limitations, we account for the impact of vaccination and different age groups on infection, hospitalization, and mortality rates, thereby developing a model that offers policymakers the ability to identify more targeted measures to protect the most vulnerable population groups. Additionally, We extend the SIRD model by introducing an additional compartment called 'Hospitalized.' This extension aims to provide public health authorities with enhanced preparedness insights based on our model's forecasts.

3   Methodology

In this section, we present: (1) the proposed extension to the SIRD epidemiological model and (2) the method used for calibrating the model parameters.

3.1   SIHRD^{a,v} Epidemiological Model

We extend the SIRD epidemiological model (Sect. 2.1) to include a 'Hospitalized (H)' compartment for individuals hospitalized as a result of infection. This enables hospitals and public health authorities to improve their preparedness efforts, such as estimating the necessary treatments and hospital bed requirements, based on the model's forecasts. Moreover, recognizing the substantial variations in infection, hospitalization, and mortality rates across different age and vaccination groups for numerous diseases, including Malaria, Influenza, and COVID-19 [19–23], we further refine each compartment by subdividing them based on individuals' age (a) and vaccination status (v). This refinement allows for a more accurate representation of the disease dynamics and assists policymakers in identifying targeted measures to protect vulnerable population groups.

Furthermore, we incorporate the potential impact of virus mutations on vaccine efficacy. Notably, SARS-CoV-2, the virus responsible for COVID-19, exhibits a high ability to evolve, as evidenced by the emergence of novel strains characterized by increased transmissibility and enhanced abilities to evade host immune responses. As an example, the Omicron variant rapidly gained global dominance in late 2021. In our analysis of COVID-19 data, we consider three distinct vaccination groups: unvaccinated individuals (uv), vaccinated individuals who have not yet received the bivalent booster (v), and vaccinated individuals who have received the bivalent booster (biv). By encompassing these different vaccination statuses, we capture the varying levels of protection conferred by the different vaccines and their potential influence on disease outcomes.

Given the evolving nature of epidemics, particularly as a result of government-imposed containment measures, seasonal trends, and evolving contagion characteristics, it is not reasonable to assume that a model with constant parameters will accurately fit all stages of the epidemic. Our model incorporates time-varying parameters allowing us to account for factors such as containment measures implemented by authorities, modifications in the epidemic characteristics, and the impact of advanced antiviral treatments. By incorporating these dynamic elements, our model provides a more comprehensive and realistic representation of the evolving nature of epidemics.

Transition dynamics between the compartments of the proposed SIHRD^{a,v} epidemiological model are shown in Fig. 1.

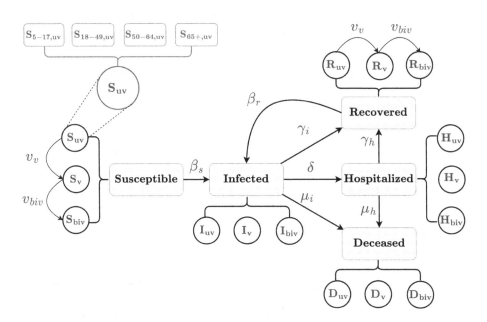

Fig. 1. SIHRD^{a,v} epidemiological model describing the transition dynamics of disease spread. The Susceptible (S), Infected (I), Hospitalized (H), Recovered (R), and Deceased (D) compartments are divided based on age (a) and vaccination groups (v). The sub-compartments are represented by subscripts denoting different age groups ($5-17$, $18-49$, $50-64$, and $65+$) and vaccination groups (uv, v, and biv). The figure also shows a detailed view of the Susceptible compartment with different age groups and the unvaccinated group. All the compartments follow the same pattern.

The evolution of the population in the compartments of the SIHRD^{a,v} model over time is given by the ordinary differential equations (ODEs) (Eq. 2). Where, N represents the size of the population under study, β denotes the infection rate, α represents the mixing coefficient to account for the imperfect mixing of population, δ signifies the rate at which infected individuals become hospitalized, γ denotes the recovery rate, and μ represents the mortality rate. Additionally, we calculate v_v; the rate at which the population is getting vaccinated and v_{biv}; the rate at which the population is getting the bivalent boosters from the vaccination data[1] collected from Centers for Disease Control and Prevention (CDC).

$$\frac{dS}{dt} = -\beta_s \frac{S}{N} I^\alpha \pm (v_v, v_{biv})S$$

$$\frac{dI}{dt} = \frac{\beta_s S + \beta_r R}{N} I^\alpha - \gamma_i I - \delta I - \mu_i I \qquad \frac{dH}{dt} = \delta I - \gamma_h H - \mu_h H \qquad (2)$$

$$\frac{dR}{dt} = -\beta_r \frac{R}{N} I^\alpha + \gamma_i I + \gamma_h H \pm (v_v, v_{biv})R \qquad \frac{dD}{dt} = \mu_i I + \mu_h H$$

The parameters β, δ, γ, and μ differ based on an individual's age and vaccination group. Furthermore, β also differs based on whether an individual has previously been infected, as prior infection provides a certain level of immunity [24–26]. Additionally, γ and μ also differ depending on whether the individual is hospitalized, as hospitalized individuals with severe symptoms have a higher likelihood of mortality.

By incorporating these varying parameters, our model captures the heterogeneity in infection, hospitalization, and mortality rates across different age and vaccination groups, leading to a more comprehensive representation of the dynamics of the epidemic.

As an example, the evolution of the population in the 65+ and unvaccinated group is given by the following ODEs (Eq. 3):

$$\frac{dS_{65+,uv}}{dt} = -\beta_{s,65+,uv} \frac{S_{65+,uv}}{N} I^\alpha - v_v S_{65+,uv}$$

$$\frac{dI_{65+,uv}}{dt} = \frac{\beta_{s,65+,uv} S_{65+,uv} + \beta_{r,65+,uv} R_{65+,uv}}{N} I^\alpha$$
$$- \gamma_{i,65+,uv} I_{65+,uv} - \delta_{65+,uv} I_{65+,uv} - \mu_{i,65+,uv} I_{65+,uv}$$

$$\frac{dH_{65+,uv}}{dt} = \delta_{65+,uv} I_{65+,uv} - \gamma_{h,65+,uv} H_{65+,uv} - \mu_{h,65+,uv} H_{65+,uv} \qquad (3)$$

$$\frac{dR_{65+,uv}}{dt} = -\beta_{r,65+,uv} \frac{R_{65+,uv}}{N} I^\alpha + \gamma_{i,65+,uv} I_{65+,uv} + \gamma_{h,65+,uv} H_{65+,uv}$$
$$- v_v R_{65+,uv}$$

$$\frac{dD_{65+,uv}}{dt} = \mu_{i,65+,uv} I_{65+,uv} + \mu_{h,65+,uv} H_{65+,uv}$$

---

[1] https://covid.cdc.gov/covid-data-tracker.

The other compartments representing different age and vaccination groups, follow similar patterns of evolution described by their respective ODEs.

3.2   Model Parameter Estimation

To estimate the model parameters α, β, δ, γ and μ we fit the solution of (2) to the COVID-19 data collected for the United States. We formulate the problem as an initial value problem (IVP) for the system of ODEs (Eg. 2) $(\mathbf{y}'(t))$ with initial values $(\mathbf{y_0})$ based on the data.

$$\mathbf{y}'_i(t) = f_i(t, y_1(t), y_2(t), ...)$$
$$\mathbf{y}(\mathbf{t_0}) = \mathbf{y_0}$$
(4)

We solve the IVP using the Explicit Runge-Kutta method of order 5 (4).

$$y_{n+1} = y_n + h \sum_{i=1}^{5} b_i k_i$$

where,
(5)

$$k_1 = f(t_n, y_n); k_2 = f(t_n + c_2 h, y_n + h(a_{21}k_1)); \dots;$$
$$k_5 = f(t_n + c_5 h, y_n + h(a_{51}k_1 + a_{52}k_2 + \dots + a_{54}k_4))$$

The estimation of the population in different compartments, based on model parameters, can be denoted by the vector $\hat{\mathbf{e}}$, while the actual values obtained from the data are denoted by the vector $\mathbf{a}$. The estimation vector $\hat{\mathbf{e}}$ is parameterized by $\{t, \alpha, \beta, \delta, \gamma, \mu\}$. We calibrate the model parameters by solving the following non-linear least squares optimization problem using the Levenberg-Marquardt algorithm [7], incorporating positivity and bound constraints on the parameters based on CDC statistics:

$$\underset{t,\alpha,\beta,\delta,\gamma,\mu}{\text{minimize}} \|\hat{\mathbf{e}} - \mathbf{a}\|_2^2$$
(6)

The Levenberg-Marquardt algorithm interpolates between the Gauss-Newton algorithm and the method of gradient descent. It is defined as follows:

$$(J^\top J + \lambda I)\delta = J^\top [y - f(\beta)]$$
(7)

Here, J represents the Jacobian matrix, $f(\beta)$ and y are vectors with i-th component $f(x_i, \beta)$ and y_i respectively. I is the identity matrix, giving as the increment δ to the estimated parameter vector β. λ is the damping parameter value, which is always positive and adjusted for each iteration. In this context, δ and β do not correspond to the epidemiological parameters but are standard notations used in the Levenberg-Marquardt algorithm.

We normalize our residuals to account for the significant differences in value ranges among compartments. For instance, the susceptible population can reach hundreds of millions, while the deceased population may be in the hundreds of thousands. Without this normalization, the Levenberg-Marquardt algorithm would place greater emphasis on data with larger ranges.

Additionally, we commute vaccination rates v_v and v_{biv} from the vaccination data[2] collected from the CDC.

Our model incorporates time-varying parameters estimated in four week intervals to accurately capture the changes associated with virus mutations, seasonal trends, advances in treatments, public health interventions, and public behavior.

4 Experiments

In this section, we present the experiments conducted to evaluate the performance and effectiveness of our proposed $SIHRD^{a,v}$ model. We estimate the epidemiological model parameters for the COVID-19 data collected for all 50 states of the United States. We perform a comparative analysis, comparing our model's predictions with the actual data. Finally, we assess a range of commonly implemented government interventions to forecast the evolution of the pandemic under different policies.

4.1 Dataset Description

We collected daily observations on cases and their outcomes, vaccine administration, and data on cases, deaths, and hospitalizations by age and vaccination groups from CDC recorded COVID-19 data for the period of April 4, 2021 till March 31, 2023[3]. Hospitalization data[4] was collected from the U.S. Department of Health and Human Services (HHS). These comprehensive datasets serve as the foundation for estimating the parameters of our $SIHRD^{a,v}$ model (Sect. 3.2), which we subsequently utilize to assess the evolution of the epidemic under various government interventions.

4.2 Model Fit and Forecasts

The model parameters are estimated at four-week intervals, covering the period from April 4, 2021, to December 31, 2022, using the COVID-19 data collected for the United States. To evaluate the model's performance, we compare its forecasts with the actual data for the period from January 1, 2023, to March 31, 2023.

Figure 2 illustrates the results of estimating the $SIHRD^{a,v}$ model parameters and comparing the model's forecasts with the actual data for the state of New

---

[2] https://covid.cdc.gov/covid-data-tracker.

[3] https://covid.cdc.gov/covid-data-tracker.

[4] https://healthdata.gov/Hospital/COVID-19-Reported-Patient-Impact-and-Hospital-Capa/g62h-syeh.

York. Specifically, we focus on the Infected and Hospitalized compartments for the 65+ and Unvaccinated group. Similarly, Fig. 3 showcases the results for the state of Pennsylvania, examining the Infected and Hospitalized compartments for the 18–49 and Vaccinated group.

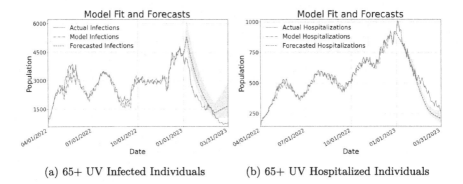

(a) 65+ UV Infected Individuals   (b) 65+ UV Hospitalized Individuals

Fig. 2. SIHRD^{a,v} Model fit and Forecasts for COVID-19 data for New York. The shaded region represents a 5% uncertainty interval.

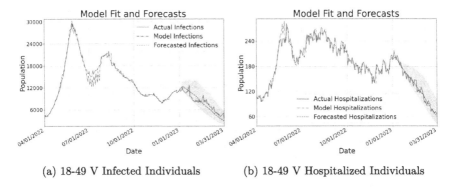

(a) 18-49 V Infected Individuals   (b) 18-49 V Hospitalized Individuals

Fig. 3. SIHRD^{a,v} Model fit and Forecasts for COVID-19 data for Pennsylvania. The shaded region represents a 5% uncertainty interval.

Through accurate forecasts, our model equips policymakers with valuable insights to develop efficient mitigation strategies. By understanding the projected trajectory of the pandemic, policymakers can make informed decisions to minimize the impact of the disease, allocate resources effectively, and implement targeted interventions tailored to specific age and vaccination groups.

4.3 Scenario Assessment

We assess the impact of three commonly applied interventions: (1) Social Distancing Mandates, (2) Mask Mandates, and (3) Social Distancing and Mask

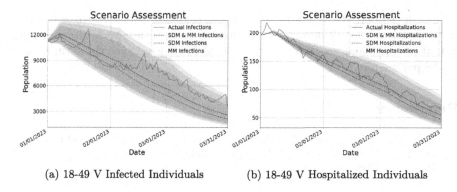

(a) 18-49 V Infected Individuals   (b) 18-49 V Hospitalized Individuals

Fig. 4. Comparison of population dynamics under different government interventions for Pennsylvania. The shaded regions represent a 5% uncertainty interval.

Mandates. Figure 4 presents the results of these scenario assessments, showcasing the predicted outcomes under each intervention.

Interestingly, at this stage of the COVID-19 pandemic, the differences among the three interventions are not substantial. There are a couple of possible explanations for this finding. First, the number of infected individuals in the total population might not be significant enough for these interventions to exhibit a noticeable distinction in their effects. Second, it is plausible that a considerable portion of the population has acquired immunity, either through vaccination or prior infection. These differences are expected to become more significant when there is a high prevalence of infected individuals in the population, as demonstrated in [5]. Estimating the model parameters based on data from earlier stages of the pandemic, such as 2020, might yield different results and counterfactual outcomes, highlighting the influence of temporal context on model performance.

Assessing the effectiveness of various interventions provides crucial insights for policymakers in their decision-making process. While the immediate impact of these interventions may not be pronounced under current circumstances, understanding their potential significance under different epidemiological conditions is essential. Policymakers can utilize these assessments to refine their strategies, adapt to changing dynamics, and implement interventions that are tailored to the specific needs and characteristics of the population.

5   Conclusion

We presented a comprehensive SIHRD^{a,v} epidemiological model for analyzing and forecasting the spread of epidemics. By incorporating time-varying parameters, age groups, vaccination status, and real-world data, our model provided accurate predictions and insights into disease dynamics. By capturing temporal variations, our model was able to better reflect the real-world scenario. Data collection from sources, such as the CDC and HHS, was crucial in ensuring model accuracy. Our SIHRD^{a,v} model serves as a valuable tool for understanding and forecasting COVID-19 dynamics, aiding policymakers, public health authorities, and researchers in developing targeted strategies. In future work, the incorporation of mobility and geographical data could further enhance the estimation of model parameters.

References

1. Brauer, F.: Compartmental models in epidemiology. In: Mathematical Epidemiology. LNM, vol. 1945, pp. 19–79. Springer, Heidelberg (2008). https://doi.org/10.1007/978-3-540-78911-6_2
2. Calafiore, G.C., Novara, C., Possieri, C.: A time-varying SIRD model for the COVID-19 contagion in Italy. Ann. Rev. Control **50**, 361–372 (2020)
3. Smith, M.C., Broniatowski, D.A.: Modeling influenza by modulating flu awareness. In: Xu, K.S., Reitter, D., Lee, D., Osgood, N. (eds.) SBP-BRiMS 2016. LNCS, vol. 9708, pp. 262–271. Springer, Cham (2016). https://doi.org/10.1007/978-3-319-39931-7_25
4. Vereshchaka, A., Kulkarni, N.: Optimization of mitigation strategies during epidemics using offline reinforcement learning. In: Thomson, R., Hussain, M.N., Dancy, C., Pyke, A. (eds.) SBP-BRiMS 2021. LNCS, vol. 12720, pp. 35–45. Springer, Cham (2021). https://doi.org/10.1007/978-3-030-80387-2_4
5. Kulkarni, N., Qiao, C., Vereshchaka, A.: Optimizing pharmaceutical and non-pharmaceutical interventions during epidemics. In: Thomson, R., Dancy, C., Pyke, A. (eds.) Social, Cultural, and Behavioral Modeling. SBP-BRiMS 2022. Lecture Notes in Computer Science. vol. 13558. Springer, Cham (2022). https://doi.org/10.1007/978-3-031-17114-7_22
6. Ranganathan, A.: The Levenberg-Marquardt algorithm. Tutoral LM Algorithm **11**(1), 101–110 (2004)
7. Gavin, H.P.: The Levenberg-Marquardt algorithm for nonlinear least squares curve-fitting problems, Department of Civil and Environmental Engineering, Duke University. vol. 19 (2019)
8. Zhang, H.F., Wang, L.H., Yin, J.P., Chen, P.H., Zhang, H.F.: Performance of the Levenberg-Marquardt neural network approach in nuclear mass prediction. J. Phys. G: Nucl. Part. Phys. **44**(4), 045110 (2017)
9. Aarnink, W., Weishaupt, A., Van Silfhout, A.: Angle-resolved x-ray photoelectron spectroscopy (ARXPS) and a modified Levenberg-Marquardt fit procedure: a new combination for modeling thin layers. Appl. Surf. Sci. **45**(1), 37–48 (1990)
10. Koh, J.M., Cheong, K.H.: Automated electron-optical system optimization through switching Levenberg-Marquardt algorithms. J. Electron Spectrosc. Relat. Phenom. **227**, 31–39 (2018)
11. Kermack, W.O., McKendrick, A.G.: A contribution to the mathematical theory of epidemics. Proc. R. Soc. London. Ser. A **115**(772), 700–721 (1927). Containing papers of a mathematical and physical character
12. Fernández-Villaverde, J., Jones, C.I.: Estimating and simulating a SIRD model of COVID-19 for many countries, states, and cities. J. Econ. Dyn. Control **140**, 104318 (2022)
13. Bailey, N.T., et al.: The mathematical theory of infectious diseases and its applications. Charles Griffin & Company Ltd, 5a Crendon Street, High Wycombe, Bucks HP13 6LE. (1975)

14. Wang, P., Jia, J.: Stationary distribution of a stochastic SIRD epidemic model of Ebola with double saturated incidence rates and vaccination. Adv. Differ. Equ. **2019**(1), 1–16 (2019)

15. Usherwood, T., LaJoie, Z., Srivastava, V.: A model and predictions for COVID-19 considering population behavior and vaccination. Sci. Rep. **11**(1), 1–11 (2021)

16. Nastasi, G., Perrone, C., Taffara, S., Vitanza, G.: A time-delayed deterministic model for the spread of COVID-19 with calibration on a real dataset. Mathematics **10**(4), 661 (2022)

17. Giordano, G., et al.: Modelling the COVID-19 epidemic and implementation of population-wide interventions in Italy. Nat. Med. **26**(6), 855–860 (2020)

18. Modeling COVID-19 scenarios for the united states. Nat. Med. **27**(1), 94–105 (2021)

19. Borchering, R.K., et al.: Modeling of future COVID-19 cases, hospitalizations, and deaths, by vaccination rates and nonpharmaceutical intervention scenarios-united states, April-September 2021. Morb. Mortal. Wkly Rep. **70**(19), 719 (2021)

20. Scobie, H.M., et al.: Monitoring incidence of COVID-19 cases, hospitalizations, and deaths, by vaccination status-13 us jurisdictions, April 4-July 17, 2021. Morb. Mortal. Wkly Rep. **70**(37), 1284 (2021)

21. O'Driscoll, M., et al.: Age-specific mortality and immunity patterns of SARS-CoV-2. Nature **590**(7844), 140–145 (2021)

22. Ma, J., Dushoff, J., Earn, D.J.: Age-specific mortality risk from pandemic influenza. J. Theor. Biol. **288**, 29–34 (2011)

23. Abdullah, S., et al.: Patterns of age-specific mortality in children in endemic areas of sub-Saharan Africa. Am. J. Trop. Med. Hyg. **77**(6), 99–105 (2007). Defining and Defeating the Intolerable Burden of Malaria III: Progress and Perspectives

24. Doolan, D.L., Dobaño, C., Baird, J.K.: Acquired immunity to Malaria. Clin. Microbiol. Rev. **22**(1), 13–36 (2009)

25. Bellan, S.E., Pulliam, J.R., Dushoff, J., Meyers, L.A.: Ebola control: effect of asymptomatic infection and acquired immunity. Lancet **384**(9953), 1499–1500 (2014)

26. Kojima, N., Klausner, J.D.: Protective immunity after recovery from SARS-CoV-2 infection. Lancet Infect. Dis. **22**(1), 12–14 (2022)

Inductive Linear Probing for Few-Shot Node Classification

Hirthik Mathavan[(✉)] [iD], Zhen Tan [iD], Nivedh Mudiam [iD], and Huan Liu [iD]

Computer Science and Engineering, Arizona State University, Tempe, AZ 85287, USA
{hmathava,ztan36,nmudiam,huanliu}@asu.edu

Abstract. Meta-learning has emerged as a powerful training strategy for few-shot node classification, demonstrating its effectiveness in the transductive setting. However, the existing literature predominantly focuses on transductive few-shot node classification, neglecting the widely studied inductive setting in the broader few-shot learning community. This oversight limits our comprehensive understanding of the performance of meta-learning based methods on graph data. In this work, we conduct an empirical study to highlight the limitations of current frameworks in the inductive few-shot node classification setting. Additionally, we propose applying a competitive baseline approach specifically tailored for inductive few-shot node classification tasks. We hope our work can provide a new path forward to better understand how the meta-learning paradigm works in the graph domain.

Keywords: Network Analysis · Few-shot Learning · Meta Learning

1 Introduction

Graphs have found extensive applications across various research fields, including social network analysis [12], bioinformatics [13], recommendation systems [11], and more. Graphs are crucial in understanding user interactions, sentiment analysis, and community detection in social media mining. For example, consider a scenario where we aim to classify user's sentiments towards a particular product or event on a social media platform. The graph can represent users as nodes and their connections as edges, capturing their relationships and interactions. By analyzing the structural properties of the graph, such as user connections, and incorporating node attributes like past sentiments or textual content, node classification algorithms can assign sentiment labels to new, unlabeled users. However, getting labeled data for node classification can take time and effort in real-world scenarios. Few-shot learning, a sub-field of machine learning, attempts to address this issue by creating a model using just a few examples. Few-shot learning has gained significant interest lately because of its capability to learn swiftly from a restricted amount of labeled data.

In recent years, meta-learning, also known as learning to learn, has emerged as a powerful technique for few-shot learning. Meta-learning involves training

© The Author(s), under exclusive license to Springer Nature Switzerland AG 2023
R. Thomson et al. (Eds.): SBP-BRiMS 2023, LNCS 14161, pp. 274–284, 2023.
https://doi.org/10.1007/978-3-031-43129-6_27

a model on a variety of tasks to learn a set of shared parameters that can be quickly adapted to new tasks with limited labeled data. In the context of graph node classification, meta-learning [5] has been used to train models that can quickly adapt to new graphs with a few labeled examples.

While meta-learning has demonstrated promising results in the field of few-shot node classification [14], most of the existing works have focused on the transductive setting, where the graph neural network (GNN) encoder is trained and evaluated on the same graph. The inductive setting, where the model is trained on a set of graphs and tested on a new, unseen graph, has received less attention in the few-shot learning community. Also, due to the message passing mechanism, where nodes exchange information with their neighboring nodes to update their own representations in GNNs, the inductive setting poses additional challenges compared to the transductive setting. Consider the example of sentiment analysis described before. In an inductive setting, we encounter new social media platforms or events where we need to classify user sentiment without access to the entire graph used during training. This reflects the reality of dealing with evolving social media platforms and ever-changing user dynamics.

Inductive few-shot learning allows us to train a model on a diverse set of graphs and test its performance on unseen graphs, mimicking the real-world scenario where we encounter novel contexts. This emphasizes the importance of studying and developing effective few-shot learning approaches in the inductive setting, enabling models to adapt and make accurate predictions in dynamic real-world environments. Therefore, this work aims to bridge this gap by providing a comprehensive study of meta-learning for few-shot node classification in the inductive setting. We empirically show that most current meta-learning frameworks cannot perform well in this setting. We propose to apply a straightforward yet effective baseline approach for inductive few-shot node classification tasks.

2   Related Work

In this section, we present an comprehensive review of the current literature concerning few-shot node classification and meta-learning, with a specific focus on the transductive setting.

2.1   Few-Shot Learning

Few-shot learning (FSL) is a machine learning paradigm that serves to address concerns of limited data by capitalizing on knowledge gained from previous training data. Some example of models that employ FSL are Model-Agnostic Meta-Learning (MAML), Prototypical Networks, and Meta-GNN.

MAML [2] tackles the few-shot learning problem by learning an optimal initialization of model parameters. It enables fast adaptation to new tasks with limited examples through a two-step process: an inner loop for task-specific updates and an outer loop for optimizing adaptation across tasks. By iteratively fine-tuning the parameters, MAML achieves effective generalization and enables efficient few-shot learning across various domains. Prototypical Networks [1] capture the essence of similarities and dissimilarities among instances through a metric-based approach by computing class prototypes based on support examples and using distance-based classification. This approach enables accurate classification in few-shot scenarios which over various domains offers a valuable approach to few-shot learning tasks. Meta-GNN [3] instead primarily addresses few-shot learning when provided with graph structured data. The model enhances the capability of GNNs to capture expressive node representations and effectively generalize to new classes or tasks with limited labeled data.

2.2 Meta Learning

In the context of few-shot node classification, meta-learning algorithms have been proposed to learn effective representations and update strategies for handling new, unseen classes with only a few labeled examples. Popular meta-learning algorithms for few-shot learning include GPN, G-Meta etc.

Graph Prototypical Network (GPN) [5,18] introduces graph prototypes, learned through iterative aggregation with GNNs, as representative embeddings from the support set. By utilizing these prototypes, GPN achieves accurate few-shot classification by computing similarity scores between query nodes and prototypes. GPN's incorporation of graph-level information and iterative aggregation enables effective generalization and robust few-shot classification on graph-structured data. G-Meta [4] combines subgraph extraction with GNNs to learn expressive node representations. It employs the MAML strategy to iteratively update and meta-update GNN parameters. This enables efficient adaptation to new tasks and improved classification on query nodes. Other models like AMM-GNN extend MAML with an attribute matching mechanism, and TENT reduces the variance among different meta-tasks for better generalization performance. Existing works primarily focus on transductive few-shot node classification, neglecting the widely studied inductive setting. We empirically evaluate meta-learning frameworks in the inductive setting to gain deeper insights into their performance on graphs.

3 Preliminaries

3.1 Problem Statement

The problem of few-shot node classification is concerned with attributed networks represented as $G = (\mathcal{V}, \mathcal{E}, X) = (A, X)$, where V is the set of nodes $v_1, v_2, \ldots, v_n$, $\mathcal{E}$ is the set of edges $e_1, e_2, \ldots, e_m$, $X = [x_1; x_2; \ldots; x_n] \in \mathbb{R}^{n \times d}$ is the matrix of node features, and $A = \{0, 1\}^{n \times n}$ is the adjacency matrix representing the network structure. Each element in A is either 0 or 1, indicating the absence or presence of an edge between nodes. The task involves a series of node classification tasks $T = \{T_i\}_{i=1}^{I}$, where T_i is a dataset for a particular task, and I is the number of such tasks. The classes of nodes available during training are referred to as base classes, while the classes during the target test phase are referred to as novel classes, and the intersection of the two sets is empty. Notably, under different settings, labels of nodes for training (i.e., C_{base}) may or may not be available during training. Conventionally, there are few labeled nodes for novel classes C_{novel} during the test phase.

Definition 1. Few-shot Node Classification (FSNC): Few-shot node classification refers to a problem in which an attributed graph $G = (A, X)$ is given, with a label space C divided into two sets, C_{base} and C_{novel}. The goal is to predict the labels of unlabeled nodes (query set Q) from C_{novel}, given only a few labeled nodes (support set S) for C_{novel}. If each task in the test set has N novel classes and K labeled nodes for each class, then this task is referred to as an N-way K-shot node classification problem.

Transductive Setting: In the transductive setting, the input graph is observed in all dataset splits, including the training, validation, and test sets (Fig. 1). The graph remains intact, and only the node labels are split for training and evaluation purposes. During training, embeddings are computed using the entire graph, and the model is trained

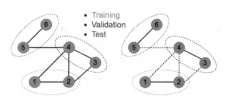

Fig. 1. Transductive/Inductive Setting

using the labels of selected nodes (e.g., node 1 and node 2). During validation, embeddings are again computed using the entire graph, and the model's performance is evaluated on the labels of other nodes (e.g., node 3 and node 4).

Inductive Setting: In the inductive setting, the graph is modified by breaking the edges between the dataset splits, resulting in different neighbor environments for nodes compared to the transductive setting (Fig. 1). For example, node 4 will no longer have an influence on the prediction of node 1. During training, embeddings are computed using the graph specific to the training split, such as the graph over node 1 and node 2. The model is trained using the labels of these selected nodes. During validation, embeddings are computed using the graph specific to the validation split, such as the graph over node 3 and node 4. The model's performance is then evaluated on the labels of these respective nodes (node 3 and node 4). This will further lead to the change of message passing, making it harder for GNNs to learn generalizable knowledge [13].

3.2 Episodic Meta-Learning for FSNC

Episodic meta-learning has emerged as an effective paradigm for addressing few-shot learning tasks, garnering substantial attention [16,17]. The underlying concept of episodic meta-learning involves training neural networks to mimic the evaluation conditions, which is believed to improve prediction performance on test tasks [16,17]. This paradigm has been successfully extended to few-shot node classification in the graph domain, as demonstrated by recent works [5,14,18]. In the context of few-shot node classification, the training phase follows a specific procedure. Meta-train tasks or episodes, denoted as T_{tr}, are generated from a base class set C_{base}, to emulate the test tasks. These episodes adhere to N-way K-shot node classification specifications. Each episode, denoted as T_t, comprises a support set S_t, and a query set Q_t, defined as follows:

$$
\begin{aligned}
T_{tr} &= \{T_t\}_{t=1}^{\mathcal{T}} = \{T_1, T_2, ..., T_{\mathcal{T}}\}, \\
T_t &= \{S_t, Q_t\}, \\
S_t &= \{(v_1, y_1), (v_2, y_2), \ldots, (v_{N \times K}, y_{N \times K})\}, \\
Q_t &= \{(v_1, y_1), (v_2, y_2), \ldots, (v_{N \times K}, y_{N \times K})\}.
\end{aligned}
\tag{1}
$$

In a typical meta-learning method, within each episode, K labeled nodes are randomly sampled from N base classes to form the support set. This support set is then used to train a GNN model, simulating the N-way K-shot node classification scenario during the test phase. Subsequently, the GNN predicts labels for a query set, which comprises nodes randomly sampled from the same classes as the support set. The optimization process involves minimizing the Cross-Entropy Loss (L_{CE}) w.r.t. the GNN encoder g_θ and the classifier f_ϕ:

$$
\theta, \psi = \arg \min_{\theta, \psi} L_{CE}(T_t; \theta, \psi).
\tag{2}
$$

Several approaches have been proposed based on this framework such as Meta-GNN [3], GPN [5], G-Meta [4] etc. Nevertheless, the evaluation of these methods has predominantly been conducted under transductive settings, neglecting the exploration of their performance in inductive settings.

3.3   Proposed Baseline

Our work is motivated by the Intransigent GNN model (I-GNN) introduced by a previous study [15,19]. The I-GNN model proposes a straightforward approach for few-shot learning that relies on reusing features instead of using complex meta-learning algorithms to achieve fast adaptation. The authors show that the I-GNN model, despite its simplicity, can achieve competitive performance compared to meta-learning based approaches. In our study, we adapt the I-GNN model to the inductive setting and propose a simple yet effective baseline for inductive few-shot node classification tasks.

The I-GNN model is designed to be inflexible and unadaptable to new tasks. The training process of I-GNN is split into two phases. In the first phase, a GNN encoder (g_θ) and a linear classifier (f_ϕ) are pre-trained on all base classes (C_{base}) using vanilla supervision through the L_{CE} loss function. A weight-decay regularization term is also applied during this phase. In the second phase, the parameter of the GNN encoder is frozen, and the classifier is discarded. When fine-tuning on a target few-shot node classification task, the pretrained GNN encoder is used to directly transfer embeddings of all nodes from the task, and a new linear classifier (f_ψ) is involved and tuned with few-shot labeled nodes from the support set (S_i) to predict labels of nodes in the query set (Q_i).

$$T'_{tr} = \cup\{T_t\}_{t=1}^{\mathcal{T}} = \cup\{T_1, T_2, ..., T_{\mathcal{T}}\}$$
$$\theta, \phi = \arg \min_{\theta,\phi} L_{CE}(T'_{tr}; \theta, \phi) + R(\theta), \qquad (3)$$

$$\psi = \arg \min_{\psi} L_{CE}(S_i; \theta, \psi) \qquad (4)$$

4 Empirical Evaluation

4.1 Experimental Settings

In this research study, various methods for few-shot node classification are evaluated through systematic experiments under the inductive setting. These methods include ProtoNet [1], MAML [2], Meta-GNN [3], G-Meta [4], GPN [5], AMM-GNN [6], and TENT [7]. The performance of these methods is compared on five real-world graph datasets: CoraFull [8], Coauthor-CS [9], Amazon-Computer [9], Cora [10], and CiteSeer [10].

Table 1. Statistics of Benchmark Datasets

Dataset	# Nodes	# Edges	# Features	$|C|$	$|C_{train}|$	$|C_{dev}|$	$|C_{test}|$
CoraFull	19,793	63,421	8,710	70	40	15	15
Coauthor-CS	18,333	81,894	6,805	15	5	5	5
Amazon-Computer	13,752	245,861	767	10	4	3	3
Cora	2,708	5,278	1,433	7	3	2	2
CiteSeer	3,327	4,552	3,703	6	2	2	2

CoraFull, Coauthor-CS, Amazon-Computer, Cora, and CiteSeer are five prevalent real-world graph datasets, each consisting of multiple node classes for training and evaluation. These datasets include citation networks, co-authorship graphs, and co-purchase graphs, and the task is to predict the category of a certain publication or paper. The number of node classes used for training, development, and testing varies depending on the dataset. Table 1 describes the statistics of the datasets.

4.2 Evaluation Protocol

This section outlines the evaluation protocol used to compare the meta-learning methods. The node label space C of an graph dataset $G = (A, X)$ is divided into $\{C_{base}, C_{novel}$ or $C_{test}\}$. C_{base} is split into C_{train} and C_{dev} (division strategy for each dataset are in Table 1). Evaluation is done by providing a GNN encoder g, a

---

Algorithm 1. UNIFIED EVALUATION PROTOCOL FOR FEW-SHOT NODE CLASSIFICATION

Input: Graph G, C_{train}, C_{dev}, C_{test}; GNN g, classifier f; parameters $EI = 10, S = 100, E = 10, M = 10000, T = 5, N = 2, 5, K = 1, 3, 5, Q = 10$

Output: f, accuracy $\mathcal{A}$, confident interval I, trained models g

    Repeat experiment for T times

1: **for** $i = 1, 2, ..., T$ **do**
2:    $j \leftarrow 1, k \leftarrow 1, a_{best} \leftarrow 0$;
3:    **while** $k \leq M$ **do**
4:       Optimize g based on the specific training strategy;   ▷ Training
5:       **if** $k \bmod EI = 0$ **then**
6:          Sample S meta-tasks from C_{dev} on G;   ▷ Validation
7:          Calculate the obtained few-shot node classification accuracy a;
8:          **if** $a > a_{best}$ **then** $a_{best} \leftarrow a, j \leftarrow 0$;
9:          **else** $j \leftarrow j + 1$;
10:        **end if**
11:      **end if**
12:      **if** $j = E$ **then** break;   ▷ Early Break
13:      **end if**
14:    **end while**
15:    Sample S meta-tasks from C_{test} on G;   ▷ Test
16:    Calculate the obtained classification accuracy a_{test};
17:    $a_r \leftarrow a_{test}, i \leftarrow i + 1$;
18: **end for**
19: Calculate averaged accuracy $\mathcal{A}$ and confident interval $\mathcal{CI}$ based on $\{a_1, a_2, ..., a_i\}$;

---

classifier, f, an epoch interval EI for validation, S sampled meta-tasks for evaluation, E epoch patience, M maximum epoch number, T experiment repeated times, and N-way K-shot, Q-query settings specification. The Algorithm 1 calculates the final FSNC accuracy $\mathcal{A}$ and confident interval $\mathcal{CI}$. The default values of all the parameters are as follows, $EI = 10; S = 100; E = 10; M = 10000; T = 5; N = \{2, 5\}; K = \{1, 3, 5\}; Q = 10$.

4.3 Comparison

In Table 2, the performance of different meta-learning methods and the proposed baseline is compared for few-shot node classification tasks. The comparison includes four distinct few-shot settings: 5-way 1-shot, 5-way 5-shot, 2-way 1-shot, and 2-way 5-shot, allowing for a comprehensive analysis. The evaluation metrics used are the average classification accuracy and the 95% confidence interval, which are computed based on multiple repetitions (T). Figure 2 presents the performance results of the CiteSeer dataset (similar trends observed in other datasets) for various N-way K-shot settings. The observations derived from the results are as follows:

- In the inductive setting, except for MAML and ProtoNet, meta-learning models exhibit a **significant performance drop** compared to the transductive setting. This decline is attributed to the challenges of generalizing knowledge from limited labeled examples to unseen data. In the transductive setting, models access the entire graph for predictions, while in the inductive setting, they must generalize to new nodes or graphs. Limited labeled data and the need for generalization contribute to lower performance in the inductive setting.
- I-GNN shows **superior performance** in the inductive setting compared to the transductive setting for certain datasets like Cora, Citeseer, and CoraFull. This can be due to its ability to capture more transferable node embedding in the inductive setting.

Table 2. Few-shot node classification results of meta-learning methods and I-GNN. Accuracy (↑) and Confidence Interval (↓) are in %. The best and second best results are bold and underlined, respectively.

Dataset	CoraFull		Coauthor-CS		Cora		Amazon-Computer		CiteSeer	
Settings	5-way 1-shot	5-way 5-shot	5-way 1-shot	5-way 5-shot	2-way 1-shot	2-way 5-shot	2-way 1-shot	2-way 5-shot	2-way 1-shot	2-way 5-shot
Inductive										
MAML	22.63 ± 1.19	27.21 ± 1.32	27.98 ± 1.42	42.12 ± 1.40	53.13 ± 2.26	57.39 ± 2.23	52.67±2.11	58.23±2.53	52.39±2.20	54.13±2.18
ProtoNet	32.43 ± 1.61	51.54 ± 1.68	32.13 ± 1.52	49.25 ± 1.50	53.04 ± 2.36	57.92 ± 2.34	61.98±2.95	70.20±2.64	52.51±2.44	55.69±2.27
Meta-GNN	34.97 ± 1.78	49.32 ± 1.99	37.78 ± 2.02	51.17 ± 1.91	52.09 ± 2.39	58.21 ± 1.52	55.47±2.43	59.12±2.55	51.18±2.04	63.68±2.65
GPN	27.90 ± 1.35	36.40 ± 1.82	35.00 ± 1.55	49.30 ± 2.63	50.00 ± 1.89	55.00 ± 1.81	49.75±0.85	54.25±2.45	52.75±1.85	59.50±2.10
AMM-GNN	36.45 ± 1.99	52.09 ± 1.90	53.30 ± 2.39	**72.64 ± 1.48**	54.36 ± 2.20	60.01 ± 2.40	51.99±1.51	52.48±1.57	52.40±2.14	54.63±2.24
G-Meta	40.76 ± 2.19	57.69 ± 1.93	46.79 ± 1.95	66.95 ± 1.43	53.78 ± 2.05	58.35 ± 2.15	52.27±1.98	61.03±2.19	52.21±2.17	54.92±2.26
TENT	38.90 ± 2.20	54.32 ± 1.65	**53.52 ± 1.73**	68.16 ± 1.18	50.40 ± 2.01	59.80 ± 2.38	**82.40±2.28**	**92.00±1.18**	57.35±2.74	64.55±2.63
I-GNN	**47.14 ± 2.08**	**59.01 ± 1.82**	37.23 ± 1.70	51.24 ± 1.42	**62.33 ± 2.67**	**70.16 ± 2.05**	59.08±2.67	68.35±2.48	**60.04±1.55**	**73.63±2.03**
Transductive										
MAML	22.63 ± 1.19	27.21 ± 1.32	27.98 ± 1.42	42.12 ± 1.40	53.13 ± 2.26	57.39 ± 2.23	52.67±2.11	58.23±2.53	52.39±2.20	54.13±2.18
ProtoNet	32.43 ± 1.61	51.54 ± 1.68	32.13 ± 1.52	49.25 ± 1.50	53.04 ± 2.36	57.92 ± 2.34	61.98±2.95	70.20±2.64	52.51±2.44	55.69±2.27
Meta-GNN	55.33 ± 2.43	70.50 ± 2.02	52.86 ± 2.14	68.59 ± 1.49	65.27 ± 2.93	72.51 ± 1.91	65.19±3.29	78.65±3.12	56.14±2.62	67.34±2.10
GPN	52.75 ± 2.32	72.82 ± 1.88	60.66 ± 2.07	**81.79 ± 1.18**	62.61 ± 2.71	76.39 ± 2.33	57.26±1.50	77.63±2.91	53.10±2.39	63.09±2.50
AMM-GNN	58.77 ± 2.49	75.61 ± 1.78	62.04 ± 2.26	81.78 ± 1.24	65.23 ± 2.67	**82.30 ± 2.07**	71.04±3.56	79.21±3.38	54.53±2.51	62.93±2.42
G-Meta	**60.44 ± 2.48**	**75.84 ± 1.70**	59.68 ± 2.16	74.18 ± 1.29	**67.03 ± 3.22**	80.05 ± 1.98	63.68±3.05	70.21±3.16	55.15±2.68	64.53±2.35
TENT	55.44 ± 2.08	70.10 ± 1.73	**63.70 ± 1.88**	76.90 ± 1.19	53.05 ± 2.78	62.15 ± 2.13	71.15±3.11	79.25±2.61	62.75±3.23	72.95±2.13
I-GNN	42.70 ± 1.92	51.46 ± 1.69	43.89 ± 1.82	55.93 ± 1.46	54.45 ± 3.13	65.18 ± 2.21	62.3±22.89	72.81±2.93	58.70±3.17	65.60±2.59

- The scores for both MAML and ProtoNet **remain the same** on all datasets because they do not utilize message-passing GNN in their approach. Since they do not leverage the graph structure and operate on a per-node basis, the performance drop observed in other meta-learning models under the inductive setting does not affect them in the same way. Therefore, their performance remains consistent between the transductive & inductive settings.
- The I-GNN model **outperforms** the meta-learning-based methods under the inductive setting, particularly on datasets like Cora, CiteSeer and Corafull, while demonstrating competitive performance on other datasets. This can be attributed to the fact that meta-learning methods typically require a large number of samples to learn effectively.

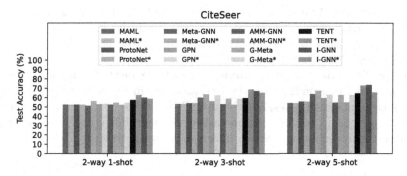

Fig. 2. Meta-Learning, I-GNN with inductive and transductive (*)

4.4 Further Analysis

To make a direct comparison between the results of meta-learning methods and I-GNN, we present additional findings in Fig. 3 and Fig. 4, which showcase the performance of all methods across different N-way K-shot settings. By analyzing these results, we can draw the following conclusions.

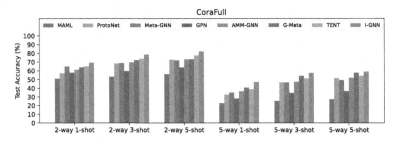

Fig. 3. N-way K-shot results of CoraFull, Meta-Learning and I-GNN.

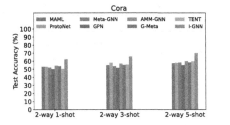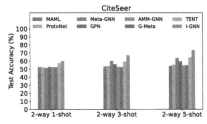

Fig. 4. N-way K-shot results of Cora and CiteSeer, Meta-Learning and I-GNN.

- As N increases, the **performance of all methods deteriorates** due to the greater variety of classes within each meta-task. This increased complexity poses challenges for classification tasks, resulting in lower performance. Figure 3 demonstrates the impact of increasing N on the classification performance using the CoraFull dataset.
- The **performance improvement** of the I-GNN method compared to meta-learning methods on the Cora dataset, as shown in Fig. 4, is notable due to its smaller number of classes, allowing I-GNN to leverage structural information for better generalization. The meta-learning methods struggle to effectively utilize the available supervision information during training.

5 Conclusion

In this paper, we investigate the performance of meta-learning methods in the inductive few-shot node classification tasks. While existing research primarily focused on the transductive setting, the inductive setting has received limited attention in the few-shot learning community. To bridge this gap, we conduct a comprehensive study of meta-learning for inductive few-shot node classification. Our empirical analysis reveals that most current meta-learning frameworks struggle in the inductive setting. To address this challenge, we propose applying a competitive baseline model called I-GNN. Experimental evaluations on five real-world datasets showcase the effectiveness of our proposed model. Our findings emphasize the need for further research in exploring the potential of meta-learning in the inductive setting, contributing to a more comprehensive understanding of few-shot node classification.

References

1. Snell, J., Swersky, K., Zemel, R.: Prototypical networks for few-shot learning. In: NeurIPS (2017)
2. Finn, C., Abbeel, P., Levine, S.: Model-agnostic meta-learning for fast adaptation of deep networks. In: ICML (2017)
3. Zhou, F., Cao, C., Zhang, K., Trajcevski, G., Zhong, T., Geng, J.: Meta-gnn: on few-shot node classification in graph meta-learning. In: CIKM (2019)
4. Huang, K., Zitnik, M.: Graph meta learning via local subgraphs. In: NeurIPS (2020)
5. Ding, K., Wang, J., Li, J., Shu, K., Liu, C., Liu, H.: Graph prototypical networks for few-shot learning on attributed networks. In: CIKM (2020)
6. Wang, N., Luo, M., Ding, K., Zhang, L., Li, J., Zheng, Q.: Graph few-shot learning with attribute matching. In: Proceedings of the 29th ACM International Conference on Information and Knowledge Management (2020)
7. Wang, S., Ding, K., Zhang, C., Chen, C., Li, J.: Task-adaptive few-shot node classification. arXiv preprint arXiv:2206.11972 (2022)
8. Bojchevski, A., Günnemann, S.: Deep gaussian embedding of graphs: unsupervised inductive learning via ranking. In: ICLR (2018)
9. Shchur, O., Mumme, M., Bojchevski, A., Günnemann, S.: Pitfalls of graph neural network evaluation. In: Relational Representation Learning Workshop, NeurIPS 2018 (2018)
10. Yang, Z., Cohen, W., Salakhudinov, R.: Revisiting semi-supervised learning with graph embeddings. In: International Conference on Machine Learning, pp. 40–48. PMLR (2016)
11. Ying, R., He, R., Chen, K., Eksombatchai, P., Hamilton, W.L., Leskovec, J.: Graph convolutional neural networks for web-scale recommender systems. In: Proceedings of the 24th ACM SIGKDD International Conference on Knowledge Discovery & Data Mining (2018)
12. Liu, P., De Sabbata, S.: Estimating locations of social media content through a graph-based link prediction. In: Proceedings of the 13th Workshop on Geographic Information Retrieval (2019)
13. Yi, H.-C., You, Z.-H., Huang, D.-S., Kwoh, C.K.: Graph representation learning in bioinformatics: trends, methods and applications. Brief. Bioinf. **23** (2021)
14. Zhou, F., Cao, C., Zhang, K., Trajcevski, G., Zhong, T., Geng, J.: Meta-GNN. In: Proceedings of the 28th ACM International Conference on Information and Knowledge Management (2019)

15. Tan, Z., Wang, S., Ding, K., Li, J., Liu, H.: Transductive linear probing: a novel framework for few-shot node classification. arXiv preprint arXiv:2212.05606 (2022)
16. Mishra, N., Rohaninejad, M., Chen, X., Abbeel, P.: A simple neural attentive meta-learner. In: ICLR (2018)
17. Ravi, S., Larochelle, H.: Optimization as a model for few-shot learning. In: International Conference on Learning Representations (2016)
18. Tan, Z., Ding, K., Guo, R., Liu, H.: Graph few-shot class-incremental learning. In: WSDM (2022)
19. Tan, Z., Ding, K., Guo, R., Liu, H.: Supervised graph contrastive learning for few-shot node classification. In: ECML-PKDD (2022)

Regression Chain Model for Predicting Epidemic Variables

Kirti Jain[1]([✉]), Vasudha Bhatnagar[1], and Sharanjit Kaur[2]

[1] Department of Computer Science, University of Delhi, Delhi, India
{kjain1,vbhatnagar}@cs.du.ac.in
[2] Acharya Narendra Dev College, University of Delhi, Delhi, India
sharanjitkaur@andc.du.ac.in

Abstract. Real-time detection and forecasting of disease dynamics is critical for healthcare authorities during epidemics. In this paper, we report a systematic investigation into the possibility of predicting three epidemic variables, viz., *peak day*, *peak infections*, and *span* of the epidemic using the Regression Chain Model.

We construct a dataset, *EpiNet*, using 35K synthetic networks of varied sizes and belonging to three network families. The dataset consists of five network features and three target variables obtained by simulating the SEIR epidemic model on the networks. We train Regression Chain Model (RCM) using four popular machine learning algorithms to predict the target variables. The model generally performs fairly well for *peak day* and *peak infections*, but the performance degrades for the *span* variable. Our preliminary investigation motivates further inquiry into the use of RCMs to replace computationally expensive epidemic simulations on larger networks.

Keywords: Contact network · Topological properties · Epidemic variables · Machine learning · Regression chain model

1 Introduction

The COVID-19 pandemic provided an unprecedented boost to research in epidemic modeling. When an epidemic spreads to a large population, early and real-time estimations of the disease infectivity are of critical importance for healthcare policy planners and administrators for managing and controlling the spread of disease. However, it is often impractical or impossible to continuously monitor the entire population to estimate the size, span, and severity of the epidemic.

1.1 Background and Motivation

Compartmental mathematical models like susceptible-infected-susceptible (SIS), susceptible-infected-recovered (SIR), susceptible-exposed-infected-recovered (SEIR), etc., have served as indispensable tools for estimating the epidemic

© The Author(s), under exclusive license to Springer Nature Switzerland AG 2023
R. Thomson et al. (Eds.): SBP-BRiMS 2023, LNCS 14161, pp. 285–294, 2023.
https://doi.org/10.1007/978-3-031-43129-6_28

dynamics for almost a century [5]. The simplifying assumption of uniform inter-actions within the population (homogeneous mixing) is a well-understood caveat of these models [1]. This assumption not only affects the quality of estimates but also overlooks the complexity of the dynamics being modeled. This limitation has promoted research related to network-based simulations for understanding disease dynamics [5]. Network-based simulations of the epidemic models deliver comparatively more realistic approximates of epidemic dynamics due to the incorporation of connectivity patterns in the population. However, the cost of network-based simulations escalates steeply with the network size. This is the prime motivation to find alternatives to expensive simulations on large networks.

During network simulation of epidemics, the notable role played by the structure and the topological properties of contact networks in the spread of contagion has been established in several studies [6,9,10,12,13]. It is reasonable to infer that topological properties of networks carry the potential for predicting the epidemic variables, viz., *peak day*, *peak cases*, and *span*. Peak cases are the maximum number of infected cases on a given day. The day when the cases are maximum is the peak day and the span denotes the time period between the first and last infected cases.

Rodrigues et al. used machine learning models to identify and rank the topological properties of the network that are crucial to estimate the outbreak size of the epidemic [11]. The major limitation of this work is the use of features of a small subset of nodes, which can be misleading due to stochasticity and non-linearity in the simulation of epidemic spread. Bucur et al. used centrality measures as features to predict outbreak sizes in networks limited to ten nodes [3]. They empirically demonstrate that it is possible to accurately predict the outbreak using network measures in isomorphic networks. However, the network size used in this study is unrealistically small, and the method does not scale-up for application in the real world. Pérez-Ortiz et al. employ thirty network properties to predict the average percentage of infected individuals using linear regression [10]. Since distance-based network properties are computationally expensive, this limits the practical applicability of this approach to large networks.

1.2 Research Contributions

A critical analysis of recent related works reveals the following research gaps: i) use of computationally expensive topological properties to predict outbreak size, ii) prediction of only one epidemic variable, iii) effectiveness reported on small networks, iv) unavailability of data to reproduce results. Our empirical study fills these gaps and contributes in the following manner. We

  i. use five inexpensive topological features of the networks that distinctively influence the pathogen spread (Sect. 2.1).
 ii. address the prediction of three epidemic variables, viz., *peak day*, *peak cases*, and *span* using the Regression Chain Model (Sect. 2.2–2.3). We empirically validate our conjecture and demonstrate the possibility of accurately predicting epidemic variables using Regression Chain Model in a restricted environment. Our results encourage further study in this direction (Sect. 3).

iii. curate a dataset, *EpiNet*, with five topological features and three target variables obtained by simulating the SEIR[1] epidemic model on synthetic networks belonging to three diverse families. The dataset will permit the reproduction of results and promote further investigation in this direction by the research community (Sect. 2.4).

Organization: Section 2 describes the methodology used in this study. Experimental settings along with the results are presented in Sect. 3. Conclusion and future work are given in Sect. 4.

2 Methodology

In this section, we describe the methodology for network construction, topological features, the epidemic spread model, and the regression chain model, along with the estimators and the metrics used to evaluate the performance. We also give details of the construction of the dataset.

2.1 Network Models and Topological Properties

We use three network models belonging to diverse families, viz., Erdos-Renyi model, Watts-Strogatz model, and Stochastic Block model to construct random, small-world, and community-based networks, respectively. These networks are extensively used for modeling social structure in epidemiology [1,8]. For each constructed network, we select those features of contact networks that impact disease dynamics most strongly. We compute the following topological properties and use them as features to train the Regression Chain Model (RCM).

i. **Average Degree:** Average degree $\bar{k}$ is the global property of the network that determines the speed of transmission of disease in the network [12]. It is computed as $\bar{k} = \frac{1}{N}\sum_i^N k_i = \frac{2M}{N}$, where k_i, M, and N denote the degree of node i, total edges, and the number of nodes, respectively. *It is established that individuals with a higher average degree have more chances of contracting/transmitting the disease* [6,12,13].

ii. **Normalized Network Density:** Density d is defined as the ratio of the number of edges M over the maximum possible number of edges in the network of N nodes, and is computed as $d = \frac{2M}{N(N-1)}$. Following [14], we compute the normalized network density as $\bar{d} = 1 + \frac{\log d}{\log N/2}$ so that the density is comparable across networks of all sizes. *It is established that networks with higher density, favor rapid transmission of the pathogen, and lead to higher number of infected individuals* [6,12,13].

iii. **Degree Variance:** This metric characterizes degree heterogeneity within a network [15]. For a graph of size N, degree variance v is computed as $v = \frac{1}{N}\sum_i^N (k_i - \bar{k})^2$. *Moreno et al. show that networks with higher heterogeneity in degree cause stronger outbreak incidence* [7]. *In such networks,*

---

[1] Note that the choice of epidemic model and parameters are disease-specific.

the infection spreads more rapidly in comparison to networks with lower variance in node degree.

iv. **Average Clustering Coefficient:** The clustering coefficient captures the degree to which its neighbors are linked to each other. For a node i with degree k_i, its clustering coefficient is defined as $c_i = \frac{2m_i}{k_i(k_i-1)}$, where m_i represents the number of links between the k_i neighbors of node i. Note that $c_i \in [0, 1]$ is a local property of the node. The average clustering coefficient of a graph $(\bar{c})$ is the global property and is computed as $\bar{c} = \frac{1}{N} \sum_{i=1}^{N} c_i$. *It is established that the clustering coefficient is an important topological characteristic that prevails in human social networks and affects pathogen transmission* [6,12,13,16].

v. **Average Shortest Path Length:** It is defined as the shortest distance averaged over all pairs of nodes in a network. Since the average shortest path length of large networks is computationally expensive, we approximate it as $\bar{p} = \exp(\frac{1}{\theta_1 \bar{d} + \theta_0})$ using normalized density $(\bar{d})$ as given by [14]. The parameters θ_1 and θ_0 are derived from the regression of $\bar{d}$ and inverse of $log\,p$, where p is the shortest path length of the sampled small networks from each network type. *It is shown that networks with short average path length exhibit fast disease spread* [6,12,13].

2.2   Epidemic Spread on Networks

Different compartmental models in epidemiology have been successfully used to model the spread of numerous contagious diseases, including COVID-19, Ebola, Chikungunya, Measles, etc. [1]. We use the susceptible-exposed-infected-recovered (SEIR) model that addresses the exposed period commonly found in most transmissible diseases. The model assumes immunity to re-infection. The population is divided into compartments, and at each time step of the dynamics, an individual can be in one of four possible states: susceptible (S), exposed (E), infected (I), or recovered (R). An infected person exposes susceptible neighbors with probability β. The exposed individuals transit to the infected state with probability α. Infected persons eventually recover with probability γ. We assume a constant population with a uniform birth and death rate for simplicity.

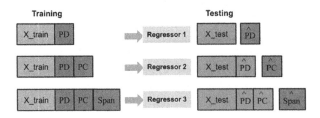

Fig. 1. RCM with the order as peak day (PD), peak cases (PC), and span (Span). Regressor 1 predicts PD, Regressor 2 predicts PC using predicted $\hat{P}D$ and Regressor 3 predicts Span using predicted $\hat{P}D$ and $\hat{P}C$.

2.3 Regression Chain Model

A Regression Chain Model (RCM) is an ensemble built over a chain of regressors to capture dependencies between the output variables (interested reader may refer to [2] for an extensive account on RCM). We set the order for the regression chain as *peak day* (PD), *peak cases* (PC), and *span* (Span). The order was empirically found to deliver the best quality predictions among all possible orders of the three target variables. Figure 1 shows the framework for the Regression Chain Modeling. We use four representative base algorithms (estimators), viz., Decision Tree, Random Forest, Kernel Ridge, and XG Boost. To quantify the assessment of prediction quality, we use two performance metrics. The first metric is the predicted coefficient of determination, $\langle R^2 \rangle$, and the second is the root mean squared error, $\langle RMSE \rangle$, both averaged over the three target variables. The higher value of $\langle R^2 \rangle$ indicates a better fit of the model with higher predictive performance, while the lower value of $\langle RMSE \rangle$ implies higher predictive accuracy.

2.4 Dataset Construction

In the absence of any real dataset for predicting epidemic variables, we curate a rich dataset called *EpiNet*, consisting of five network properties (features) and three epidemic variables (targets to be predicted) for networks with varying sizes (N) and average degrees ($\bar{k}$). We generate 15K small networks ($N \in [20K\text{--}60K]$), 10K medium networks ($N \in [60K\text{--}150K]$), and 10K large networks ($N \in [150K\text{--}300K]$), with equal number of instances belonging to three network families[2] - Random, Small-world, and Community-based networks. We set average degrees for all generated networks in the range [6, 40], as observed in real-life social networks from the SNAP[3] library.

We compute five topological properties mentioned in Sect. 2.1 for each constructed network. Subsequently, we simulate epidemic spread using the SEIR model for specific parameters and note three dependent epidemic variables, viz., peak day (in year), the fraction of infections on the peak day, and the span of the epidemic (in year). To mitigate the effect of stochasticity, we average epidemic variables over *ten* simulation runs of the SEIR spreading process on each network. Hence, we get a feature vector of five network properties and three target variables for each network. Based on the network sizes, we split *EpiNet* into three partitions, corresponding to small networks (D-SN), medium networks (D-MN), and large networks (D-LN). The dataset can be used for further investigation by the research community and is available on GitHub[4].

3 Experiments and Results

This section presents the details of the experiments carried out to examine the feasibility of using RCM as a substitute for expensive simulation of epidemic

---

[2] We omit scale-free networks as they are inappropriate to study epidemic spread [4].
[3] https://snap.stanford.edu/data/#socnets.
[4] https://github.com/kirtiJain25/EpiNet.

spread on networks for a given set of epidemic parameters. Following Pérez-Ortiz et al., we use $\beta = 0.155, \alpha = 1/5.2$ and $\gamma = 1/12.39$, for SEIR epidemic simulation and obtain the target variables in the training set. In line with our objective, we formulate the following research questions.

i. *How do the network properties influence epidemic variables for the specific epidemic model and epidemic parameters?* (Sect. 3.1)
ii. *Is it possible to predict three epidemic variables with reasonable accuracy using Regression Chain Model?* (Sect. 3.2)
iii. *How sensitive is the performance of RCM to network size?* (Sect. 3.3)

We create the networks using the Igraph library and simulate the SEIR epidemic model in Python (64bits, v3.7.2) on an Intel(R) Core(TM) i7 CPU @1.80 GHz with 16 GB RAM. We train RCM using the Scikit-learn library.

3.1 Influence of Network Properties on Epidemic Variables

We study the relationship between three selected network topological characteristics, viz., average degree, average clustering coefficient, and average shortest path length with three epidemic variables through scatter plots (Fig. 2).

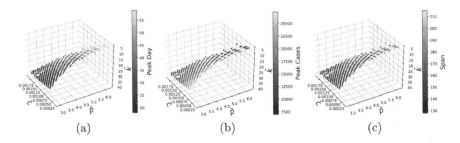

(a)   (b)   (c)

Fig. 2. Effect of three selected network properties, viz., average degree ($\bar{k}$), average clustering coefficient ($\bar{c}$), and average shortest path length ($\bar{p}$) on three epidemic variables, viz., peak day, peak cases, and epidemic span in Figs (a)–(c) respectively.

It is clear from figures (a)–(c) that topological properties have a profound impact on the three epidemic variables. Networks with low average degrees, high average path lengths, and small clustering coefficients have delayed *peak days* with reduced *peak infections* and larger *spans*. In addition, low path length and high average degree fuel the epidemic spread leading to early *peak day* and higher *peak infections*. We also observe that *peak day* has a negative relationship with *peak cases*. Early peak day results in a higher number of cases, and vice versa. On the other hand *peak day* and *span* are positively correlated. The earlier the peak day, the shorter the epidemic duration.

Thus, it is sufficient to conclude that network properties influence epidemic dynamics and are potent to be used as features to predict three epidemic variables.

3.2   Model Validation

We use ten-fold cross-validation method to assess the competence of regression chain models for all three training sets, followed by testing the model on unseen data. Figure 3 shows the heatmap of $\langle R^2 \rangle$ scores and $\langle RMSE \rangle$ values of the RCM using four base algorithms, (a) Decision Tree, (b) Random Forest, (c) Kernel Ridge, and (d) XG Boost. The lower and upper triangles of each cell show $\langle R^2 \rangle$ scores and $\langle RMSE \rangle$ values respectively.

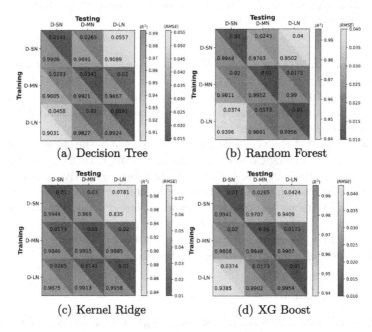

Fig. 3. Predictive performance of the RCMs trained and tested on topological properties of small (D-SN), medium (D-MN), and large (D-LN) networks. The lower and upper triangles in each cell denote $\langle R^2 \rangle$ scores and $\langle RMSE \rangle$ values, respectively.

It is clear that the cross-validated performance for all regressors (diagonal cells in the heatmap) is high. The non-diagonal cells in the heatmap correspond to the performance of the model on unseen data. We observe that performance degrades marginally, i.e. low $\langle R^2 \rangle$ scores and high $\langle RMSE \rangle$, for the model trained on D-SN and tested on D-LN, and vice versa. The overall high $\langle R^2 \rangle$ scores and low $\langle RMSE \rangle$, in all cases and for all base algorithms demonstrate the competence of the RCMs for predicting epidemic variables in networks.

Figure 4 shows the R^2 and RMSE scores of the model trained on D-SN and tested on D-MN and D-LN. Each metric is computed for three epidemic variables individually. We show the cross-validated performance on D-SN (dark blue colored bars in Fig. 4). We observe high R^2 and low RMSE scores for *peak day*

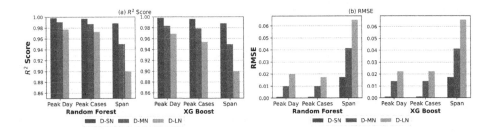

Fig. 4. Predictive performance of three epidemic variables by RCMs trained on D-SN and tested on D-SN, D-MN, and D-LN.

and *peak cases*, while the predicted R^2 score for *span* variable is comparatively low with high RMSE value for medium (D-MN) and large (D-LN) networks (Figs. 4(a) and (b)). This pulls down the overall $\langle R^2 \rangle$ score and raises the $\langle RMSE \rangle$ value when the model trained on small networks is used to predict epidemic variables for larger networks.

We conjecture that the prediction error for the *span* of the epidemic arises due to non-linearity in topological features with increasing network size. Our results motivate further study in this direction to improve predictive models so that all three epidemic variables are predicted accurately without performing costly epidemic simulations on the contact networks. *Nevertheless, it is reasonable to conclude from this experiment that RCMs are capable of predicting epidemic variables with high accuracy on similar-sized networks and may deliver slightly degraded performance on different-sized networks.*

3.3 Sensitivity Analysis

Having observed that the model performance degrades for networks of dis-similar sizes, we examine the sensitivity of the predicted variable to the size of the network. We train the model using D-SN, pool D-MN, and D-LN data sets, and test the model on the pooled test set. We group instances into eight batches (Fig. 5) with approximately equal numbers of records per batch and report batch-wise R^2 scores and RMSE values for Random Forest-based RCM.

We observe a marginal decrease in R^2 scores and a marginal increase in RMSE values for *peak days* and *peak cases* with increasing network size. However, the R^2 score degrades notably for the *span* variable for larger networks ($N > 120K$). This observation ratifies our earlier observation that the regression chain model trained using topological properties of small networks (D-SN) is capable of reliably predicting *peak day* and *peak cases* for medium and large networks, thereby saving computational expense incurred by epidemic simulations. However, the prediction of the *span* variable is not accurate. This is because the model is unable to capture the relationship between the network features and the duration of the epidemic spread.

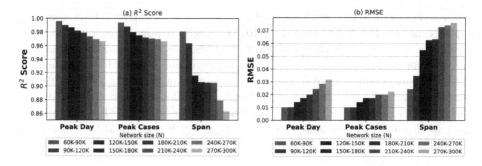

Fig. 5. Predictive accuracy for three epidemic variables for Random Forest-based regressor trained on small networks (D-SN) and tested on larger networks (D-MN and D-LN) grouped by network size.

We conclude that basic network characteristics are insufficient for accurately predicting the span variable, and further study is required to understand the role of other topological properties on the span of the epidemic.

3.4 Discussion

Our empirical study shows promising results and indicates that inexpensive models trained on small networks can predict two epidemic variables with reasonably high accuracy. Prediction of the third variable (epidemic span) is challenging. However, certain important issues must be noted before using this approach.

i. Sensitivity of the method to population size necessitates curating training data sets for different network sizes for accurate predictions. Models trained on networks of vastly dissimilar sizes may deliver inaccurate predictions.
ii. Since the constructed data set is disease-specific, the training set needs to be curated using the appropriate epidemic spread model and its parameters.
iii. Several observable and unobserved variables influence the epidemic spread in the complex landscape of social, political, and economic realities in the real world. To account for these factors, the simulation of the epidemic spread needs to be tweaked accordingly. Our simulations in this study do not account for any external factor and hence offer the *best-case* results.

We believe that this work will help advance the study of recognizing machine learning models that proxy for expensive network-based simulation.

4 Conclusion

In this research, we examine the possibility of predicting three epidemic variables using the regression chain model (RCM). We curate a rich data set called *EpiNet*, consisting of five network properties (features) and three epidemic variables (targets captured using SEIR epidemic model) for 35K networks of varying types and sizes. The dataset is split into three partitions (small - D-SN, medium - D-MN, large - D-LN), and we train RCM using four popular regressors.

Our results establish the possibility of predicting three epidemic variables, viz. *peak day*, *peak cases*, and *span*, using a Regression Chain Model trained on the topological properties of the underlying contact networks as a substitute for costly epidemic simulations on large networks. Detailed analysis of the predicted variables reveals that prediction accuracy for the *span* variable is lower compared to that of *peak days* and *peak cases*. Further study is warranted to understand the additional topological characteristics required for its accurate prediction.

References

1. Barabási, A.L.: Network Science Book. Cambridge University Press, Cambridge (2014). http://barabasi.com/networksciencebook
2. Borchani, H., Varando, G., Bielza, C., Larranaga, P.: A survey on multi-output regression. Wiley Interdisc. Rev. Data Mining Knowl. Disc. **5**(5), 216–233 (2015)
3. Bucur, D., Holme, P.: Beyond ranking nodes: predicting epidemic outbreak sizes by network centralities. PLoS Comput. Biol. **16**(7), 1–20 (2020)
4. Du, M.: Contact tracing as a measure to combat covid-19 and other infectious diseases. Am. J. Infect. Control **50**(6), 638–644 (2022)
5. Hethcote, H.W.: The basic epidemiology models: models, expressions for R0, parameter estimation and applications, vol. 16, chap. 1, pp. 1–61. World Scientific Publishing, Singapore (2008)
6. Keeling, M.J., Eames, K.T.D.: Networks and epidemic models. J. Royal Soc. Interface **2**(4), 295–307 (2005)
7. Moreno, Y., Pastor-Satorras, R., Vespignani, A.: Epidemic outbreaks in complex heterogeneous networks. Eur. Phys. J. B-Cond. Matter Complex Syst. **26**(4), 521–529 (2002)
8. Newman, M.E.J.: The structure and function of complex networks. SIAM Rev. **45**, 167–256 (2003)
9. Newman, M.E.: Spread of epidemic disease on networks. Phys. Rev. E **66**(1), 016128 (2002)
10. Pérez-Ortiz, et al.: Network topological determinants of pathogen spread. Sci. Rep. **12**(1), 1–13 (2022)
11. Rodrigues, F.A., et al.: A machine learning approach to predicting dynamical observables from network structure. arXiv preprint arXiv:1910.00544 (2019)
12. Shirley, M.D., Rushton, S.P.: The impacts of network topology on disease spread. Ecol. Complex. **2**(3), 287–299 (2005)
13. Small, M., Cavanagh, D.: Modelling strong control measures for epidemic propagation with networks - a COVID-19 case study. IEEE Access **8**, 109719–109731 (2020)
14. Smith, R.D.: Average path length in complex networks: patterns and predictions. arXiv preprint arXiv:0710.2947 (2007)
15. Snijders, T.A.: The degree variance: an index of graph heterogeneity. Social Netw. **3**(3), 163–174 (1981)
16. Volz, E.M., et al.: Effects of heterogeneous and clustered contact patterns on infectious disease dynamics. PLoS Comput. Biol. **7**, e1002042 (2011)

Physical Distancing and Mask Wearing Behavior Dataset Generator from CCTV Footages Using YOLOv8

Roland P. Abao$^{(\boxtimes)}$ ⓘ, Maria Regina Justina E. Estuar ⓘ,
and Patricia Angela R. Abu ⓘ

Department of Information System and Computer Science, Ateneo de Manila
University, Quezon City, Philippines
roland.p.abao@obf.ateneo.edu, {restuar,pabu}@ateneo.edu

Abstract. Computer simulations using agent-based approach aimed at
modeling human behavior require a robust dataset derived from actual
observation to serve as ground truth. This paper details an approach for
developing a movement behavior dataset generator from CCTV footages
with respect to two health-related behaviors: face mask wearing and
physical distancing, while addressing the privacy concerns of confidential CCTV data. A two-stage YOLOv8-based cascaded approach was
implemented for object tracking and detection. The first stage involves
tracking of individuals in the video feed to determine physical distancing
behavior using the pre-trained YOLOv8 xLarge model paired with Bot-
SORT multi-object tracker and OpenCV Perspective-n-Point pose estimation. The second stage involves determining the mask wearing behavior of the tracked individuals using the best-performing model among
the five YOLOv8 models (nano, small, medium, large, and xLarge), each
trained for 50 epochs on a custom CCTV dataset. Results show that the
custom-trained xLarge model performed the best on the mask detection
task with the following metric scores: mAP50 $= 0.94$; mAP50-95 $= 0.63$;
and F1 $= 0.872$. The faces of all the tracked individuals are blurred-out
in the resulting video frames to preserve the privacy of the CCTV data.
Finally, the developed system is able to generate the corresponding mask-
distancing behavior dataset and annotated output videos from the input
CCTV raw footages.

Keywords: YOLOv8 · dataset generator · CCTV data · physical
distancing · mask wearing

1 Background

On the onset of the COVID-19 pandemic, physical distancing and mask wearing were among the non-pharmaceutical health interventions recommended by

Ateneo Social Computing Laboratory and DOST-Engineering Research and Development for Technology (ERDT).

public health authorities to slow down the spread of the disease among the population. Today, COVID-19 is no longer considered a public health emergency of international concern [10]. However, for low-middle income countries (LMIC), limited health capacity still necessitates observation of mask wearing and social distancing. Living with the virus entails, first and foremost, having a substantial portion of the population vaccinated, and keeping up with the voluntary preventive health behaviors such as mask wearing and physical distancing specially in enclosed indoor spaces [4].

There is a need to capture public health behavior to develop a model that will be able to predict movement based on compliance or non-compliance to public health standards. Computer simulations aimed at modeling complex human behavior, as well as other domains of interest, require a robust dataset for the development and validation of a simulation model. Specifically, data-driven agent-based models (ABM) [7] requires having actual ground truth data for modeling. Observation and capturing of real world data is used to model and infer behavioral patterns of the agents and other latent processes based on the actual observed data. The reliability of ABMs can only be as good as the assumptions and calibrations performed in developing the model [2].

In the context of modeling physical distancing and mask wearing behavior, a good source of data are the actual behavior captured from closed circuit television (CCTV) cameras installed in informal public places. In this environment, individuals may be influenced by each other to follow or not the preventive health behaviors. Processing of CCTV data to extract relevant information may be accomplished by deep learning-based object detection algorithms such as the various variants of either two-stage detectors (e.g. R-CNN, fast R-CNN, and faster R-CNN) or single stage detectors (e.g. SSD and YOLO) [11]. Among the different object detection algorithms, the YOLO (You Only Look Once) framework has been used more often by researchers for its exceptional balance of fast inference speed and high-accuracy detection, enabling a real-time and accurate identification of objects in the video frames [5,8,11]. Since its inception in 2015, the original YOLO model had undergone numerous modifications, each of which built upon the prior versions to address previous flaws and improve detection performance. As of the time of writing, YOLOv8 [3] released last January of 2023 is the state-of-the-art version of YOLO, which has the highest detection accuracy as compared to its predecessors.

Processing of CCTV data to extract relevant information in relation to the mask wearing and physical distancing behavior, poses some challenges in relation to object detection. First, the object detection model should have the ability to detect the mask wearing status of each person in the video frame, as well as compute for the distance of one person to another. Existing pre-trained object detection models, including the latest YOLOv8 model, are able to detect multiple persons in the frame already with good accuracy. However, modification (fine-tuning) of the model is still required to detect specific use cases, such as the mask and distancing behavior of the person. Secondly, CCTV cameras, which are almost always placed on a fixed elevated location, produce video frames at a high angle or bird's-eye point-of-view (POV). Existing object detection models specifically designed to detect persons and faces in the frame, however, are usually trained on a datasets (e.g. PASCAL VOC and COCO datasets) mostly composed of images with over-the-shoulder or at the eye-level POV. Detection models trained from eye-level POV datasets may not necessarily perform very well on the actual captured CCTV feeds with high-angle POV because of the POV difference in the training process and the actual use case. Thirdly, CCTV data contains confidential information, prominently the facial features of the person in the frame. Data privacy measures should be carefully implemented to preserve the identity privacy of the individuals present in the resulting video feeds. Accounting for these challenges in extracting behavioral information from confidential CCTV data, this paper describes and presents current findings in the development a movement behavior dataset generator -with respect to physical distancing and mask wearing- from CCTV feeds, while preserving the identity privacy of the individuals in the resulting video frames. Specifically, the study addresses the following research questions:

1. How can the real-world physical distance be computed from the detected individuals in the CCTV footages?
2. How do the different YOLOv8 models (nano, small, medium, large, and xLarge) perform in the face mask detection task of CCTV footage?
3. When processing confidential CCTV data, how can the privacy of the individuals be preserved on the resulting output video frames?

Additionally, the following are the major contributions of the study:

- Five variants of face mask detection models based on YOLOv8 fine-tuned for high-angle point-of-view video frames such as in CCTV footages.
- A behavior dataset generator system employing a two-stage YOLOv8-based cascaded approach of processing CCTV data, whilst preserving the identity privacy of the individuals in the resulting video frames.

2 Methodology

This section describes the procedure in the development of a dataset generator for physical distancing and mask wearing behavior captured from CCTV footages. The first step involves collecting CCTV footages as presented in Sect. 2.1. A two-stage YOLOv8-based cascaded approach for object tracking and detection was then implemented in the study. The first stage, as detailed in Sect. 2.2, involves tracking the position of the individuals in the frame to determine their physical distancing behavior using a pre-trained YOLOv8 model. The second stage of the cascade involves using a custom-trained YOLOv8-based model for detecting the mask wearing behavior of the individuals, as detailed in Sect. 2.3. Finally, Sect. 2.4 presents the method for integrating the two stages -object tracking and mask detection- to generate a behavior dataset, as well as the method to preserve the privacy identity of individuals in the resulting CCTV frames.

2.1   Collecting Data from CCTV Footages

The trial data was obtained from CCTV footages located in informal public places within a university setting. All CCTV cameras were accompanied with signage informing the public of surveillance recording using CCTV camera. A total of 100 recorded footages were obtained from October 1 to 30, 2022. A non-disclosure agreement was executed by the concerned parties prior to receipt of the confidential CCTV data. The selected informal locations included study areas, activity areas, and walkways where individuals have a considerable leeway to practice wearing of face mask and observe physical distance from each other.

2.2   First Stage: Object Tracking of All Individuals in the Frame

In this stage, the study utilized the pre-trained YOLOv8 xLarge model, which has the highest mean average precision score ($mAP_{50-95} = 53.9$) among all the pre-trained YOLOv8 models [3], for detecting the individual person (class $= 0$) in the CCTV frames. For object tracking, the default Bot-SORT multi-object tracker algorithm [1] built-into the YOLO package was used. The object tracker produces a bounding box and assigns a unique ID for each tracked individual in the frame. The X and Y coordinates of the middle-bottom part of the bounding box, representing the point of the ground where the person is currently standing (or sitting), was used as the anchor point of the individual in the camera frame, which should be converted into the real-world XYZ coordinates (where $Z = 0$). Translating between the camera frame coordinates and the real-world coordinates was made possible by identifying 6 known 3-dimensional (3d) fixed points in the real-world and pinpointing its corresponding 2-dimensional (2d) points on the on the camera frame. The identified 3d to 2d translation points were applied to OpenCV's perspective-n-point pose estimation algorithm [6] to get the vector rotation and translation matrix. The dot product of the camera matrix properties and the concatenated rotation & translation matrices would result to the image projection matrix and its inverse projection, which can then be used to compute for the desired real-world coordinates given the input camera frame coordinates.

2.3   Second Stage: Mask Detection

For the mask detection stage, a custom mask dataset to fine-tune the pre-trained YOLOv8 models was first created consisting of a total of 500 random snapshots from the gathered CCTV data. All faces in the snapshots were manually annotated with either noMask (class $= 0$) or withMask (class $= 1$) using LABE-LIMG, an open-source image annotation tool. The custom dataset was split into a 80%-10%-10% distribution for the train, validation, and test set respectively. The custom mask dataset was then used to produce five YOLOv8-based mask detection models, namely: mask_YOLOv8n (nano), mask_YOLOv8s (small), mask_YOLOv8m (medium), mask_YOLOv8l (large) and mask_YOLOv8x (extra large). The model fine-tuning was implemented with the following parameters:

Algorithm 1: Pseudo-code of the two-stage YOLOv8-based cascaded approach for object tracking and detection in generating a behavior dataset from CCTV data

---

Data: $CCTV\_footage$, $3d\_to\_2d\_translation\_pts$
Result: $output\_csv\_file$ /* Behavior dataset CSV file */
            $output\_video\_file$ /* Annotated output video file */

1   $time \leftarrow 0$
2   $df \leftarrow pandas.DataFrame()$
     /* First model: object tracking of individuals in the frames   */
3   $model \leftarrow YOLO(yolov8x)$
4   $frame\_results \leftarrow model.track(CCTV\_footage,\ tracker = botsort)$
5   **for** $result$ **in** $frame\_results$ **do**
6      $realXY \leftarrow \{\}$
      /* Second model: mask detection of all faces in the frame   */
7      $mask\_model \leftarrow YOLO(best\_mask\_model)$
8      $mask\_results \leftarrow mask\_model.predict(result.orig\_img)$
      /* Pixelate all detected faces to preserve identity privacy */
9      $resulting\_frame \leftarrow result.orig\_img.copy()$
10   **for** box_face **in** $mask\_results.boxes$ **do**
11       $x1, y1, x2, y2 \leftarrow box\_face.xyxy$
12       $resulting\_frame[y1 : y2, x1 : x2] \leftarrow pixelate\_face(box\_face.xyxy)$
13   **end**
      /* Identifying the mask wearing status of the individuals   */
14   **for** box **in** $result.boxes$ **do**
15       $mask\_stat,\ mask\_conf,\ max\_area \leftarrow null, null, min\_integer$
16   **for** box_face **in** $mask\_results.boxes$ **do**
17   **if** $area\_intersection(box.xyxy,\ box\_face.xyxy) > max\_area$ **then**
18           $mask\_stat,\ mask\_conf \leftarrow box\_face.cls,\ box\_face.conf$
19           $max\_area \leftarrow area\_intersection(box.xyxy,\ box\_face.xyxy)$
20   **end**
21   **end**
22       $id \leftarrow box.id$
23       $realXY[id] \leftarrow get\_world\_XY(box.xyxy,\ 3d\_to\_2d\_translation\_pts)$
24   **append** $[time, id, box.xyxy, mask\_stat, mask\_conf, realXY[id]]$ to df
25   **end**
      /* Computing for the Euclidean distance of the individuals   */
26   **for** row **in** $df[df.time == time].iterrows()$ **do**
27       $closest\_dist \leftarrow max\_integer$
28   **for** j **in** $realXY$ **where** $row.id \neq j$ **do**
29         $d \leftarrow math.dist(realXY[row.id],\ realXY[j])$
30   **if** $d \leq closest\_dist$ **then**
31           $closest\_dist \leftarrow d$
32   **end**
33   **end**
34   **annotate** $resulting\_frame$ **with** $row.id, row.mask\_stat, closest\_dist$
35   **end**
36     $time \leftarrow time + 1$
37   **write** $resulting\_frame$ to $output\_video\_file$
38 **end**
39 **save** df **as** $output\_csv\_file$

---

epoch of 50, batch size of 8, image size of 640 pixels, SGD as the optimizer, and an initial & final learning rate of 0.01. After the training process, the best-performing mask detection model based on the F1, mAP50, and mAP50-95 metric scores was used as the final model for mask detection in the study.

2.4   Combining the Two Stages and Generating the Behavior Dataset

Generating the desired behavior dataset was achieved by cascading the two YOLOv8-based models: the pre-trained YOLOv8 xLarge object (person) tracking model and the custom-trained YOLOv8-based mask detection model. Both models are able to produce a bounding box enclosing the tracked individuals and detected faces in the CCTV frames respectively. For each bounding box enclosing a person, its corresponding mask wearing behavior is determined by the bounding box of the face with the largest area of intersection with itself (see LINES 14–21 of ALGORITHM 1 for the pseudo-code implementation). For determining the physical distancing behavior, the euclidean distance for each individual relative to its nearest neighbor in the frame was computed with the help of the DIST() function from the python math library (see LINES 26–35 of ALGORITHM 1 for the pseudo-code implementation). All the detected faces in the input frames using the best-performing mask model were pixelated (blurred-out) to a point where the facial features are no longer distinguishable, as denoted in LINES 9–13 of ALGORITHM 1 to preserve the identity privacy of individuals in the resulting video frames. The blurring-out process was carried out by dividing the cropped image of the face into multiple blocks and filling each block with a color value based on the mean of all the pixel colors within the block. Finally, the output behavior dataset for the CCTV footage was generated as a CSV file with the following column information: **time** indicating the time stamp of the frame; **id** referring to the unique ID of the tracked individual; **box.xyxy** conveying the points of the bounding box enclosing the tracked individual in the frame; **mask_stat** & **mask_conf** indicating the mask wearing behavior and the classification confidence score; and **real_XY** referring to the real-world x & y coordinates of the individual.

3   Results and Discussion

The generated custom mask dataset consisting of a total of 500 random snapshots from the CCTV data was split into 80% (train set) - 10% (val set) - and 10% (test set) distribution. The 400 images in the train set have a total of 2,034 instances (39% unmasked & 61% masked) of labeled mask wearing behavior. The validation and test sets have 248 (39% unmasked & 61% masked) and 286 (44% unmasked & 56% masked) instances respectively. Results showed that there are more instances of faces labeled as masked than unmasked across the train-val-test distribution, reflecting the general mask behavior of the test university constituents at the time period the CCTV data was gathered. Although there is a slight imbalance of the unmasked-masked class on our custom dataset, a small imbalance of this scale is still acceptable not to cause any possible problems in the learning process of the models [9].

After using the custom mask dataset to fine-tune the five scaled versions of YOLOv8 models (nano, small, medium, large, and xLarge), the training process produced the resulting graphs shown in Fig. 1 consisting of the training losses and performance metrics for the five models over 50 epochs. From the loss graphs (train cls_loss & val cls_loss), there exists a small spike in the loss values at around the 33rd epoch, prominently on the nano and xLarge models. This gives a hint that the models may already have achieved its best version prior to epoch 33. Nevertheless, the general downward trend of the training and validation losses indicate that the models are not over-fitting our custom dataset, which is a good indication that the models may also perform well even for the other unseen dataset. For the precision, recall, and mean average precision scores (mAP50 & mAP50-95), the graphs show a steady increase in values during the early epochs and eventually saturating at a certain point, indicating that the models may not necessarily benefit from training beyond the 50th epoch anymore. The nano version achieved its best model at the 50th epoch, the small model on the 45th epoch, medium at the 31st epoch, large at the 25th epoch, and xLarge at 28th epoch.

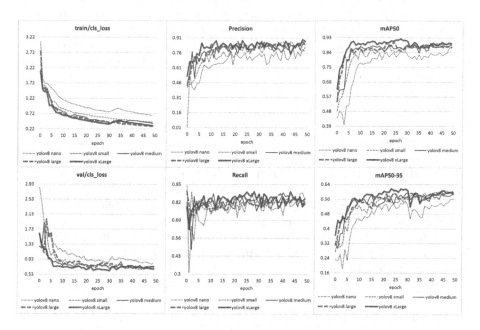

Fig. 1. Training results of the different YOLOv8 models fine-tuned on the custom mask dataset for 50 epoch.

Table 1. Summary of the developed fine-tuned YOLOv8-based mask detection models.

	Params (M)	Speed* (ms/img)	Precision	Recall	F1	mAP50	mAP50-95
nano	3.006	15.5	0.808	0.811	0.809	0.864	0.548
small	11.126	18.1	0.866	0.850	0.858	0.925	0.617
medium	25.841	19.5	0.876	0.849	0.862	0.918	0.625
large	43.608	25.3	0.878	0.808	0.842	0.898	0.609
xLarge	68.125	30.2	0.904	0.843	**0.872**	**0.940**	**0.630**

*(Speed = preprocess + inference + postprocess) using NVIDIA 4060 mobile GPU

For the detailed training (fine-tuning) results, Table 1 shows the summary properties of the five fine-tuned YOLOv8 models for mask detection. Results showed that the xLarge is the best-performing model with F1 = 0.872, mAP50 = 0.940 and mAP50-95 = 0.630. On the other hand, results showed that the nano model is the least performing model with F1 = 0.809, mAP50 = 0.864 and mAP50-95 = 0.548. However, when considering the average speed of the models, the nano variant, with an average speed of 15.5 ms/img, is the fastest among the five models. As expected the xLarge variant, with an average speed of 30.2 ms/img, performs the slowest. These results suggest that the nano model should be used if faster detection time is at most priority. If however, the accuracy rate (in terms of detection & classification) is more desired over than speed, the xLarge model is more suited for such kind of use case. The small (F1 = 0.858 & speed = 18.1), medium (F1 = 0.862 & speed = 19.5), and large (F1 = 0.842 & speed = 25.3) models are also good options when considering a balance between speed and accuracy.

After implementing the two-stage YOLOv8-based cascaded object tracking and detection pseudo-code shown in ALGORITHM 1 into a Python code, the study is able to process the raw CCTV footages producing the desired behavior dataset and output video files. Figure 2 shows an example snapshot of the resulting video frame generated by the system, as well as its corresponding entry in the output behavior dataset CSV file. The tracked individuals in the video frames were annotated with a bounding box (colored as red if unmasked and colored as blue if masked) enclosing the person. The unique ID, mask behavior, and the physical distance value (in meters) of the person to its nearest neighbor were also included in the annotated frame. Additionally, all faces in the resulting frames were adequately blurred-out (pixelated) enough to be unrecognizable, preserving the identity privacy of the individuals. The system was also able to extract other relevant information from the image frame and generate those data in the corresponding resulting behavior dataset.

Though the developed system is able to achieve its goal to generate the mask and distancing behavior of the individuals from the input CCTV frames, the study still comes with some limitations. Since the mask detection model was trained (fine-tuned) on a custom dataset taken from the CCTV footages of a test university, the model may only work best on the conditions akin to the extrinsic appearance of the individuals in training dataset. For example, almost all individuals in the training dataset were only using common medical and non-medical face masks when wearing one, thus other clothing (e.g. hijab and other similar clothing) and some accessories (e.g. head cap, costumes covering the head, etc.) may affect drastically the detection performance of the models of the study. The way how the individuals are captured in the CCTV camera image frame also poses another limitation on the developed system. Some image frames are not able to capture the whole body of the individuals in the frame (see the bounding box of individuals in Fig. 2 as an example). The system can only assume the real world position of the individual base from the resulting bounding box generated by the object (person) detection model, resulting in an inaccurate 2d to 3d coordinate translation result when only a portion of the individual's body is detected in the image frame.

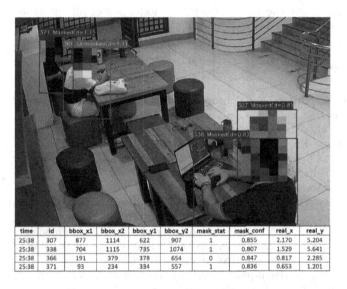

time	id	bbox_x1	bbox_x2	bbox_y1	bbox_y2	mask_stat	mask_conf	real_x	real_y
25:38	307	877	1114	622	907	1	0.855	2.170	5.204
25:38	338	704	1115	735	1074	1	0.807	1.529	5.641
25:38	366	191	379	378	654	0	0.847	0.817	2.285
25:38	371	93	234	334	557	1	0.836	0.653	1.201

Fig. 2. Example snapshot of the resulting video frame and its corresponding entry in the generated behavior dataset.

4   Conclusion

A detailed implementation of a two-stage YOLOv8-based cascaded approach was used in generating a movement behavior dataset with respect to two behaviors: physical distancing and mask wearing. The first stage involved tracking of individuals in the video feed to determine their physical distancing behavior based on the euclidean distance. The second stage involves determining the mask wearing behavior of the tracked individuals using the custom-trained YOLOv8 model. From the five different YOLOv8 models (nano, small, medium, large, and xLarge), the xLarge mask detection variant has the highest performance scores while the nano variant has the lowest scores. In terms of speed, however, the nano variant is the fastest while the xLarge variant is the slowest among the five models. The medium variant presents a good balance between accuracy and speed. Finally, a blurring-out method was employed to make the faces in the resulting video frames unrecognizable, thus preserving the identity privacy of the individuals. The source code used in the study and the fine-tuned models are available at https://github.com/rpabao/dataset-generator-YOLOv8.

Future works of the study include validating the actual accuracy of the fine-tuned mask detection and distancing models on real world data (and other external dataset), to establish the soundness of the behavior dataset generator developed in the study. Another possible route would be to use other deep learning algorithms (e.g. R-CNNs, SSD, and other versions of YOLO) for object detection and compare its performance to the fine-tuned YOLOv8-based models in this study, to get the best performing model out of all the object detection algorithms available.

References

1. Aharon, N., Orfaig, R., Bobrovsky, B.Z.: Bot-sort: robust associations multi-pedestrian tracking. arXiv preprint arXiv:2206.14651 (2022)
2. An, L., et al.: Challenges, tasks, and opportunities in modeling agent-based complex systems. Ecol. Model. **457**, 109685 (2021)
3. Jocher, G., Chaurasia, A., Qiu, J.: Yolo by ultralytics (2023). https://github.com/ultralytics/ultralytics. Accessed 26 Mar 2023
4. Kasai, T.: Adapting to life with covid-19 and staying safe (2021). https://www.who.int/westernpacific/news-room/commentaries/detail-hq/adapting-to-life-with-covid-19-and-staying-safe. Accessed May 2023
5. Nazir, A., Wani, M.A.: You only look once-object detection models: a review. In: 2023 10th International Conference on Computing for Sustainable Global Development (INDIACom), pp. 1088–1095. IEEE (2023)
6. OpenCV: Perspective-n-point (pnp) pose computation (nd). https://docs.opencv.org/4.x/d5/d1f/calib3d_solvePnP.html. Accessed 26 Mar 2023
7. Ravaioli, G., Domingos, T., Teixeira, R.F.M.: A framework for data-driven agent-based modelling of agricultural land use. Land **12**(4) (2023). https://doi.org/10.3390/land12040756
8. Terven, J., Cordova-Esparza, D.: A comprehensive review of yolo: from yolov1 to yolov8 and beyond. arXiv preprint arXiv:2304.00501 (2023)
9. Weiss, G.M.: Foundations of imbalanced learning. In: Imbalanced Learning: Foundations, Algorithms, and Applications, pp. 13–41 (2013)
10. World Health Organization: Statement on the fifteenth meeting of the ihr (2005) emergency committee on the covid-19 pandemic (2023). https://www.who.int/news/item/05-05-2023-statement-on-the-fifteenth-meeting-of-the-international-health-regulations-(2005)-emergency-committee-regarding-the-coronavirus-disease-(covid-19)-pandemic. Accessed 10 May 2023
11. Xiao, Y., et al.: A review of object detection based on deep learning. Multimedia Tools Appl. **79**, 23729–23791 (2020). https://doi.org/10.1007/s11042-020-08976-6

Moving Beyond Stance Detection in Cross-Cutting Communication Analysis

Rezvaneh Rezapour[1(✉)], Daniela Delgado Ramos[2], Sullam Jeoung[2], and Jana Diesner[2]

[1] Drexel University, Philadelphia, USA
sr3563@drexel.edu
[2] University of Illinois Urbana-Champaign, Urbana, USA

Abstract. In today's social media landscape, personal opinions on controversial topics are widespread. While some platforms provide structured environments for discussing such matters, fostering cross-cutting communication among individuals, understanding how people engage in these discussions remains a challenge. This study aims to understand the dynamics of discussing controversial topics, focusing specifically on the topic of abortion. Using an aspect-based approach, we employ BERT-based topic modeling and attention mechanisms to identify key aspects of debates. Through clustering, we identify highly polarizing aspects and examine the contextual nuances and sentiment surrounding them. Our methodology enhances our understanding of cross-cutting communication on controversial topics and offers an in-depth analysis of consensus and disagreement among participants. Our study contributes to the field of stance analysis, revealing opportunities for mutual understanding and uncovering diverse perspectives on controversial issues. (**Warning:** *this paper contains content that may be triggering.*)

Keywords: Stance Analysis · Contextual analysis · Cross-cutting communication

1 Introduction

The use of social media platforms such as Facebook and Twitter has become a prevalent aspect of modern communication, enabling individuals to engage in activities such as sharing information and ideas and staying up-to-date with current events[20,29]. These platforms are also used to discuss controversial issues, particularly subjective topics such as social, cultural, or political matters [1]. These discussions can be emotionally charged, characterized by polarizing views, and involve divisive language that may impede constructive dialogues [14]. Controversial issues are often rooted in subjective perceptions of cultural, societal,

---

D. D. Ramos and S. Jeoung—Equal contribution.

R. Thomson et al. (Eds.): SBP-BRiMS 2023, LNCS 14161, pp. 305–315, 2023.
https://doi.org/10.1007/978-3-031-43129-6_30

and political characteristics and differences [23,24]. These differences in perceptions and beliefs, particularly among groups with diverging interests or values, can lead to debates and conflicts that may be challenging to resolve [21].

Social media platforms also opened up new avenues for recognizing opposing viewpoints and facilitating constructive dialogue among individuals and groups with different perspectives. Specifically, the study of stance, i.e., individuals' perspectives or attitudes towards a particular topic or issue [27], and cross-cutting communication, i.e., the exchange of ideas between individuals or groups with opposing viewpoints, has become a critical area of research in this regard [16]. Stance analysis has been used to extract people's positions on specific topics. Using linguistic features and machine learning algorithms, researchers are able to determine whether individuals are in favor of or against a particular topic [17]. Beyond classification, examining the reasoning and evidence used by participants to support their positions, as well as analyzing the social and contextual factors that influence their viewpoints is also crucial [6,22]. Analysis of cross-cutting communication can provide insights into the elements, or *frames* [10], that facilitate or impede productive discourse in social media.

This study aims to understand online cross-cutting communication by analyzing how diverse viewpoints are addressed when controversial topics are discussed. Our goal is to understand what aspects (i.e., topics or themes characterizing the perceptions and opinions of the speaker) influence the dynamics between participants holding opposing viewpoints in (online) debates. We employ attention-based and topic-modeling techniques to identify key aspects of arguments and thoroughly analyze the sentiment and context surrounding each aspect discussed to provide deeper insights into these dynamics. As a case study, we analyze online debates on the highly polarizing topic of abortion using a debate dataset [9]. By examining discussions surrounding conflicting aspects such as the embryo, the rights of pregnant people, and governmental regulation, we explore the factors or aspects driving cross-cutting communication.

Our findings show how language use among different groups of people with opposite stances mirrors their opinions on abortion, and how they leverage various aspects to reinforce their respective perspectives. Our analysis shows that the debates around abortion are multi-faceted, extending beyond morals and covering scientific, ethical, legal, and social issues. We also found instances of agreement among groups regarding aspects and their context for discussing themes. Contextual analysis unveiled differences in word choice, perspectives, and priorities between the groups. Our sentiment analysis highlighted the highly emotional nature of the abortion discussion. Understanding the complexity of abortion discourse has important implications for public opinion. By grasping these causes, we can develop strategies to encourage empathy, respect, and informed decision-making, fostering constructive dialogue among diverse perspectives.

2   Related Work

Stance detection plays a crucial role in understanding individuals' perspectives on various issues [27]. Previous studies on stance have explored different methods such as sentiment analysis and topic modeling to enhance the prediction of stance. Stance detection data has been collected from different online platforms, including online discussion boards [9], Twitter [17], and websites [19]. Moham- mad et al. [17] introduced the SemEval 2016 shared task, which focuses on stance detection using Twitter as a data source. They released a baseline model and dataset for analyzing stance in user-generated texts. Zhang et al. [28] utilized the SemEval dataset to train a bidirectional LSTM model incorporating semantic and emotional valences as features. Other methods, such as topic modeling [26] and morality analysis [22,23] have been employed to identify underlying themes and reasoning behind a stance. BERT and other deep learning methods have been leveraged to achieve higher accuracy in stance detection [25].

Detecting stance is challenging due to the subjective nature of individuals' and groups' diverse linguistic expressions [23]. Extensive research is needed for reliable algorithms and models for accurate stance detection. Understanding the context is crucial as it impacts text interpretation, including cross-cutting communication and framing [10,11]. Previous studies explored framing detection challenges [4,7]. Dahlberg and Sahlgren investigated framing of "outsiders" in Swedish politics using Random Indexing [7]. This work builds upon previous work in these areas and employs automated identification of topics, sentiment, and sentence structures for the analysis of cross-cutting communication.

3   Methodology

3.1   Data

In this study, we used a dataset collected by Durmus and Cardie [9], which contains 78,376 debates from a debate website (debate.org) across 23 topics, including politics, religion, technology, music, and travel. Each debate consists of multiple rounds where opposing sides present their arguments, and other users in the community can provide comments and votes. In total, the dataset includes 606,102 comments and 199,210 votes. The dataset includes expressions of users' viewpoints and stances on several controversial issues, including abortion. From the original dataset, we selected the debates containing the term 'abortion' in the titles (N = 1950 debates). In the selected debates, each debater self-identified their stance on abortion, either as in Favor (PRO) or Against (CON). Then, we thoroughly examined each debate's title, description, and conversation to iden- tify whether the stance label and debaters' positions were consistent. For exam- ple, we discovered that debates with titles like "Banning abortion" or "abortion is a form of murder" are congruent with the stance suggested by their titles (e.g., in Favor of banning abortion), indicating a stance against abortion. Con- sequently, we adjusted the position for these cases to ensure accuracy throughout our dataset. To clean the text, we removed citations or quotations, which are typically used to provide references to previous arguments or external sources.

3.2   Aspect Extraction

When deliberating on a specific matter (e.g., abortion), people typically refer to various aspects of such matter in order to strengthen their position as in favor or against it. Based on this premise, our objective is to extract the relevant aspects of such discussions and examine how individuals frame their arguments in debates on a controversial topic, as well as the similarities and differences in the context surrounding these aspects. As existing research [13,22] suggests that nouns play a critical role in representing aspects in written texts, we tagged, extracted, and lemmatized words tagged as nouns. To select the most salient aspects, we employed two methods. First, we applied the Top2Vec technique [2] to extract aspects from the debates. Using BERT [8] *distilbert-base-nli-mean-tokens*, we embedded the arguments in the corpus, applied dimensional reduction, and clustering, and used TF-IDF to sort and extract the most salient words within each cluster. This resulted in a list of 500 aspects. In addition, we complemented the topic modeling with attention-based aspect extraction. Each word in a corpus w is associated with a feature vector $e_w \in \mathbb{R}^d$. Concatenating the feature vectors resulted in a word embedding matrix $\mathbf{E} \in \mathbb{R}^{V \times d}$, where V is the vocabulary size. The goal with this step is to learn the embeddings of the aspects, specifically in matrix $\mathbf{T} \in \mathbb{R}^{K \times d}$, where K is the number of predefined aspects, $K < V$. Defining a sentence embedding z_s, which is the weighted sum of word embeddings, we get $z_s = \sum_{i=1}^{n} a_i e_{w_i}$. The a_i is computed by an attention model, which is interpreted as the weight of the word. The model is trained to minimize the reconstruction error, hence maximizing the representative aspects for the given dataset. The two lists were merged, resulting in a final list of 723 aspects.

3.3   Aspect Clustering

To examine which aspects are overrepresented in the in Favor and Against stances, respectively, we calculated their z-score [18]. To do so, we compute the relative importance of the words in each stance. We denote the odds of word w as $\Omega_w = \pi_w/(1 - \pi_w)$, where π refers to the likelihood from the Dirichlet prior of word w. We denote $\Omega_w^{(i)}$ to indicate the odds of word w with respect to stance i. The relative odds $\delta_w^{(i)}$ is defined as

$$\delta_w^{(i)} = \log(\Omega_w^{(i)}/\Omega_w)$$

The z-score of word w, z_w is the log-odds-ratio, taking the variance into account.

$$z_w = \delta_w^{(i-j)}/\sqrt{\sigma^2(\delta_w^{(i-j)})}$$

where σ^2 is the variance of $/\delta_w$ and $\delta_w^{(i-j)}$ refers to $\delta_w^{(i)}$-$\delta_w^{(j)}$, i and j indicating two different stances. The data were grouped into two categories; in Favor and Against of abortion, and z-scores were computed per aspect. The results are shown in Fig. 1. The z-score on the y-axis indicates the significance of the aspect used in each stance: a higher positive z-score signifies greater importance for in Favor, while a lower negative z-score suggests greater significance for the Against. The x-axis represents the logarithm of word frequency, with higher x-values indicating a higher frequency of word occurrences.

Once the z-score of aspects were calculated, we clustered them using frequency and z-score. Four aspect clusters were identified (Fig. 1): Cluster 1 contains infrequent, polarizing aspects; Cluster 2 are common, less polarizing aspects; and Cluster 3 are highly polarized, frequently occurring aspects. Other aspects in Cluster −1 (light blue-shaded) were excluded from further analysis due to their low polarization and infrequency in our data.

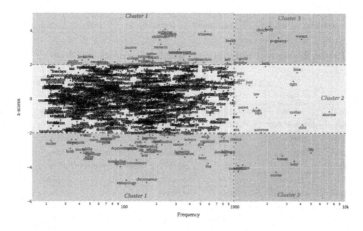

Fig. 1. Z-scores on the y-axis of the graph represent the polarity of aspects. Positive z-scores indicate aspects that are emphasized in favor of abortion, while negative z-scores indicate aspects that are emphasized in opposition to the topic. (Color figure online)

3.4   Contextual Analysis

We analyzed the context and perspectives surrounding each aspect to understand the views of those discussing in Favor and Against abortion. For this purpose, we leverage vector representations to capture both contextual meaning and semantic relationships of words [12]. Specifically, we utilized pre-trained WORD2VEC word embeddings, which were fine-tuned using our dataset, and categorized by stance and aspect. By comparing nearest neighbors based on cosine similarity, we identified differences and similarities in discourse.

3.5   Sentiment Analysis

To conduct sentiment analysis, we used a multilingual *XLM-RoBERTa-based* model trained on around 198M tweets and fine-tuned for sentiment analysis [3]. For the analysis of sentiment for each aspect, we first split the debates into sentences and tagged each sentence with its respective sentiment score. We then extracted the sentences including each aspect, clustered the sentences based on stance, and calculated the mean of the sentiment score for each aspect per stance.

Table 1. Nearest neighboring words of aspects in Cluster3

Life		Woman		Baby		Human	
Favor	Against	Favor	Against	Favor	Against	Favor	Against
necessity	opportunity	toll	tragedy	mum	deformity	component	insect
survival	instant	resort	partner	sympathy	mom	preborn	beings
certainty	extension	tragedy	trauma	bond	deed	essence	therefore
convenience	saying	pregnancy	sterilization	plenty	dad	sustenance	qualifie
accord	hypocrite	hangar	feminist	help	babys	insect	ant
equation	deprivation	affect	addition	inconvenience	saying	tapeworm	pig
judgement	born	vessel	grayer	attachment	inconvenience	amoeba	controversy
supersede	another	birthing	distress	newborn	trouble	makeup	non
stake	believer	criminalization	precaution	amanda	let	lifeform	mosquito
live	infringement	refusal	likelihood	hardship	child	mammal	mammal

Table 2. Nearest neighboring words of aspects in Cluster2

Abortion		Child		Fetus		Mother	
Favor	Against	Favor	Against	Favor	Against	Favor	Against
alley	illegal	neglect	irresponsibility	essence	synonym	inability	fate
illegal	alley	inability	custody	requirement	complexity	necessity	mom
conviction	candidate	upbringing	lifestyle	outside	blob	intervention	deed
back	run	turmoil	visitation	container	qualifie	labor	dad
illegality	childbirth	bond	dad	surrounding	misconception	inconvenience	equation
allowance	clue	trouble	expense	breathing	lab	jeopardy	endangerment
pregnancy	professional	difficulty	trouble	embryos	born	difficulty	inconvenience
addition	addition	stigma	inconvenience	womb	equivalent	stake	comfort
endanger	illegalization	misery	guardianship	extension	mammal	wish	grayer
practice	stat	hardship	foolishness	lifeform	appendage	hardship	permission

Table 3. Nearest neighboring words of aspects in Cluster1

Health		Self		Killing		Development	
Favor	Against	Favor	Against	Favor	Against	Favor	Against
risk	wichita	concept	confidence	justifiable	mercy	essence	invino
concern	department	presence	defense	manslaughter	verb	tadpole	puberty
risks	hazard	communication	presence	intent	unlawful	size	adolescence
terminal	complication	preservation	preservation	premeditation	premeditation	stage	adulthood
illness	percentage	contra	war	taking	innocence	embryogenesis	vertebrate
abnormality	danger	variation	fdh	reducto	murder	mechanism	size
criminalization	threat	awareness	awareness	murder	intention	inch	stage
professional	services	capacity	aggression	noun	absence	capability	acronym
public	professional	ownership	ownership	malice	malice	feature	appearance
wellbeing	mental	message	interest	kill	USFG	fish	sled

4 Results and Discussion

Our z-score analysis of aspects revealed three significant clusters associated with the broad topic of abortion. As shown in Fig. 1, Cluster −1 and Cluster 1 contain the majority of infrequent aspects. Cluster 2 contains less polarizing but more frequent aspects, and Cluster 1 and Cluster 3 contain the most polarized aspects, with the latter ones being more frequently used in our dataset. In Cluster 3, polarized words such as "woman", "choice", and "body" occur more frequently in Favor of than Against abortion, while words like "adoption" and "murder" occur in debates Against abortion. Cluster 2 contains words like "law", "right", "abortion" and "pain" indicating more regulatory and social aspects. Cluster 1 includes biological and physiological aspects, such as terms referring to embryology and fetal development stages. These clusters highlight the complex scientific, ethical, legal/regulatory, societal, and medical aspects of the abortion topic. The analysis of the top 10 aspects in each cluster reveals different patterns in debates in Favor and Against abortion. More specifically, in Cluster 2, "fetus" appeared more frequently in debates in Favor of abortion, while the use of "child" as an aspect was more prevalent in Against discussions. In Cluster 3, "baby" as an aspect is more frequently discussed in Against abortion. As previous studies have shown, different words can have different meanings to people with different ideologies [15]. Our findings show that word choices and language use can potentially be influenced by how people perceive abortion.

Analyzing words within each cluster reveals differences in word choice, perspectives, and priorities between users in Favor of and Against abortion. Tables 1, 2, and 3 show the closest (semantically and contextually) words to the extracted aspects in our dataset. In cluster 3, the contexts around the aspect of "woman" in Favor of abortion are words like "toll" and "affect"; suggesting a focus on the negative consequences that can result from denying women access to abortion; e.g., *"pregnancy takes its toll on the body and changes it forever."* Words like "pregnancy," "vessel," and "birthing" emphasize the physical and medical aspects of pregnancy. The inclusion of "criminalization" and "refusal" suggests a concern about legal and societal barriers to accessing abortion. On the other hand, the against debates include words like "partner," "trauma," and "distress," which suggest a focus on the emotional experiences of women who are against abortion. Words like "sterilization," "precaution," and "likelihood" suggest a focus on the prevention of unwanted pregnancy e.g., *"If abortion were made punishable by incarceration of the doctor and sterilization of the patient, then there would be few unwanted pregnancies."* The inclusion of "feminist" can suggest a perception of abortion as a divisive feminist issue. The word "tragedy" appears in both lists but carries different contexts, reflecting the tragedy of forced pregnancy for pro-choice individuals and the tragedy of ending a life for anti-abortion proponents. In the debates against abortion, "tragedy" is used in contexts such as *"abortion is a cruel tragedy but it is also a choice that should never be made"*, while in favor debates we have *"What I am asking for here is consistency. Are the embryo and the skin cell BOTH potential human beings? Or are NEITHER human beings? [...] If both are potential human beings then every skin cell that falls off your body is a tragedy equal to embryonic abortion."*

For the aspect of "baby" in Cluster 3, debates in Favor of abortion contain words such as "mum," "sympathy," "bond," "attachment," and "newborn"; suggesting positive associations with babies and motherhood. These words can indicate that those who are in favor of abortion may still value the idea of motherhood and the emotional bonds that can be formed between a parent and child, e.g., *"Abortion isn't that bad a thing. [...] And whilst killing a newborn is wrong, yes killing a fetus isn't. They both don't have feelings, but it's the parental attachment. You know that first minute you look into your newborn babies eyes. You don't have that with a Fetus. That's why I guess it's ok."* Cluster 2 represents a less polarizing perspective in the discussions on abortion. Specifically, when examining the aspect of "fetus," arguments both against and in favor of abortion converge on the question of how to define a fetus and whether it should be regarded as a human being e.g., *"UNBORN CHILD: A synonym for "fetus" that describes a macro-cellular organism between the stages of conception and birth."* At the same time, pro-abortion arguments also engage in the discussion with words like "essence," "requirement," "embryos," or "lifeform," e.g., *"[...]the fetus cannot be considered as a seperate lifeform. Abortion isn't equivalent to killing a newborn, because the fetus isn't capable of living an independent life."* While the stances on the fetus definition also differ, the discussion still points in the same direction. In Cluster 1, the closest words around the topic of *killing* in debates in favor of abortion are "justifiable," "manslaughter," and "reducto" suggesting that those in favor of abortion argue that it is justifiable to end a pregnancy, e.g., *"having abortion in the first trimester, is not considered manslaughter. "* On the other hand, the closest words around the aspect of "killing" in debates against abortion are "unlawful," "innocence," and "absence" suggesting that those arguing against abortion believe that ending a pregnancy is an unlawful and immoral act, e.g., *"My reasoning for this is that it is the unlawful killing of human being."*

The sentiment analysis reveals a prevalence of negative sentiment in discussions regarding abortion. In all three clusters, discussions around the aspects both in Favor and Against exhibit negative sentiment. Cluster 2 demonstrates the highest negativity, with an average sentiment of -0.58 for in Favor of abortion and -0.45 for against it. Following closely is cluster 3, with an average sentiment of -0.50 in favor and -0.45 against. To gain deeper insights into the nature of these discussions, we employed LIWC [5] to analyze the tone and emotion surrounding each aspect. We discovered that in debates Against abortion, discussions contained a greater presence of sad emotions compared to those in Favor. Furthermore, when debating "adoption", individuals Against abortion expressed more positive emotions. On the other hand, debates against abortion exhibited higher levels of anxiety when discussing topics such as the "baby" and "rape". Overall, we found that discussions on polarized aspects can reveal the underlying values and beliefs of each side. The context surrounding words used in debates can also help to understand the nature and themes of discussions. In general, analyzing the language used in polarized discussions can help to understand the perspectives, priorities, and beliefs of each side, and highlight areas of common ground or differences that need to be addressed.

5 Conclusion

In this study, we deployed an aspect-based approach to analyze cross-cutting communication. As a case study, we examined debates around a highly polarized topic (abortion). Using a combination of BERT-based topic modeling and an attention mechanism, we extracted the most salient aspects (frames) representing the main themes or topics of the discussion around abortion. This approach enabled a deeper understanding of the discussion around a bigger topic and facilitated a high-dimensional analysis of the areas of agreement and disagreement among people that address the same topic (abortion) but from opposing viewpoints. By employing our method, we were able to delve into some intricate details of the themes and aspects related to cross-cutting discussions of a specific topic and exploring the multifaceted nature of the debate. This approach contributes to a nuanced and comprehensive analysis of complex and controversial topics, providing valuable insights into the various dimensions and perspectives involved. Social media is a popular venue for discussions on a wide range of topics, and while it has the potential to bring people together, it can also lead to polarization, echo chambers, and segregation. Understanding the aspects of these discussions is crucial for users or participants to connect, communicate, and gain a better understanding of each other's perspectives. Our study is limited as we only focus on one dataset from one platform and one polarizing topic. In addition, the context in which the discussions take place, including the platform used, and the audiences involved play a significant role in determining the stances taken. Future research will extend the findings presented here to analyze the stance toward more issues from a more diverse set of social contexts and cultures.

Acknowledgment. This work was supported in part by the Cline Center for Advanced Social Research at the University of Illinois Urbana-Champaign, including a Linowes Fellowship.

References

1. Addawood, A., Rezapour, R., Abdar, O., Diesner, J.: Telling apart tweets associated with controversial versus non-controversial topics. In: Proceedings of the Second Workshop on NLP and Computational Social Science, Vancouver, Canada, August 2017, pp. 32–41. Association for Computational Linguistics (2017). https://doi.org/10.18653/v1/W17-2905. https://aclanthology.org/W17-2905
2. Angelov, D.: Top2Vec: distributed representations of topics. CoRR abs/2008.09470 (2020). https://arxiv.org/abs/2008.09470
3. Barbieri, F., Espinosa Anke, L., Camacho-Collados, J.: XLM-T: a multilingual language model toolkit for Twitter. arXiv e-prints arXiv:2104.12250 (2021)
4. Baumer, E., Elovic, E., Qin, Y., Polletta, F., Gay, G.: Testing and comparing computational approaches for identifying the language of framing in political news. In: Proceedings of the 2015 Conference of the North American Chapter of the Association for Computational Linguistics, pp. 1472–1482 (2015)

5. Boyd, R.L., Ashokkumar, A., Seraj, S., Pennebaker, J.W.: The development and psychometric properties of LIWC-22, pp. 1–47. University of Texas at Austin, Austin, TX (2022)
6. Card, D., Gross, J.H., Boydstun, A., Smith, N.A.: Analyzing framing through the casts of characters in the news. In: Proceedings of the 2016 Conference on Empirical Methods in Natural Language Processing, pp. 1410–1420 (2016)
7. Dahlberg, S., Sahlgren, M.: Issue framing and language use in the Swedish blogosphere. In: From Text to Political Positions: Text Analysis Across Disciplines, pp. 71–92 . John Benjamins, Amsterdam (2014)
8. Devlin, J., Chang, M.W., Lee, K., Toutanova, K.: BERT: pre-training of deep bidirectional transformers for language understanding. In: Proceedings of the 2019 Conference of the North American Chapter of the Association for Computational Linguistics: Human Language Technologies, Volume 1 (Long and Short Papers), Minneapolis, Minnesota, June 2019, pp. 4171–4186. Association for Computational Linguistics (2019). https://doi.org/10.18653/v1/N19-1423. https://aclanthology.org/N19-1423
9. Durmus, E., Cardie, C.: A corpus for modeling user and language effects in argumentation on online debating. In: Proceedings of the 57th Annual Meeting of the Association for Computational Linguistics, pp. 602–607 (2019)
10. Entman, R.M.: Framing: toward clarification of a fractured paradigm. J. Commun. **43**(4), 51–58 (1993)
11. Entman, R.M.: Framing bias: media in the distribution of power. J. Commun. **57**(1), 163–173 (2007)
12. Gonen, H., Jawahar, G., Seddah, D., Goldberg, Y.: Simple, interpretable and stable method for detecting words with usage change across corpora. arXiv preprint arXiv:2112.14330 (2021)
13. Gupta, V., Lehal, G.S.: A survey of text summarization extractive techniques. J. Emerg. Technol. Web Intell. **2**(3), 258–268 (2010)
14. Jones, D.A.: The polarizing effect of new media messages. Int. J. Pub. Opin. Res. **14**(2), 158–174 (2002)
15. KhudaBukhsh, A.R., Sarkar, R., Kamlet, M.S., Mitchell, T.: We don't speak the same language: interpreting polarization through machine translation. In: Proceedings of the AAAI Conference on Artificial Intelligence, vol. 35, pp. 14893–14901 (2021)
16. Matthes, J., Knoll, J., Valenzuela, S., Hopmann, D.N., Von Sikorski, C.: A meta-analysis of the effects of cross-cutting exposure on political participation. Polit. Commun. **36**(4), 523–542 (2019)
17. Mohammad, S., Kiritchenko, S., Sobhani, P., Zhu, X., Cherry, C.: SemEval-2016 Task 6: detecting stance in tweets. In: Proceedings of the 10th International Workshop on Semantic Evaluation, SemEval-2016, pp. 31–41 (2016)
18. Monroe, B.L., Colaresi, M.P., Quinn, K.M.: Fightin' words: lexical feature selection and evaluation for identifying the content of political conflict. Polit. Anal. **16**(4), 372–403 (2008)
19. Murakami, A., Raymond, R.: Support or oppose?: classifying positions in online debates from reply activities and opinion expressions. In: Proceedings of the 23rd International Conference on Computational Linguistics: Posters, pp. 869–875. Association for Computational Linguistics (2010)
20. Newman, N.: Mainstream media and the distribution of news in the age of social media (2011)
21. Prior, M.: Media and political polarization. Annu. Rev. Polit. Sci. **16**, 101–127 (2013)

22. Rezapour, R., Dinh, L., Diesner, J.: Incorporating the measurement of moral foundations theory into analyzing stances on controversial topics. In: Proceedings of the 32nd ACM Conference on Hypertext and Social Media, pp. 177–188 (2021)
23. Rezapour, R., Shah, S.H., Diesner, J.: Enhancing the measurement of social effects by capturing morality. In: Proceedings of the 10th Workshop on Computational Approaches to Subjectivity, Sentiment and Social Media Analysis, pp. 35–45 (2019)
24. Roccas, S., Brewer, M.B.: Social identity complexity. Pers. Soc. Psychol. Rev. **6**(2), 88–106 (2002)
25. Siddiqua, U.A., Chy, A.N., Aono, M.: Tweet stance detection using an attention based neural ensemble model. In: Proceedings of the 2019 Conference of the North American Chapter of the Association for Computational Linguistics: Human Language Technologies, Volume 1 (Long and Short Papers), pp. 1868–1873 (2019)
26. Skeppstedt, M., Stede, M., Kerren, A.: Stance-taking in topics extracted from vaccine-related tweets and discussion forum posts. In: Proceedings of the 2018 EMNLP Workshop SMM4H: The 3rd Social Media Mining for Health Applications Workshop & Shared Task, pp. 5–8 (2018)
27. Somasundaran, S., Wiebe, J.: Recognizing stances in online debates. In: Proceedings of the Joint Conference of the 47th Annual Meeting of the ACL and the 4th International Joint Conference on Natural Language Processing of the AFNLP, Suntec, Singapore, August 2009, pp. 226–234 (2009)
28. Zhang, B., Yang, M., Li, X., Ye, Y., Xu, X., Dai, K.: Enhancing cross-target stance detection with transferable semantic-emotion knowledge. In: Proceedings of the 58th Annual Meeting of the Association for Computational Linguistics, pp. 3188–3197 (2020)
29. Gil de Zúñiga, H., Jung, N., Valenzuela, S.: Social media use for news and individuals' social capital, civic engagement and political participation. J. Comput. Mediat. Commun. **17**(3), 319–336 (2012)

Pandemic Personas: Analyzing Identity Signals in COVID-19 Discourse on Twitter

Scott Leo Renshaw$^{(\boxtimes)}$ (ID), Samantha C. Phillips (ID), Michael Miller Yoder (ID), and Kathleen M. Carley (ID)

IDeaS Center, Software and Societal Systems Department, Carnegie Mellon University, 5000 Forbes Ave., Pittsburgh 15213, USA
{srenshaw,samanthp,mamille3,carley}@cs.cmu.edu

Abstract. The first year of the COVID-19 pandemic coincided with significant social and political changes. This article presents an exploratory analysis of Twitter users' self-representations in the context of COVID-19. While some identities remained stable throughout the year, others appear to have been influenced by external events such as the Black Lives Matter protests and the U.S. presidential election. We also examine how users' political identities are expressed in tweets, finding that right-leaning users are more likely to mention left-leaning identities, although both conservative and liberal-leaning users showed increased out-group focus in the period leading up to and after the election. Our study provides a comprehensive overview of user personas in a critical discourse topic, shedding light on identity signaling on social media. We end by identifying several clear opportunities for future research on social identity self-presentation on social media.

Keywords: Identities · Social media · COVID-19

1 Introduction

With in-person modes of socialization curbed by stay-at-home mandates, social distancing, and masking, many people turned to online social networks to connect with their extended networks and gain up-to-date information about the COVID-19 pandemic in 2020 [13].

One of the unique aspects of social media is that it allows users to curate personas of their "digital" selves, in which they can adorn their user biographies,

This work was supported in part by the Knight Foundation and the Office of Naval Research grant MURI: Persuasion, Identity, & Morality in Social-Cyber Environments, N00014-21-12749. Additional support was provided by the Center for Computational Analysis of Social and Organizational Systems (CASOS) at Carnegie Mellon University. The views and conclusions contained in this document are those of the authors and should not be interpreted as representing the official policies, either expressed or implied, of the Knight Foundation, Office of Naval Research, or the U.S. Government.

© The Author(s), under exclusive license to Springer Nature Switzerland AG 2023
R. Thomson et al. (Eds.): SBP-BRiMS 2023, LNCS 14161, pp. 316–325, 2023.
https://doi.org/10.1007/978-3-031-43129-6_31

usernames, and messages with identity signals.[1] Therefore, it is crucial to analyze how users present their identity signals on social media, especially in the context of a major historical event like the COVID-19 pandemic. This analysis helps us understand how information is shared, received, and imbued with these identity signals and how that impacts discourse spaces [12].

This article examines how Twitter users' identities and corresponding tweets in the COVID-19 communication space evolved throughout 2020. By analyzing identity trends in the COVID-19 discourse space, we first identify stable and volatile identities across 2020; second, we identify how offline events, like the Black Lives Matter protests and the 2020 U.S. Presidential Election, align with changes in user identity frequency across 2020; and third, we delve into the use of the COVID-19 communication space for political in-group and out-group identity signaling, investigating the communication styles of conservative and liberal-leaning users on Twitter.

By examining the identity signals of users on Twitter across 2020, we can glean insights into how offline events and social dynamics intersect with the ongoing COVID-19 discourse on social media.

2   Prior Work

Identities are often conceptualized as social signaling, or, in sociologist Erving Goffman's terms, as a "performance" that varies across time and situation [4]. This perspective fits the nature of social media and Twitter particularly well, where users choose what words to put in their biographies to present themselves – a "front" stage performance. Sociolinguists studying how identity is expressed in language often adopt this "performative" perspective instead of a view of identity as stable, internal, or psychological [2]. We also draw on social identity theory [11], which recognizes that different identities may be more or less relevant depending on the situation and that identities are developed through in-group membership in the context of the out-group. We adopt this focus on in- and out-groups for political identities in this dataset.

Other work on identity on social media has also drawn on Goffman's dramaturgical metaphor of self-presentation, including Bullingham and Vasconcelos [3], who find that online presentation on Second Life often mirrors offline self-presentation with selective edits. Bazarova and Choi [1] apply a functional approach in modeling the reasons for self-disclosure online, finding in a small study that disclosure in public posts was motivated by self-expression and social validation.

In the context of Twitter, there have been several approaches that have considered elements of user biographies. Among studies that apply quantitative and computational methods to examine self-presented identities in social media biographies, Pathak et al. [8] and Yoder et al. [14] have utilized bottom-up

---

[1] We construe the presentation of selves through user biography information as a kind of digital social signal - a sort of badge or indicator of being part of, related to, or even against particular kinds of social categories or movements.

extraction approaches to collecting and measuring identities present in Twitter and Tumblr datasets. Magelinski and Carley [7] developed a measure of identity prototypicality for online communities – utilizing several features including hashtags, mentions, emojis, and identity phrases in user bios.

Our study focuses on the use of identity signals in user biographies within the COVID-19 communication space – a historic human event that generated much conversation online, affecting humans across the globe. These identity signals over the first year of COVID-19 should lend themselves to how users position and present their "digital" selves during this event, how it may relate to other events that occurred during that year, and how political in-group and out-group signaling was co-opted COVID-19 discourse during 2020, a presidential election year.

3   Methods

3.1   Data Collection

Our dataset was collected using a streaming keyword search via Twitter v1 API between January 1, 2020, and December 31, 2020. We collected tweets that contained at least one of the following terms: coronavirus, Wuhan virus, Wuhanvirus, 2019nCoV, NCoV, NCoV2019, covid-19, covid19, covid 19. We utilized two samples of this data to focus on two distinct areas: identity use frequency in user biographies across each month of 2020 and political in-group/out-group identity use in biographies and tweets.

To collect identity usage in biographies and tweets, we leveraged term matches with an extensive lexicon of English identity terms in the NetMapper software[2] [5]. Keeping all tweets from users with at least one identity in our lexicon yielded a full dataset of 349 million tweets, 14 million unique users, and 20 million unique biographies

3.2   Characterizing Bio Identities

To identify general trends in identity use beyond specific terms, we manually examined the 1000 most frequent identities in biographies over the entire year and identified several key categories that we thought would provide useful and interesting contrasts. This included 40 political identity terms (e.g., MAGA, American, Republican, Patriot), 22 religious identities (e.g., Christian, Muslim, Jesus, Sinner, Worshipper), 7 racial identities (e.g., White, Black, POC, Hispanic), and 4 gender/sexuality identity terms (Lesbian, Transgender, Gay, Queer).

To analyze fluctuations in identity use, we calculated ranked orderings of identities by frequency per month. Identities with low average change in rank between months are the most stable in their participation in the COVID-19 communication space, whereas looking at those with the greatest absolute value

---

[2] https://netanomics.com/netmapper/.

change among months reveals the identities that quickly enter or depart from this discourse.

3.3   Political Bio & Tweet In/Out-Group Analysis

We analyze in-group and out-group political signaling by comparing political identity use in tweets from users with political identities in their bios. We selected popular political identities that had a clear left- or right-leaning orientation in an American context. The right-leaning identities selected were "MAGA", "patriot", "republican", and "conservative". For the left-leaning group, we grouped the following cues: "democrat", "liberal", "socialist", and "leftist".

To estimate the prevalence of in-group and out-group political signaling over time, we calculate the number of unique users that have a right- or left-leaning identity in their bio and signal a right- or left-leaning identity in their tweets for each quarter in 2020.

4   Results

4.1   Identity Use Across 2020

We identified a total of 10,109 unique identities from users participating in the COVID-19 communication space on Twitter, including pronouns, family roles, political, racial, national, professional titles, and other affiliations. The raw frequency count of identities was highly skewed to a relatively small number of identities – the median frequency of appearance was 207, the 3rd quartile was 1694, with a max value of 10,748,463 – therefore, to separate signal from noise we generated sub-categories among the top 1000 most frequent terms.

In Table 1 we identify the most stable identities, ordered by the top 10 stable identity signals, and identities that had the largest single-month rank change, organized by descending order of absolute value change. Signals found to be the most stable have the lowest number of changes in rank across the year, which can be interpreted as users of a particular identity remaining engaged with the COVID-19 communication space on Twitter across the year. There appears to be a strong mix of both American political categories, particularly right-leaning identities ("American," "MAGA," and "Republican," "Patriot" in the top 15), as well as Christian religious identities ("God," "Savior," "Jesus," "Catholic," "Christian"). The only non-political/religious signaled categories are "White," a racial category, and Queer, a gender/sexual identity category.

For the top 10 most significant changes between months, a positive value indicates a decrease in the rank – this can be interpreted as a decrease in the overall participation of this identifier in the COVID-19 communication space. Conversely, a negative value indicates an increase in rank from the previous month. The top 5 terms that lost rank were "Fascists," "Fascist," "Republicans," "ANTIFA," and "Constitutionalists." Each of these rank losses occurred in the month of March, meaning that there was an overall reduction of these identity group's participation (relative to all other identities being used) during the first month that the initial Emergency Mandate was issued in the United States. Interestingly, because of how prominent the "Republican" identity is in our dataset, it is simultaneously a category that is among the most 15 top stable terms we have identified, but also had one of the largest single-month changes among all the terms – this also means that it was able to steadily recover its participation in the topic space post-March.

Interestingly, three of these identities see a large resurgence during the month of June. We find that "ANTIFA," and "Fascist/Fascists" gain a large rank increase, with "ANTIFA" in particular having the largest single-month increase in the entire subset of terms of interest, gaining nearly 500 in ranking between those months. Alongside these terms, the identity signal for "POC" (people of color) and "racist" also had a large increase in their ranking in June – this appears to correspond with the BLM movement protests after the death of George Floyd in late May 2020.

Table 1. Stable and Volatile Identities: Average change for most stable identities and most significant month-to-month changes for volatile identities.

Most Stable				Most Significant Changes			
Identity	Avg Change	Top Rank	Category	Identity	Month	Change	Category
american	0	34	Polit.	antifa	Jun	−489	Polit.
god	0.27	13	Relig.	fascists	Mar	269	Polit.
president	−0.27	45	Polit.	fascist	Mar	257	Polit.
savior	−0.36	440	Relig.	republicans	Mar	244	Polit.
white	0.36	112	Racial	poc	Jun	−239.50	Racial
jesus	−0.45	49	Relig.	fascists	Jun	−236.50	Polit.
maga	0.64	16	Polit.	antifa	Mar	222	Polit.
catholic	−0.73	153	Relig.	constitutionalist	Mar	206	Polit.
christian	0.91	33	Relig.	presidential	Feb	−194	Polit.
queer	1.00	136	Gender	racist	Jun	−180	Polit.

4.2   Average Rank over Time

To get an idea of how our general sub-categories vary over the course of the year, Fig. 1 shows the average rank of all terms within these groupings. Political identity signals hit the lowest value in March but regained rank leading up to and past the 2020 Presidential Election. Racial and gender/sexual identity categories have what appears to be a correlated co-movement in June, again coinciding with the BLM movement. Religious identities seem to have a fairly stable participation rank across the year.

4.3   Political Identities in User Bios and Tweets

Now that we have looked at general trends in the data for identities used, we now explore relationships between in-group and out-group identity signaling within

Fig. 1. Average rank of all terms within a given category.

the COVID-19 communication space. Overall, we found that 249,009 users with a right-leaning identity used a political in-group signal in their tweet(s), however, a much larger proportion of the unique users (873,420) with a right-leaning identity in their bio used a left-leaning identity cue in their tweet(s). Compare this to the tendency for left-leaning identities, where we found 131,209 left-leaning bio identity users signaled right-leaning identities in their tweet(s), with a slightly larger number of left-leaning users (151,761) referencing right-leaning identities.

To get a better idea of the general tendency for in-group/out-group signaling by these political-leaning accounts on Twitter, in Fig. 2 we show the proportion of users with right-leaning bio identities using right/left-leaning identities in their messages in reference to left-leaning bio identities using right/left-leaning signals across 2020, by quarter. To determine if the differences between in-group and out-group tendencies were significantly different between groups, we conducted 4 chi-square tests of independence for each quarter, which shows us that indeed the lean of the identity cues in the tweets is dependent on the lean of the identity cues in associated bios throughout the year (Q1: $\chi^2 = 8{,}754$, p-value < 0.01, Q2: $\chi^2 = 30{,}763$, p-value < 0.01, Q3: $\chi^2 = 43{,}754$, p-value < 0.01, Q4: $\chi^2 = 22{,}929$, p-value < 0.01).

This analysis reveals a general tendency for right-leaning users to signal to the out-group more than their group during 2020. Left-leaning accounts seem to have a more mixed tendency, with earlier in the year signaling more toward their group, and leading up to and during election months focusing more attention on out-groups, nearly matching the right-leaning tendency.

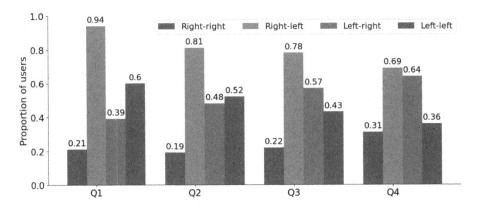

Fig. 2. Proportion of users with right-leaning bio identity that use a right/left-leaning identity in their tweet(s) and proportion of users with left-leaning bio identity that use a right/left-leaning identity in their tweet(s).

5 Discussion

Within the communication landscape of COVID-19 on Twitter in 2020, we find that users signaled a wide variety of identities, but these presentations of identity were fairly centralized around key identities, such as religious, racial, and political/national categories. In particular, American, religious (mainly Christian), white, and Republican were the most stable in participation. Changes in identities mentioned in tweets were expected, however, systematic fluctuations of users' *biography* in COVID-19 participation were unexpected. It may indicate that there are systematic movements of users expressing similar identities that are organically "stumbling" into the communication space or an impact of users strategically changing their bios to reflect current trends while discussing COVID-19.

Through our overview, we observed three major events that appear to have affected the COVID-19 communication space: the initial stay-at-home order in the United States, the Black Lives Matter protests after George Floyd's death, and the American Presidential Election. During these events, certain identities decreased their overall participation (relative to other identities) in the COVID-19 discourse, while others saw a resurgence. For instance, Republican, Constitutionalists, Liberal, and ANTIFA-identifying users decreased their participation in March 2020. In contrast, ANTIFA, POC, and references to fascism saw an increase in usage during the first month of the BLM protests. Additionally, gender and racial identity signals appeared to have correlated co-movement around this period, with increases in their overall participation in the post-June period.

The impact of these events, moreover, appears to catalyze not only upticks of certain kinds of identities but also appear to relate to the emergence of new popular identity signals, as evidenced by several high-ranking emergent or relatively novel identity terms – for instance, during the Black Lives Matter movement in

2020, terms like "ANTIFA," "POC," and "fascist(s)" saw increased participation in the general COVID-19 conversation on Twitter.

The shifts seen in user identity over this period could be potentially attributed to two key factors: 1) users adopting specific identities that subsequently surface in the COVID-19 topic space, or 2) users signaling certain identities to more intensely engage in the pandemic conversation due to perceived benefits – such as reaching a wider audience for their cause or discussing the intersection of inequalities that events like the pandemic would naturally exacerbate. These shifts reflect users leveraging and asserting a persona, with motivations surely varying from expressing demographic categories to bots and other bad actors hoping to hijack persona categories that may provide benefits of authenticity and authority [10].

Our more in-depth analysis of political in-group and out-group dynamics revealed divergent communication styles between conservative-leaning and liberal-leaning bio-tweet pairings. Conservative-leaning users exhibited a general tendency toward focusing on their liberal counterparts than their own, whereas liberal-leaning users started more self-focused in early 2020 (perhaps relating to the 2020 Democratic Primary Convention) shifting to similar levels of out-group focus by the end of 2020, around the 2020 Presidential Election.

Here again, we see that exogenous events appear to play a particularly interesting role for in-group/out-group bio-tweet identity signals. During the pre-election period, the behavior of liberal-leaning users focusing on the out-group ramped up as it was at its height during and right after the election, almost mirroring their conservative counterparts. Similarly, because of the potential impact of the Democratic Primary Election occurring early in 2020, we may not necessarily be seeing a general tendency for conservative and liberal individuals, but patterns that are referential to ongoing exogenous events and how these external events can steer co-opting of major, non-political, current events like COVID-19 to win favors within the group and benefit their political party broadly. Alternatively, out-group animosity and negative emotion have been found to predict higher engagement on social media, which is particularly motivating during contentious periods like elections, and could also be salient in driving these in-group/out-group tendencies [6,9].

6   Conclusion

In this paper, we presented an exploratory deep-dive into identity terms used in biographies and tweets within COVID-19 discourse on Twitter throughout 2020. Broadly our overview revealed that several identities stayed consistently engaged in the space across the year, while also showing that some identity signals drop in and out of the discourse in dramatic ways. These changing identities often seemed to co-occur with exogenous events, particularly ones that involved significant social and political tensions, including the BLM protests and the U.S. presidential election.

Our focus on in-group and out-group references of political identities in user bios and tweets showed a general tendency for conservatives to mention out-group (liberal) political identities across the year more often than their liberal counterparts. However, liberal-leaning users across the year increasingly mention out-group political identities in tweets as they get closer to the presidential election, again suggesting the importance of exogenous events in identity signaling in COVID-19 discourse. Further, we found general evidence of identities being formed in the absence of or opponent of certain groups or concepts – namely, being anti - "racist" or "fascist."

Broadly, we see these findings as the first step in establishing a larger body of research focused on how social actors (e.g., real people, bots, malicious agents) utilize social identities as masks to portray themselves in online social spaces. While we have focused here on the COVID-19 communication space, the issue of COVID-19 as the major topic of interest during a particular year means understanding how secondary issues and/or events may attempt to participate generally in that space, either as communicators representing a certain point of view or, perhaps, to garner greater attention for their cause of interest. We end our deep-dive of looking at the stability of identities, emerging identities signals, and political in-group out-group use of a critical and highly salient, non-political, issue (COVID-19) by sketching out future lines of research in this area.

6.1   Future Work

Future work should include the following areas of investigation:

1. **Disentangling process of increased participation by some identities:** One important aspect to explore is whether users are changing their bios to add new signals to the ongoing discourse or if there is an influx of new users who have "stumbled" into the space.
2. **Potential co-opting of major salient topics like COVID-19:** Another area of interest is determining if users are attempting to divert attention to their political or social issues by leveraging the broad attention garnered by major topics like COVID-19 – "wave-catching."
3. **Role of hashtags in biographies:** Investigate the use of hashtags in biographies to understand how they indicate or signal affiliation to certain groups and movements. Unlike hashtags in tweets, hashtags in biographies may provide direct entree to communication spaces but are not indexed in the hashtag communication channel. It also may indicate topics of discourse a user engages with and provides a signal to others on which topics to interact with them.
4. **Stabilization of novel terms:** Analyze discourse across multiple years to determine whether novel terms eventually stabilize or if they are primarily event-driven and have a short half-life.
5. **Role of social influence in identity adoption:** Examine the role of moral language and emotionality in communication between users in predicting shifts in identity use.

6.2 Limitations

There are many ways individuals may choose to express or represent their identities. Using a dictionary approach limits our ability to capture novel representations of identities, misspellings, and so on. We measure the broad use of identity terms without delving into the context of identity mentions beyond qualitative examples. Future work should take into account how the use of identity terms varies, for example mentioning an out-group negatively in one's bio ("anti-racist," for example). Finally, we focus on U.S. national events, but more analysis of the role of local and international events would benefit the completeness of this work.

References

1. Bazarova, N.N., Choi, Y.H.: Self-disclosure in social media: extending the functional approach to disclosure motivations and characteristics on social network sites. J. Commun. **64**(4), 635–657 (2014)
2. Bucholtz, M., Hall, K.: Identity and interaction: a sociocultural linguistic approach. Discourse Stud. **7**(4–5), 585–614 (2005)
3. Bullingham, L., Vasconcelos, A.C.: 'The presentation of self in the online world': Goffman and the study of online identities. J. Inf. Sci. **39**(1), 101–112 (2013). https://doi.org/10.1177/0165551512470051
4. Goffman, E.: The Presentation of Self in Everyday Life. Doubleday, Oxford, England (1959)
5. Joseph, K., Wei, W., Benigni, M., Carley, K.M.: A social-event based approach to sentiment analysis of identities and behaviors in text. J. Math. Sociol. **40**(3), 137–166 (2016)
6. Jost, J.T., Baldassarri, D.S., Druckman, J.N.: Cognitive-motivational mechanisms of political polarization in social-communicative contexts. Nat. Rev. Psychol. **1**(10), 560–576 (2022)
7. Magelinski, T., Carley, K.M.: Identity-based attribute prototypes distinguish communities on Twitter. arXiv preprint arXiv:2302.10172 (2023)
8. Pathak, A., Madani, N., Joseph, K.: A method to analyze multiple social identities in Twitter bios. Proc. ACM Hum. Comput. Interact. **5**(CSCW2), 1–35 (2021)
9. Rathje, S., Van Bavel, J.J., Van Der Linden, S.: Out-group animosity drives engagement on social media. Proc. Natl. Acad. Sci. **118**(26), e2024292118 (2021)
10. Shahi, G.K., Dirkson, A., Majchrzak, T.A.: An exploratory study of Covid-19 misinformation on Twitter. Online Soc. Netw. Media **22**, 100104 (2021)
11. Tajfel, H.: Social identity and intergroup behaviour. Soc. Sci. Inf. **13**(2), 65–93 (1974)
12. Taylor, S.J., Muchnik, L., Kumar, M., Aral, S.: Identity effects in social media. Nat. Hum. Behav. **7**(1), 27–37 (2023)
13. Tsao, S.F., Chen, H., Tisseverasinghe, T., Yang, Y., Li, L., Butt, Z.A.: What social media told us in the time of COVID-19: a scoping review. Lancet Digit. Health **3**(3), e175–e194 (2021)
14. Yoder, M.M., et al.: Phans, Stans and Cishets: self-presentation effects on content propagation in Tumblr. In: 12th ACM Conference on Web Science, pp. 39–48 (2020)

Poster Abstract

Selected Poster Presentation Abstracts

Robert Thomson[✉]

United States Military Academy, West Point, NY 10996, USA
robert.thomson@westpoint.edu

Abstract. This year the SBP-BRiMS Conference had 73 Full Paper submissions, of which 31 were accepted for publication. We also were able to accept 42 non-archival submissions to a poster session. The present paper presents an overview of 18 selected poster abstracts that were presented at the conference. These posters include social network analysis, agent-based modeling, cognitive modeling, and human simulation studies; covering the domains of health and pandemic modeling, social-cyber maneuvers, foundation models, and human decision-making. This overview shows the range of topics at the conference this year.

Keywords: Social cybersecurity · Human modeling · Social media network analysis

1 Selected Poster Abstracts

The present section describes 18 selected posters from the SBP-BRiMS 2023 Annual Conference. These posters were reviewed by an average of 3 non-program committee members and present an overview of the kinds of research presented at the conference.

Jinhang Jiang, Mei Feng, Kiran Kumar Bandeli and Karthik Srinivasan. *Forecasting Future Economic Uncertainty with Sentiments Embedded in Social Media.* This paper suggests a method for predicting the Economic Uncertainty Index in the Equity Market by analyzing social media data from Reddit. The proposed framework includes various custom pre-processing and analytics techniques, such as BERTopic for identifying latent topics and regularized linear models. Using this framework, this study conducts a sophisticated descriptive and explanatory analysis of a vast collection of Reddit posts about personal finance. The research provides valuable insights into Reddit's discussion topics and their ability to forecast economic uncertainty accurately. Overall, this study highlights the potential of explainable deep learning and social media data to enhance economic decision-making and forecasting.

Guocheng Feng, Huaiyu Cai and Wei Quan. *Exploring the Emotional and Mental Well-Being of Individuals with Long COVID Through*

Twitter Analysis. The COVID-19 pandemic has led to the emergence of Long COVID, a cluster of symptoms that persist after infection. Long COVID patients may also experience mental health challenges, making it essential to understand individuals' emotional and mental well-being. This study aims to gain a deeper understanding of Long COVID individuals' emotional and mental well-being, identify the topics that most concern them, and explore potential correlations between their emotions and social media activity. Specifically, we classify tweets into four categories based on the content, detect the presence of six basic emotions, and extract prevalent topics. Our analyses reveal that negative emotions dominated throughout the study period, with two peaks during critical periods, such as the outbreak of new COVID variants. The findings of this study have implications for policy and measures for addressing the mental health challenges of individuals with Long COVID and provide a foundation for future work.

Matthew Hicks and Kathleen M. Carley. *BEND Battle: An Agent Based Simulation of Social-Cyber Maneuvers.* BEND Battle is an agent-based model that simulates social media users conducting BEND social-cyber maneuvers in order to influence each other on a single topic. It provides a visualization of two sides conducting maneuvers against each other and the effects of those maneuvers. Previous BEND maneuver simulations are grounded in medium-specific implementations. This simulation attempts to normalize confounding factors and focus on how BEND maneuvers interact on a level-playing field. Results suggest that explain and negate maneuvers provide a broad counter to a wide range of opposing BEND maneuvers.

David Beskow, Haley Seaward and Jasmine Talley. *ArxNet Model and Data: Building Social Networks from Image Archives.* A corresponding explosion in digital images has accompanied the rapid adoption of mobile technology around the world. People and their activities are routinely captured in digital image and video files. By their very nature, these images and videos often portray social and professional connections. Individuals in the same picture are often connected in some meaningful way. Our research seeks to identify and model social connections found in images using modern face detection technology and social network analysis. The proposed methods are then demonstrated on the public image repository associated with the 2022 Emmy's Award Presentation.

Gayane Grigoryan, Andrew Collins, Wimarsha Jayanetti and Dean Chatfield. *Game-Theoretic Coalition Formation Using Genetic Algorithm and Agent-Based Modeling.* Understanding the behavior of coalition formation is of significant importance. This paper investigates coalition formation in agentbased modeling using a genetic algorithm. Agent-based simulations are utilized to gather data on coalition formation by employing an inverse generative approach. The study encompasses a wide range of game sizes and examines various characteristics of coalition formation. Specifically, it analyzes the average number of coalition suggestions per game size and the time required to achieve the final coalition structure based on respective fitness functions. By delving into these aspects, this research contributes to a deeper understanding of coalition formation dynamics in agent-based simulation models.

Christine Lepird and Dr. Kathleen M. Carley. *Automated Pink Slime Detection from Social Network Features.* A network of over a thousand seemingly local news sites was found to be controlled by a handful of partisan national organizations who have plans for expansion to new regions. These polarized sites (referred to as "pink slime") can undermine trust in the local news in susceptible communities. While over a thousand of these sites have been labeled and identified, discovering new sites remains challenging; the current process of discovering new sites involves a tedious IP address lookup process. This research proposes a methodology for detecting emerging pink slime sites through analyzing the network of Facebook pages that share content linking to these sites (as well as others of known credibility) and assigning credibility scores to the news domains in order to quickly flag any suspicious activity. It allows researchers to efficiently surveil the social media landscape to find new sources of pink slime as they emerge. This paper demonstrates the importance of the new features through machine learning validation and then applies the methods to a recent dataset for the discovery of new pink slime sites.

Nahiyan Bin Noor, Niloofar Yousefi, Billy Spann and Nitin Agarwal. *Toxicity in Reddit Discussion Threads: Impacts and Predictive Insights.* Harmful content, including abusive language, disrespect, and hate speech, is a growing concern on social media platforms. Despite efforts to tackle this issue, completely preventing the impact of harmful content on individuals and communities remains challenging. This paper utilizes Reddit data using a tree structure to study the impact of toxic content on communities. Machine learning algorithms are employed to classify the toxicity of each leaf node based on its parent, grandparent nodes, and the overall average toxicity of the tree. Our methodology aids policymakers in identifying early warning signs of toxicity and guiding harmful comments toward less toxic avenues. The research offers a comprehensive analysis of social media platform toxicity, facilitating a better understanding of variations across platforms and the impact of toxic content on different communities. Our findings offer valuable insights into the prevalence and impact of toxic content on social media platforms, and our approach can be used in future studies to provide a more nuanced understanding of this complex issue.

Harsh Katakwar, Cleotilde Gonzalez and Varun Dutt. *Attackers Have Some Prior Beliefs: Understanding Cognitive Factors of Confirmation Bias on Adversarial Decisions.* Cyberattacks are hazardous, and honeypot deception has been shown to be successful in combating them. Due to involvement of multiple factors in cyber situation, the adversary is likely to suffer from various cognitive biases. One of the many cognitive biases that affect adversarial decisions in cyberspace is confirmation bias. However, little is known about the cognitive mechanisms that drive confirmation bias in adversarial decision-making. To test for confirmation bias, one hundred and twenty participants were recruited via a crowdsourcing website and were randomly assigned to one of two between-subjects conditions in a deception-based cybersecurity simulation. Results revealed the presence of confirmation bias in adversarial decisions. Thereafter, a cognitive Instance-based Learning model was built involving recency, frequency, and cognitive noise to understand the reasons behind the reliance on confirmation bias. Results revealed that participants showed reliance on recent events and high cognitive noise in their decisions. We highlight the implications of our findings for cyber decisions in the presence of deception in the real world.

Catherine King, Peter Carragher and Kathleen M. Carley. *Citation Network Analysis of Misinformation Interventions.* In recent years, social media misinformation has become a growing problem around the world. Researchers in a variety of fields have been investigating the most effective and acceptable ways to counter fake news online. This review paper analyzes the citation network of relevant papers to provide policy makers and academics with a contextual background on the state of the literature in this field. This analysis found that there are many under and over-studied interventions in the literature, and the literature focuses almost exclusively on the effectiveness of countermeasures without considering the critical metric of user acceptance. Additionally, publication venues in this field have mostly remained disjointed and focused on specific disciplines, lacking collaborative or cross-disciplinary work.

Obianuju Okeke, Mert Can Cakmak, Ugochukwu Onyepunuka, Billy Spann and Nitin Agarwal. *Evaluating Emotion and Morality Bias in YouTube's Recommendation Algorithm for the China-Uyghur Crisis.* This study introduces a drift analysis methodology to explore patterns in YouTube's recommendation algorithm concerning the China- Uyghur crisis. Recognizing the influence of this conflict on a global discourse, we inspect the bias within YouTube's algorithm that can potentially affect information propagation about this crisis. Utilizing a dataset gathered from multiple layers of video recommendations, we apply emotion analysis and morality assessment to examine the algorithm's behavior. Our findings from the emotion analysis indicate a trend towards more positive emotions and a decline in negative emotions as users progress through recommended videos. Moreover, our morality assessment, based on the Moral Foundations Theory, indicates a decrease in moral vices and an emergent preference for certain moral virtues in the recommended content. This comprehensive analysis offers insights into how YouTube's recommendation algorithm might influence the perception of highly polarizing global issues, such as the China-Uyghur crisis.

Steven Albert. *What Can We Learn from Simple Computational Models of Health Behavior Change?* Simple computational models designed to appropriately represent hypothesized mechanisms of health behavior change may be valuable given the challenges of behavior modeling. We developed a simple approach drawn from research on the psychology of health to assesses adoption of health-protective behavior according to (i) its efficacy in reducing disease risk, (ii) the influence of others who themselves differ in the likelihood of behavior change, and (iii) the threshold of perceived threat needed for behavior change. Simulations vary the efficacy of behavior adoption for reducing disease risk, the influence of others on threat perception, and the level of threat associated with behavior change. Face validity of model dynamics is confirmed by the greater number of behavior adopters when expected efficacy of disease reduction is higher. Likewise, a greater number adopt protective behavior when threat thresholds are lower within each disease risk reduction scenario. Social influence had its greatest effect in scenarios of low expected efficacy of disease risk reduction with a high threshold for behavior change. Only in this scenario did high protective behavior influence from social contacts substantially increase behavior adoption. Observational and experimental efforts will allow a test of this finding and parameters for more sophisticated models.

Chi-Shiun Tsai. *Staying Connected & Healthy: Exploring the Impact of the COVID-19 Pandemic on Friendship Networks and Well-Being of High School Students.* This study examines the impact of school closures during the COVID-19 pandemic on the friendship networks of high school students and their well-being. This study utilizes survey data collected from 536 high school students between October 2021 to June 2022. By using mixed-effects models, the results indicate that students reported feeling less happy during the school closures. Moreover, this study introduces several new network metrics, including Differences in Closeness and Differences in Fondness. The former between the ego and alter's perspectives is positively associated with happiness and health. In addition, the clustering coefficient at the individual level shows a positive correlation with happiness. This study highlights the significance of considering the closeness aspect of relationships and the asymmetries in perceived relationships to better understand the well-being of high school students during school closures.

Siyu Wu, Amir Bagherzadeh, Frank Ritter and Farnaz Tehranchi. *Cognition Models Bake-Off: Lessons Learned from Creating Long Running Cognitive Models.* We design and develop models capable of emulating human-like game playing behaviors through interaction with unaltered interface. Each model is designed to represent a specific type of human behavior observed during gameplay. These models were constructed using the ACT-R cognitive architecture and incorporate perceptual-motor knowledge of the virtual environment. Our implementation follows an "eyes and hands" approach, incorporating enhanced components for visual perception and robotic hand actions at the bitmap level. During the experimentation phase, one of the cognitive models ran for a prolonged period, spanning approximately 4 h as it drove a simulated bus from Tucson to Las Vegas. In contrast, the other model demonstrated a shorter gameplay duration. We employed two different design approaches for these models, incorporating varied knowledge representations and actions. These approaches were either matched or unmatched with respect to the fine details of human behavior, such as the number of course corrections, average speed, and learning rate. Upon analyzing the performance of these models, it was discovered that the longer-running model did not exhibit the same level of fatigue as is typically observed in human subjects. Additionally, the shorter-running model and its behavior in this task revealed significant limitations in both the ACT-R framework and the extended hands and eyes framework. These limitations specifically pertained to the efficiency of visual pattern matching in a dynamic environment and the correspondence time of ACT-R and its extended eyes and hands components. We have documented these limitations and have already begun implementing them for further investigation.

Iuliia Alieva, Ian Kloo and Kathleen M. Carley. *Analyzing Russia's Propaganda Discourse on Twitter Using Network Analysis: A Case Study of the 2022 Russian Invasion of Ukraine.* This paper examines Russia's propaganda discourse on Twitter during the 2022 Russian invasion of Ukraine. The study employs network analysis to identify key communities and narratives associated with the prevalent and damaging narrative of "fascism/Nazism" in discussions related to the invasion. By combining network science approaches and community clustering, the research identifies influential actors and examines the most impactful messages in spreading this disinformation narrative. The study focuses on the role of social media in the Russian-Ukrainian conflict, where both sides leverage online platforms to shape public opinion and manipulate geopolitical dynamics. Russian-affiliated accounts, state-funded media outlets, bots, and trolls propagate narratives that align with their motives, while Ukrainian social media seeks support from Western nations and highlights their own military efforts. The related works section explores the tactics and impact of Russian propaganda, including the use of social media platforms, algorithms, and automation. It also discusses the narratives employed by Russian state propaganda and their persuasive techniques. The English and Russian Twitter data is analyzed using a mixed-method pipeline, including network analysis and community clustering. The ORA software tool is used for network analytics to identify key actors and communities within Twitter data. Overall, this research contributes to the understanding of propaganda dissemination on social media platforms and provides insights into the narratives and communities involved in spreading disinformation during the 2022 Russian invasion of Ukraine.

Shadi Shajari, Mustafa Alassad and Nitin Agarwal. *Detecting Suspicious Commenter Mob Behaviors on YouTube Using Graph2Vec.* YouTube, a widely popular online platform, has transformed the dynamics of content consumption and interaction for users worldwide. With its extensive range of content creators and viewers, YouTube serves as a hub for video sharing, entertainment, and information dissemination. However, the exponential growth of users and their active engagement on the platform has raised concerns regarding suspicious commenter behaviors, particularly in the comment section. This paper presents a social network analysis-based methodology for detecting suspicious commenter mob-like behaviors among YouTube channels and the similarities therein. The method aims to characterize channels based on the level of such behavior and identify common patterns across them. To evaluate the effectiveness of the proposed model, we conducted an analysis of 20 YouTube channels, consisting of 7,782 videos, 294,199 commenters, and 596,982 comments. These channels were specifically selected for propagating false views about the U.S. Military. The analysis revealed significant similarities among the channels, shedding light on the prevalence of suspicious commenter behavior. By understanding these similarities, we contribute to a better understanding of the dynamics of suspicious behavior on YouTube channels, which can inform strategies for addressing and mitigating such behavior.

Mayor Inna Gurung, Hayder Al Rubaye, Nitin Agarwal and Ahmed Al-Taweel. *Analyzing Narrative Evolution About South China Sea Dispute on YouTube: An Exploratory Study Using GPT-3.* YouTube has emerged as the leading platform for video sharing, catering to billions of users worldwide. However, most research conducted on YouTube primarily focuses on analyzing metadata, overlooking the actual content of the videos. In order to bridge this gap, our research delves into the narratives embedded in 9,000 YouTube videos pertaining to the South China Sea Dispute, utilizing Generative Pretrained Transformer (GPT-3). Additionally, we have developed a visualization tool that facilitates the analysis process by visually depicting the interconnections between keywords and narratives. We found that narratives related to the South China Sea Dispute started gaining attention in 2016 due to China's actions in constructing and militarizing artificial islands. Moreover, in 2022, there was a staggering five-fold increase in misinformation-riddled anti-western narratives from January to December, portraying the United States as attempting to exploit the South China Sea dispute. This suggests manipulating the information environment through narrative amplification maneuvers in content surrounding the South China Sea dispute on YouTube. Furthermore, our examination of the dominant cluster of narratives revealed additional narratives being injected into the information environment about the Russia-Ukraine war and the COVID-19 pandemic.

Al Zadid Sultan Bin Habib, Md Asif Bin Syed, Md Tanvirul Islam and Donald A. Adjeroh. *Cardiovascular Disease Risk Prediction via Social Media.* Researchers use Twitter and sentiment analysis to predict Cardiovascular Disease (CVD) risk. We developed a new dictionary of CVD-related keywords by analyzing emotions expressed in tweets. Tweets from eighteen US states, including the Appalachian region, were collected. Using the VADER model for sentiment analysis, users were classified as potentially at CVD risk. Machine Learning (ML) models were employed to classify individuals' CVD risk and applied to a CDC dataset with demographic information to make the comparison. Performance evaluation metrics such as Test Accuracy, Precision, Recall, F1 score, Mathew's Correlation Coefficient (MCC), and

Cohen's Kappa (CK) score were considered. Results demonstrated that analyzing tweets' emotions surpassed the predictive power of demographic data alone, enabling the identification of individuals at potential risk of developing CVD. This research highlights the potential of Natural Language Processing (NLP) and ML techniques in using tweets to identify individuals with CVD risks, providing an alternative approach to traditional demographic information for public health monitoring.

Dian Hu, Manisha Salinas, Daniel Lee, Sandra Bisbee, Folakemi Odedina and Hongfang Liu. *What Happened After the Great Resignation? an Observational Study of a Health Research Center Using Social Network Analysis as Part of a Community Outreach and Engagement Program Evaluation.* "The Great Resignation" has become a concern for many in healthcare since the pandemic. Inspired by the literature on social network analysis, we applied SNA techniques to analyze the impact of the Great Resignation on a large health research center. We found that although the great resignation has caused evident turbulence among inter-program and inter-scholar collaborations, most of those who did not resign were able to adjust and establish new connections within the center. This study has reaffirmed the strength of SNA in understanding organizational structures in health institutions and has demonstrated SNA's potential in real-time and interactive program evaluation.

Author Index

A

Abao, Roland P.   295
Abdo, Jacques Bou   199
Abu, Patricia Angela R.   295
Alizadeh, Meysam   253
Amro, Falah   219
Arya, Pratyush   231

B

Bhatnagar, Vasudha   285
Biswas, Ahana   33
Buettner Jr., Raymond R.   179

C

Cadavid, Lorena   209
Calay, Tristan J.   199
Carley, Kathleen M.   23, 95, 105, 115, 139,
    159, 169, 209, 316
Carley, L. Richard   169
Carragher, Peter   159
Chabot, Eugene   179
Chang, Rong-Ching   44, 85
Chen, Kai   44

D

Dabas, Mahavir   189
Dela Cruz, Jose Mari Luis M.   64
Delgado Ramos, Daniela   305
Diesner, Jana   305
Domson, Obed   241
Dutt, Varun   189

E

Estuar, Maria Regina Justina E.   64, 295

F

Franco, Carlos J.   209
Frangia, William   149
Frydenlund, Erika   126, 241

G

Golbeck, Jennifer   3
Gupta, Aadhar   189

H

He, Zihao   44
Hossain, Liaquat   199
Hu, Wenjia   105
Huseynli, Guljannat   126

J

Jacobs, Charity S.   95, 115
Jain, Kirti   285
Jeoung, Sullam   305
Jin, Zhifei   105

K

Kaur, Sharanjit   285
Kloo, Ian   23
Klutse, Edinam Kofi   74
Kulkarni, Nitin   263

L

Lerman, Kristina   44, 85
Lin, Yu-Ru   33
Liu, Huan   274
Llinas, Brian   126
Llinas, Humberto   126
Lothian, Nick   54
Luo, Jiebo   74
Lyu, Hanjia   74

M

Mathavan, Hirthik   274
May, Jonathan   44, 85
Mortimore, David   179
Mudiam, Nivedh   274
Mulahuwaish, Aos   199

R. Thomson et al. (Eds.): SBP-BRiMS 2023, LNCS 14161, pp. 337–338, 2023.
https://doi.org/10.1007/978-3-031-43129-6

N
Ng, Lynnette Hui Xian   115, 159
Niven, Tim   33
Nuamah-Amoabeng, Samuel   74

P
Padilla, Jose J.   126, 241
Palacio, Katherine   126
Pentland, Alex 'Sandy'   54
Phillips, Samantha C.   95, 316
Purohit, Hemant   219

Q
Qiao, Chunming   263
Qin, Xuanlong   12
Qolomany, Basheer   199

R
Renshaw, Scott Leo   316
Rezapour, Rezvaneh   305

S
Salazar-Serna, Kathleen   209
Shapiro, Jacob N.   253

Sharma, Sakshi   189
Shin, Jeongkeun   169
Song, Guohui   241
South, Tobin   54
Srivastava, Nisheeth   231

T
Tam, Tony   12
Tan, Zhen   274
Thomson, Robert   149, 329

U
Uttrani, Shashank   189
Uyheng, Joshua   95

V
Vereshchaka, Alina   263

W
Williams, Evan M.   139

Y
Yabe, Takahiro   54
Yoder, Michael Miller   316